P9-ANZ-970

SCHAUM'S®
outlines

Elementary Mathematics

──────────────────────────────── *Second Edition*

Barnett Rich, Ph.D.

Revised by

Philip A. Schmidt, Ph.D.
Professor of Secondary Education
SUNY College at New Paltz
New Paltz, New York

Schaum's Outline Series

New York   Chicago   San Francisco   Lisbon   London   Madrid
Mexico City   Milan   New Delhi   San Juan   Seoul
Singapore   Sydney   Toronto

DR. BARNETT RICH held a doctor of philosophy degree (Ph.D.) from Columbia University and a doctor of jurisprudence (J.D.) from New York University. He began his professional career at Townsend Harris Hall High School of New York City and was one of the prominent organizers of the High School of Music and Art where he served as the Administrative Assistant. Later he taught at CUNY and Columbia University and held the post of Chairman of Mathematics at Brooklyn Technical High School for 14 years. Among his many achievements are the 6 degrees that he earned and the 23 books that he wrote, among them Schaum's Outlines of *Elementary Algebra*, *Modern Elementary Algebra*, and *Geometry*.

DR. PHILIP A. SCHMIDT has a B.S. from Brooklyn College (with a major in mathematics), an M.A. in mathematics, and a Ph.D. in mathematics education from Syracuse University. He is currently Professor of Secondary Education at SUNY College at New Paltz. He is the author of *3000 Solved Problems in Precalculus* and *2500 Solved Problems in College Algebra and Trigonometry* and coauthor of *Schaum's Outline of College Mathematics* as well as numerous journal articles.

Schaum's Outline of Review of
ELEMENTARY MATHEMATICS

1 2 3 4 5 6 7 8 9 10   CUS/CUS   1 9 8 7 6 5 4 3 2 1

ISBN   978-0-07-176254-0
MHID   0-07-176254-x

Preface

For the third time in my career, I have been given the responsibility for revising one of the late Dr. Barnett Rich's exemplary texts. In 1987, I revised *Geometry*, and in 1993 I revised *Elementary Algebra*. Although this text (*Elementary Mathematics*) is quite distinct from those earlier revisions, Dr. Rich's keen insights into developmental mathematics are once again clearly present in this text.

This text is unique in that it begins with a most elementary view of arithmetic, but progresses through a thorough review of arithmetic, algebra, and geometry, finally introducing the reader to trigonometry. All important principles are stated in clear, crisp language, leading students who may have less than a substantial background in elementary arithmetic and algebra through a thorough set of tutorials in secondary level mathematics.

I have reorganized portions of the book, and I have introduced modern terminology and the calculator throughout. It is my hope that I retained Dr. Rich's fine sense of pedagogy as I have made these revisions.

As always, I thank the fine editorial staff at McGraw-Hill for their assistance, especially Maureen Walker. My warmest thanks are offered to Jean (Mrs. Barnett) Rich for her friendship and warm support throughout this revision and prior work as well. Finally, I dedicate this book to my son, Reed Schmidt, whose love of mathematics continuously encourages me as a mathematics educator.

DR PHILIP A. SCHMIDT

SUNY at New Paltz
New Paltz, New York
January, 1997

Preface to the First Edition

This survey of pre-college mathematics is designed for those who want intensive help in learning arithmetic, geometry, and algebra, or who need a concentrated review of these crucial subjects. It provides maximum assistance, far beyond the traditional book in elementary mathematics, for the following reasons:

(1) Each important rule, formula, and principle is stated in simple language, and is immediately applied to one or more sets of solved problems.

(2) Each procedure is developed step-by-step, with each step applied to problems alongside the procedure.

(3) Each set of solved problems is used to clarify and illustrate a rule or principle. The particular character of each such set is indicated by a title. *There are 2500 carefully selected and fully solved problems.*

(4) Each set of supplementary problems provides further application of a rule or principle. A guide number with each such set refers a student needing help to the related set of solved problems. *There are 3200 supplementary problems.* Each of these contains its required answer and, where needed, further aids to solution.

The author wishes to acknowledge the cooperation of Mr. Thomas J. Dembofsky and his staff.

BARNETT RICH

Contents

Chapter 1

Fundamentals of Arithmetic:
Number

1. NUMBER

Numbers and Sets of Numbers

Number is a fundamental idea in mathematics. A number may be expressed by a symbol called a *numeral*. A number may be named by a word.

Thus, the number 5 may be written as a numeral "5" or named by the word "five."

A *set of numbers* is a collection of identifiable numbers. Each number in a set is a member or element of the set. In mathematics, the study of sets is fundamental to the study of other branches of the subject.

To specify a set of numbers by *roster*, list the numbers inside braces. A capital letter may be used to refer to a set. Thus, use $S = \{1, 2, 3, 4, 5\}$ to specify the set of the first five counting numbers.

Finite and Infinite Sets

A *finite set* is one having a limited number of members. The members of a finite set can be counted. Thus, $S = \{1, 2, 3, 4, 5\}$ is a finite set. Also, the set of the first million numbers is a finite set.

An *infinite set* is one which is not a finite set. Since there is no limit to the number of counting numbers, the set of counting numbers is an infinite set. Three dots (. . .) are needed to list an infinite set. Read the three dots as "and so on" or "and so on in the same pattern."

Thus, the infinite set of counting numbers may be listed as $C = \{1, 2, 3, 4, 5, \ldots\}$.

The symbol of three dots (. . .) may also be used to list a finite set having a great number of members. Thus, the set of the first hundred counting numbers may be listed as

$$S = \{1, 2, 3, \ldots, 98, 99, 100\}$$

Natural Numbers

Any counting number is called a natural number since counting can be done using one's natural fingers. In fact, the symbols 1, 2, 3, 4, 5, 6, 7, 8, 9, 0 are called *digits*, from a Latin word meaning finger. Any natural number of our decimal system can be expressed using only these ten digits.

Thus, the digits 1 and 7 are used to express the natural or counting numbers 17 and 71.

The *successor* of a natural number is the next greater natural number. Thus, the successor of 99 is 100.

Whole Numbers

A *whole number* is either 0 or a natural number. Zero (0) is neither a counting number nor a natural number.

Odd and Even Whole Numbers

The *set of even numbers* consists of 0, 2, 4, 6, 8, and all whole numbers whose last digit is one of these.

The *set of odd numbers* consists of 1, 3, 5, 7, 9, and all whole numbers whose last digit is one of these. A whole number is either an odd number or an even number.

For example, 1352 is an even number and 2461 is an odd number.

1

Specifying Infinite Sets of Numbers

The infinite sets of natural numbers, whole numbers, even numbers, and odd numbers can be specified as follows:

$$\text{Set of natural numbers:} \qquad N = \{1, 2, 3, 4, 5, \ldots\}$$
$$\text{Set of whole numbers:} \qquad W = \{0, 1, 2, 3, 4, 5, \ldots\}$$
$$\text{Set of even whole numbers:} \qquad E = \{0, 2, 4, 6, 8, 10, \ldots\}$$
$$\text{Set of odd whole numbers:} \qquad Q = \{1, 3, 5, 7, 9, 11, \ldots\}$$

Two-Digit and Three-Digit Numbers

Many of our problems involve numbers having two or three digits. A number having two digits is a *two-digit number* as long as the first digit as not a zero digit. A number having three digits is a *three-digit number* as long as the first digit is a nonzero digit.

Thus, 19 is a two-digit number while 09 is not. Also, 139 is a three-digit number while 039 and 009 are not.

1.1   NAMING MISSING NUMBERS

Name the missing numbers in each set.

(a) $\{5, 6, 7, \ldots, 11, 12, 13\}$ (d) $\{10, 20, 30, \ldots, 60, 70, 80\}$
(b) $\{3, 5, 7, \ldots, 15, 17, 19\}$ (e) $\{22, 33, 44, \ldots, 77, 88, 99\}$
(c) $\{20, 22, 24, \ldots, 34, 36, 38\}$ (f) $\{151, 252, 353, \ldots, 757, 858, 959\}$

Illustrative Solution (c) The numbers in the set beginning with 20 and ending with 38 are even numbers. Hence, the missing numbers are 26, 28, 30, and 32. *Ans.*

Ans. (a) 8, 9, 10 (b) 9, 11, 13 (d) 40, 50 (e) 55, 66 (f) 454, 555, 656

1.2   DETERMINING WHETHER A NUMBER IS ODD OR EVEN

Determine whether each number is odd or even.

(a) 25 (b) 52 (c) 136 (d) 163 (e) 135792 (f) 246801

Solutions

The odd numbers are those whose last digit is odd. Hence the odd numbers are 25 in (a), 163 in (d), and 246801 in (f).

The even numbers are those whose last digit is even. Hence the even numbers are 52 in (b), 136 in (c), and 135792 in (e).

1.3   LISTING FINITE SETS

List each set: (a) the set of whole numbers less than 5, (b) the set of natural numbers less than 5, (c) the set of odd numbers less than 13, (d) the set of even numbers greater than 7 and less than 15, (e) the set of two-digit numbers less than 50 with both digits the same, (f) the set of two-digit numbers the sum of whose digits is 4.

Illustrative Solution (f) The two-digit numbers whose digits have a sum of 4 are 13, 22, 31, and 40. Do not include 04 since the first digit is 0. Think of 04 as 4, a single-digit number.

Ans. (a) $\{0, 1, 2, 3, 4\}$ (c) $\{1, 3, 5, 7, 9, 11\}$ (e) $\{11, 22, 33, 44\}$
        (b) $\{1, 2, 3, 4\}$ (d) $\{8, 10, 12, 14\}$

1.4   NAMING NUMBERS

Name each number: (a) the least natural number, (b) the greatest natural number, (c) the least two-digit odd whole number, (d) the greatest two-digit even whole number, (e) the least three-digit odd whole number all of whose digits are the same, (f) the greatest three-digit even whole number all of whose digits are different.

Illustrative Solution   (*b*) There is no greatest natural number. The set of natural numbers is an infinite set without limit to the number of members.

Ans.   (*a*) 1   (*c*) 11   (*d*) 98   (*e*) 111   (*f*) 986

1.5   LISTING FINITE AND INFINITE SETS USING THREE DOTS

Using three dots, list each set: (*a*) the set of natural numbers greater than 10,   (*b*) the set of whole numbers between 25 and 75,   (*c*) the set of odd numbers greater than 10,   (*d*) the set of even numbers less than 100,   (*e*) the set of whole numbers less than 100 whose last digit is 5,   (*f*) the set of three-digit whole numbers.

Illustrative Solution   (*b*) List a large finite set by naming the first three and the last three members of the set; thus, {26, 27, 28, ..., 72, 73, 74}. Note the three dots representing all the numbers that are not listed.

Ans.   (*a*) {11, 12, 13, ...}   (*d*) {0, 2, 4, ..., 94, 96, 98}   (*f*) {100, 101, 102, ..., 997, 998, 999}
         (*c*) {11, 13, 15, ...}   (*e*) {5, 15, 25, ..., 75, 85, 95}

1.6   NAMING SUCCESSORS OF NATURAL OR WHOLE NUMBERS

Name the successor of each.

(*a*) 99   (*b*) 909   (*c*) 990   (*d*) 9009   (*e*) 9090   (*f*) 9999

Illustrative Solution   (*e*) To obtain the successor of a number, add 1 to the number. Hence, the successor of 9090 is 9090 + 1, or 9091.   *Ans.*

Ans.   (*a*) 100   (*b*) 910   (*c*) 991   (*d*) 9010   (*f*) 10,000

2.   DECIMAL SYSTEM OF NUMERATION

Examine the numbers in Table 1-1 and note, as we go from ones to billions, how the place value of a digit becomes ten times as great from any place to the place immediately to the left.

Table 1-1.   Place Values

Number	Billions	Hundred Millions	Ten Millions	Millions	Hundred Thousands	Ten Thousands	Thousands	Hundreds	Tens	Units or Ones
(*a*)										3
(*b*)									4	0
(*c*)								5	0	0
(*d*)								5	4	3
(*e*)				5	0	0	4	0	3	0
(*f*)	5	0	0	4	0	0	0	3	0	0

Discussion of Numbers in Table 1-1

(*a*) The number 3 is a one-digit number. The only digit of a one-digit number is a *units digit*. The units digit 3 has a *face value* of 3.

(*b*) The number 40 is a two-digit number. In a two-digit number, the first digit is a *tens digit* and the last digit is a units digit. The tens digit 4 has a value of 40. The units digit 0 enables us to place a nonzero digit such as 4 in tens place.

(*c*) The number 500 is a three-digit number. In a three-digit number, the first digit is a *hundreds digit*, the second digit is a tens digit, and the third digit is a units digit. The hundreds digit 5 has a value of 500. The zero digits in tens and units places enable us to place 5 in hundreds place. *Think of a zero digit as the absence of any value.*

(*d*) The number 543 is a three-digit number. Since the hundreds digit is 5, the tens digit is 4, and the units digit is 3, the value of 543 is 5 hundreds + 4 tens + 3 units; that is $500 + 40 + 3$. The expression $500 + 40 + 3$ is the *expanded form* of 543.

(*e*) The number 5,004,030 has three nonzero digits, 5, 4, and 3. Since the millions digit is 5, the thousands digit is 4, and the tens digit is 3, the value of 5,004,030 is 5 millions + 4 thousands + 3 tens; that is, $5,000,000 + 4,000 + 30$. The expanded form, $5,000,000 + 4,000 + 30$, shows that the number should be read as "five million, four thousand, thirty." "Thirty" is a modification of "three tens." Note, as shown here, that *commas may be used to mark off three-digit groups from right to left.* Marking off three-digit groups in this way simplifies the reading and also the writing of large numbers.

(*f*) The number 5,004,000,300 has three nonzero digits, 5, 4, and 3. Since the billions digit is 5, the millions digit is 4, and the hundreds digit is 3, the value of 5,004,000,300 is 5 billions + 4 millions + 3 hundreds. The expanded form, $5,000,000,000 + 4,000,000 + 300$, shows that the number should be read as "five billion, four million, three hundred." Note the use of commas to mark off three-digit groups from right to left.

Expressing Numbers in Expanded Form

A number is expressed in expanded form as follows: (1) express the value of each nonzero digit separately, and (2) place plus signs between these values.

Thus, as above, express 543 in expanded form as $500 + 40 + 3$. Also, express 5,004,030 in expanded form as $5,000,000 + 4,000 + 30$.

Naming the Digits of Two-Digit and Three-Digit Numbers

A two-digit number consists of a tens digit followed by a units digit.

Thus, the digits of 45 are the tens digit 4 and the units digit 5. Read 45 as "forty five." "Forty" is a modification of "four tens."

A three-digit number consists of a hundreds digit, a tens digit, and lastly, a units digit.

Thus, the digits of 135 are the hundreds digit 1, the tens digit 3, and lastly, the units digit 5. Read 135 as "one hundred, thirty five."

Reading and Marking Off Numbers Having More Than Three Digits

Note the way in which numbers (*e*) and (*f*) are marked off in groups of three digits from right to left. Separating large numbers in groups of three digits in this way simplifies both the reading and the writing of these numbers.

Thus, 123,000 is read as "one hundred twenty three *thousand*" and 123,000,000 is read as "one hundred twenty three *million*."

Important Notes:

(1) Do not use "and" when reading whole numbers.

(2) Note the use of the singular. Read 500 as "five hundred," not "five hundreds."

2.1   Naming Nonzero Digits According to Place

Name each nonzero digit according to its place in the number.

(a) 57   (b) 570   (c) 5,700   (d) 489   (e) 48,090   (f) 4,080,900

Illustrative Solution   (b) Since $570 = 500 + 70$, 5 is hundreds digit and 7 is tens digit.

Ans.   (a) 5 is tens digit, 7 is units digit;   (c) 5 is thousands digit, 7 is hundreds digit;   (d) 4 is hundreds digit, 8 is tens digit, 9 is units digit;   (e) 4 is ten thousands digits, 8 is thousands digit, 9 is tens digit; (f) 4 is millions digit, 8 is ten thousands digit, 9 is hundreds digit.

2.2   Reading Numbers

Read each: (a) 67,   (b) 76,   (c) 607,   (d) 7,600,   (e) 60,070,   (f) 3,004,050.

Illustrative Solution   (c) Since $607 = 600 + 7$, read this as "six hundred seven."

Ans.   (a) sixty seven   (d) seven thousand, six hundred   (f) three million, four thousand, fifty
        (b) seventy six   (e) sixty thousand, seventy

2.3   Writing Numbers in Expanded Form

Write each in expanded form:

(a) 38   (b) 83   (c) 308   (d) 8,003   (e) 459   (f) 40,509

Illustrative Solution   (c) Expand using only the nonzero digits, 3 and 8. Hence, $308 = 300 + 8$.

Ans.   (a) $30 + 8$ (b) $80 + 3$ (d) $8,000 + 3$ (e) $400 + 50 + 9$ (f) $40,000 + 500 + 9$

2.4   Converting Numbers From Verbal to Digit Form

Write each in digit form:
(a) twenty   (d) two hundred forty thousand
(b) two hundred four   (e) twenty million four hundred thousand
(c) four thousand, two hundred   (f) four hundred billion, two hundred million, forty two

Illustrative Solution   (d) Write 240 for two hundred forty, then add three zeros for thousand.
Ans.   240,000.

Ans.   (a) 20   (b) 204   (c) 4,200   (e) 20,400,000   (f) 400,200,000,042

3.   NUMBER LINE

Constructing a Number Line

A *number line* is constructed by dividing a line into equal segments, as in Fig. 1-1. The arrowhead in Fig. 1-1 indicates that the number line extends to the right without end.

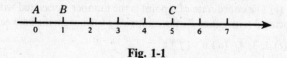

Fig. 1-1

Pairing Points and Numbers on a Number Line

As in Fig. 1-1, the points of division of a number line may be paired with successive whole numbers and identified by capital letters. For each pair of points and numbers, the number is the *coordinate* of the point and the point is the *graph* of the number.

Thus, in Fig. 1-1, 0 is the coordinate of point *A*, and *A* is the graph of 0. Also, 5 is the coordinate of point *C*, and *C* is the graph of 5.

Models of a Number Line

A good *model of a number line* may be the edge of a ruler, a yardstick, or a tape measure. In Fig. 1-2, the edge of a ruler in centimeters serves as a model of a number line.

Number Line on a Ruler

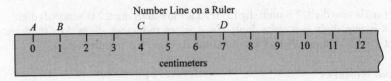

Fig. 1-2

The unit of the number line, Fig. 1-2, is one centimeter, the length of one of the equal segments. Note in Fig. 1-2 that 4 is the coordinate of point *C*, and point *D* is the graph of 7.

Graphing a Set of Numbers on a Number Line

As in Fig. 1-3, a set of numbers is graphed on a number line by making heavy the points that are the graphs of the numbers in the set.

To Graph a Set of Numbers on a Number Line

Graph the set of even numbers between 1 and 9.

PROCEDURE	SOLUTION
1. Construct a number line having a convenient unit:	1.
2. Find the numbers to be graphed:	2. The even numbers between 1 and 9 are 2, 4, 6, and 8.
3. Make heavy the points to be graphed:	3. Graph {2, 4, 6, 8}

Fig. 1-3

3.1 NAMING THE COORDINATES OF GIVEN POINTS

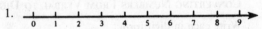

Name the coordinate of: (*a*) point *B*, (*b*) point *F*, (*c*) numbered points between *B* and *F*, (*d*) point halfway between *D* and *J*, (*e*) midpoint of segment between *E* and *G*, (*f*) point 3 units to the right of *F*.

Illustrative Solution (*e*) The coordinate of a point is the number associated with it. The required coordinate is 5, the number associated with point *F*, the midpoint of the segment between *E* and *G*.

Ans. (*a*) 1   (*b*) 5   (*c*) 2, 3, 4   (*d*) 6   (*f*) 8

3.2 NAMING THE GRAPHS OF GIVEN COORDINATES

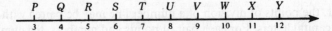

Name the graph of: (*a*) 6, (*b*) 10, (*c*) the odd coordinates, (*d*) the last two even coordinates, (*e*) coordinates grearer than 4 and smaller than 8, (*f*) the successor of 9.

Illustrative Solution   (*d*) The graph of a coordinate is the point associated with the coordinate. The last two even coordinates are 10 and 12. The graphs or points associated with 10 and 12 are *W* and *Y* respectively.

Ans.   (*a*) *S*   (*b*) *W*   (*c*) *P, R, T, V, X*   (*e*) *R, S, T*   (*f*) *W*

3.3   GRAPHING SETS OF NUMBERS

Graph: (*a*) the set of the first three whole numbers;   (*b*) the set of the last digits of odd numbers;   (*c*) the set of even numbers between 3 and 9;   (*d*) the set of odd numbers greater than 0 and less than 6;   (*e*) the set of the successors of 1, 4, and 7;   (*f*) the set of the numbers whose successors are 2, 7, and 9.

Solutions

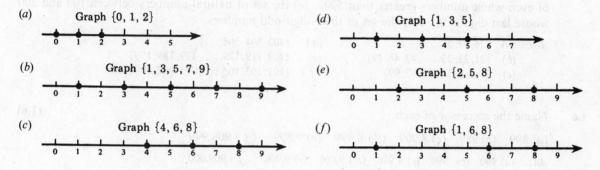

Supplementary Problems

The numbers in parentheses at the right of each set of problems indicate where to find the same type of example in this chapter. For example, (1.1) indicates that Set 1.1 in this chapter involves problems of the same type. If help is needed, a student should review the set to which reference is made.

1.1.   Name the missing numbers in each set.   **(1.1)**

    (*a*) $\{9, 10, 11, \ldots, 15, 16, 17\}$ (*d*) $\{15, 25, 35, \ldots, 65, 75, 85\}$
    (*b*) $\{4, 6, 8, \ldots, 16, 18, 20\}$ (*e*) $\{25, 30, 35, \ldots, 70, 75, 80\}$
    (*c*) $\{99, 101, 103, \ldots, 113, 115, 117\}$ (*f*) $\{939, 838, 737, \ldots, 333, 232, 131\}$

Ans.   (*a*)   12, 13, 14   (*c*)   105,107,109,111   (*e*)   40, 45, 50, 55, 60, 65
       (*b*)   10, 12, 14   (*d*)   45, 55   (*f*)   636, 535, 434

1.2.   Determine whether each number is odd or even.   **(1.2)**

    (*a*) 32   (*b*) 67   (*c*) 245   (*d*) 138   (*e*) 123456   (*f*) 234567

Ans.   (*a*) even   (*b*) odd   (*c*) odd   (*d*) even   (*e*) even   (*f*) odd

1.3.   List each set: (*a*) the set of whole numbers between 10 and 15,   (*b*) the set of odd   **(1.3)**
whole numbers less than 30 and greater than 20,   (*c*) the set of two-digit even numbers less than 20,   (*d*) the set of two-digit odd numbers both of whose digits are the same,   (*e*) the set of two-digit numbers less than 50, the sum of whose digits is 5,   (*f*) the set of three-digit numbers, each of which has the digits 1, 2, and 3.

Ans.   (*a*) $\{11, 12, 13, 14\}$ (*c*) $\{10, 12, 14, 16, 18\}$ (*e*) $\{14, 23, 32, 41\}$
       (*b*) $\{21, 23, 25, 27, 29\}$ (*d*) $\{11, 33, 55, 77, 99\}$ (*f*) $\{123, 132, 213, 231, 312, 321\}$

1.4. Name each number: (*a*) the least whole number,   (*b*) the greatest whole number,   **(1.4)**
(*c*) the least two-digit odd whole number each of whose digits is greater than 3,   (*d*) the greatest
two-digit even whole number both of whose digits are the same,   (*e*) the least three-digit odd
whole number all of whose digits are different,   (*f*) the greatest three-digit even whole number
having the digits 1, 2, and 3.

> *Ans.* (*a*) 0   (*b*) Such a number does not exist.   (*c*) 45   (*d*) 88   (*e*) 103   (*f*) 312

1.5. Using three dots, list: (*a*) the set of whole numbers greater than 25,   (*b*) the set of   **(1.5)**
natural numbers between 20 and 50,   (*c*) the set of odd whole numbers less than 100,   (*d*) the set
of even whole numbers greater than 500,   (*e*) the set of natural numbers between 100 and 200
whose last digit is 9,   (*f*) the set of three-digit odd numbers.

> *Ans.* (*a*) {26, 27, 28, ...}   (*d*) {502, 504, 506, ...}
>   (*b*) {21, 22, 23, ..., 47, 48, 49}   (*e*) {109, 119, 129, ..., 179, 189, 199}
>   (*c*) {1, 3, 5, ..., 95, 97, 99}   (*f*) {101, 103, 105, ..., 995, 997, 999}

1.6. Name the successor of each.   **(1.6)**

(*a*) 899   (*b*) 989   (*c*) 8,909   (*d*) 8,999   (*e*) 9,899   (*f*) 908,999

> *Ans.* (*a*) 900   (*b*) 990   (*c*) 8,910   (*d*) 9,000   (*e*) 9,900   (*f*) 909,000

1.7. Name each nonzero digit according to its place in the number.   **(2.1)**

(*a*) 85   (*b*) 850   (*c*) 8,500   (*d*) 691   (*e*) 90,016   (*f*) 90,100,060

> *Ans.* (*a*) 8 is tens digit, 5 is units digit
>   (*b*) 8 is hundreds digit, 5 is tens digit
>   (*c*) 8 is thousands digit, 5 is hundreds digit
>   (*d*) 6 is hundreds digit, 9 is tens digit, 1 is units digit
>   (*e*) 9 is ten thousands digit, 1 is tens digit, 6 is units digit
>   (*f*) 9 is ten millions digit, 1 is hundred thousands digit, 6 is tens digit

1.8. Read each.   **(2.2)**

(*a*) 92   (*b*) 209   (*c*) 2,900   (*d*) 678   (*e*) 600,807   (*f*) 60,800,700

> *Ans.* (*a*) ninety two   (*d*) six hundred seventy eight
>   (*b*) two hundred nine   (*e*) six hundred thousand, eight hundred seven
>   (*c*) two thousand, nine hundred   (*f*) sixty million, eight hundred thousand, seven hundred

1.9. Write each in expanded form.   **(2.3)**

(*a*) 82   (*b*) 208   (*c*) 2,800   (*d*) 359   (*e*) 300,509   (*f*) 30,500,900

> *Ans.* (*a*) 80 + 2   (*c*) 2,000 + 800   (*e*) 300,000 + 500 + 9
>   (*b*) 200 + 8   (*d*) 300 + 50 + 9   (*f*) 30,000,000 + 500,000 + 900

1.10. Write each in digit form.   **(2.4)**

(*a*) eighty   (*d*) three hundred forty two
(*b*) eight hundred five   (*e*) four thousand, two hundred three
(*c*) five thousand eight   (*f*) forty million, twenty thousand, thirty

> *Ans.* (*a*) 80   (*b*) 805   (*c*) 5,008   (*d*) 342   (*e*) 4,203   (*f*) 40,020,030

1.11. Name the coordinate of each point specified below.   **(3.1)**

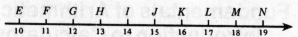

(a) point H,   (b) point M,   (c) numbered points between I and L,   (d) point halfway between E and M,   (e) point 4 units to the left of N,   (f) point 5 units to the right of point F.

Ans.   (a) 13   (b) 18   (c) 15, 16   (d) 14   (e) 15   (f) 16

1.12. Name the graph of each coordinate specified below.   **(3.2)**

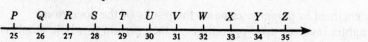

(a) 29,   (b) 35,   (c) the first three even coordinates,   (d) the last two odd coordinates,   (e) the coordinates greater than 26 and less than 30,   (f) the successor of the first coordinate having both odd digits.

Ans.   (a) T (b) Z (c) Q, S, U (d) X, Z (e) R, S, T (f) W

1.13. Graph: (a) the set of the first four natural numbers,   (b) the set of the last digits of   **(3.3)** whole even numbers,   (c) the set of odd whole numbers between 4 and 14,   (d) the set of even whole numbers greater than 21 and less than 29,   (e) the set of successsors of 13, 18, and 19,   (f) the set of numbers whose successors are 24, 29, and 32.

Ans.

(a)   Graph {1, 2, 3, 4}

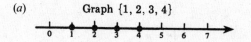

(d)   Graph {22, 24, 26, 28}

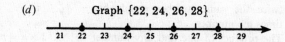

(b)   Graph {0, 2, 4, 6, 8}

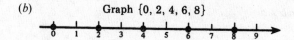

(e)   Graph {14, 19, 20}

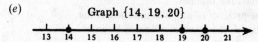

(c)   Graph {5, 7, 9, 11, 13}

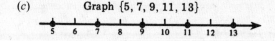

(f)   Graph {23, 28, 31}

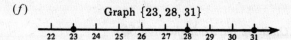

Chapter 2

Fundamentals of Arithmetic and Introduction to Calculators

1. FUNDAMENTAL OPERATIONS OF ARITHMETIC

The four fundamental operations of arithmetic are

1. ADDITION (sum)	3. MULTIPLICATION (product)
2. SUBTRACTION (difference)	4. DIVISION (quotient)

The words enclosed in parentheses name the results of the operations.

This chapter involves these operations as they pertain to whole numbers.

Using a Letter to Represent a Number

A letter may be used to represent any number in a set of numbers. Thus $n + 5$ may be used to represent "any number increased by 5." The set of numbers may be the set of whole numbers or any other set of numbers.

The use of letters enables us to shorten verbal statements considerably, as in each of the following:

Verbal Statements	Equivalent Statements	
	USING $\times$	OMITTING $\times$
1. The sum of three times a number and twice the same number equals five times the number.	$3 \times n + 2 \times n = 5 \times n$	$3n + 2n = 5n$
2. The product of one and any number equals the number.	$1 \times n = n$	$1n = n$
3. The product of any two numbers is the same as the product obtained by interchanging the numbers.	$a \times b = b \times a$	$ab = ba$

Omitting the Multiplication Sign

The multiplication sign may be omitted between a number and a letter, or between two letters.

Thus, in the first equivalent statement to the right above, $3n$ was used for "three times a number." Also, ab in the last equivalent statement represents "the product of any two numbers" The multiplication sign cannot be omitted between numbers being multiplied. We cannot write 3×4 as 34!

1.1 STATING PRODUCTS WITHOUT MULTIPLICATION SIGNS

State each product without multiplication signs.

(a) $10 \times n$ (c) $b \times c$ (e) $4 \times 2 \times z$
(b) $1 \times n$ (d) $7 \times a \times b$ (f) $2 \times 3 \times a \times g$

Illustrative Solution (f) Use 6, the product of 2 and 3. Omit the multiplication sign between 6 and a, and between a and g. *Ans.* $6ag$.

Ans. (a) $10n$ (b) $1n$ or n (c) bc (d) $7ab$ (e) $8z$

1.2   CHANGING VERBAL STATEMENTS TO EQUIVALENT STATEMENTS

Change each verbal statement to an equivalent statement using appropriate letters.

(*a*)   Ten times a number is equal to eight times the number, increased by twice the number.
(*b*)   The sum of a number and twice the number equals three times the number.
(*c*)   Five times a number less twice the number equals three times the number.
(*d*)   The perimeter of a square is four times the length of one of its sides.
(*e*)   The selling price of an article is the sum of the cost and the profit.
(*f*)   The difference between ten times a number and the number is nine times the number.

Illustrative Solution   (*d*) Use p for "perimeter of a square" and s for "length of one of its sides."
Ans. $p = 4s$.

Ans.   (*a*) $10n = 8n + 2n$ (*c*) $5n - 2n = 3n$ (*f*) $10n - n = 9n$
(*b*) $n + 2n = 3n$ (*e*) $s = c + p$

2.   FUNDAMENTAL PROPERTIES OF ADDITION

Terms Used in Addition: Addends, Sum

Addends are numbers being added. Their *sum* is the answer obtained. Thus, in $2 + 3 = 5$, the addends are 2 and 3. Their sum is 5.

Commutative Law of Addition

> **Commutative Law of Addition**
> Interchanging addends does not change their sum.

Thus, $2 + 3 = 3 + 2$. In general, for any two numbers a and b, $a + b = b + a$.
Using a cash register, if first \$5.37 is rung up, and then \$4.25, the sum displayed must be the same as that which would be displayed if \$4.25 were the first to be rung up followed by \$5.37. Here, $\$5.37 + \$4.25 = \$4.25 + \$5.37$.

Additive Identity Property: Additive Identity

> **Additive Identity Property**
> Adding zero to any number results in the same number.

Thus, $543 + 0 = 543$. In general, for any number n, $n + 0 = n$.
Zero (0) is the *additive identity*, since adding zero to any number results in identically the same number.

Using a Number Line to Perform Addition

To Add Two Numbers Using a Number Line

Using a number line, add: (*a*) $3 + 4$ (*b*) $4 + 3$

PROCEDURE   SOLUTIONS

1. Begin at point whose coordinate is the first number   Begin at 3, Fig. 2-1(*a*)   Begin at 4, Fig. 2-1(*b*).

2. Go to the right a number of units equal to the second number: Go right 4 units.   Go right 3 units.

3. The sum is the coordinate of the point reached: Reach 7, the sum.   Reach 7, the sum.
Ans. 7   *Ans.* 7

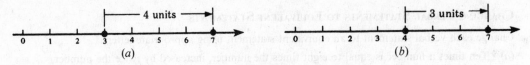

Fig. 2-1

Note that Fig. 2-1 illustrates the Commutative Law of Addition by showing that $3 + 4 = 4 + 3$.

Constructing an Addition Table of Basic Sums

Table 2-1.   Addition Table for Digits from 2 to 9

Addends	2	3	4	5	6	7	8	9
2	4	5	6	7	8	9	10	11
3	5	6	7	8	9	10	11	12
4	6	7	8	9	10	11	12	13
5	7	8	9	10	11	12	13	14
6	8	9	10	11	12	13	14	15
7	9	10	11	12	13	14	15	16
8	10	11	12	13	14	15	16	17
9	11	12	13	14	15	16	17	18

The addends in the addition table are the digits 2, 3, 4, 5, 6, 7, 8, and 9. The digit 0 is not included as an addend since adding 0 to any number results in the same number. The digit 1 is not used as an addend since it is easy to remember that adding 1 to any number results in the successor of that number.

A *basic sum* is the sum of two digits. The addition table contains 64 basic sums. Thus, adding digits 9 and 7, or 7 and 9, we obtain the basic sum 16.

Basic Sums in the Table

1. The basic sums in the shaded boxes of Table 2-1 are the *doubles* of the digits. Thus, the shaded 6 equals $3 + 3$; that is, 6 is the double of 3.

2. Going from left to right in any row, or from top to bottom in any column, each new number is one more than the previous number. Hence, each new number is the *successor* of the previous number. Thus, after 4, by adding 1 each time, we obtain 5, 6, 7, 8, 9, 10, and 11.

3. For each basic sum to the left of a shaded box, there is the same basic sum above the shaded box. Thus, the basic sum 5, one box to the left of 6, is the same as the basic sum 5 which is one box above 6. This illustrates the Commutative Law of Addition since one of the 5s is $2 + 3$ and the other 5 is $3 + 2$.

Memorizing the Basic Sums

Each basic sum should be memorized. To this end, each of the 8 doubles must be memorized and also each of the 28 basic sums either to the left of the shaded boxes or to the right of them. When you have memorized one group of 28 basic sums, you can recall the other group of 28 by means of the Commutative Law of Addition.

Associative Law of Addition

> **Associative Law of Addition**
> The way in which numbers are added in groups of two does not change their sum.

Thus, the sum $50 + 10 + 2$ may be found by using two different methods. In each of these methods, a different pair of addends is used. As shown in the following illustrative problem, parentheses are used to pair addends.

To Add Three Numbers

Find the sum: $50 + 10 + 2$.

METHOD 1
1. Pair the first two numbers.

$$(50 + 10) + 2$$

2. Add the pair in parentheses.

$$60 + 2$$

3. Add the resulting two numbers.

$$62 \quad Ans.$$

METHOD 2
1. Pair the last two numbers

$$50 + (10 + 2)$$

2. Add the pair in parentheses.

$$50 + 12$$

3. Add the resulting two numbers.

$$62 \quad Ans.$$

Since the results are the same, $(50 + 10) + 2 = 50 + (10 + 2)$.

In general: $a + b + c = (a + b) + c = a + (b + c)$ for any three numbers a, b, and c.

As in the following example, a sum may be simplified by using both the Commutative and Associative Laws of Addition:

$$25 + (37 + 75) = 25 + (75 + 37) \quad \text{Commutative Law of Addition}$$
$$= (25 + 75) + 37 \quad \text{Associative Law of Addition}$$
$$= 100 + 37 = 137 \quad Ans.$$

2.1   APPLYING THE COMMUTATIVE LAW OF ADDITION

State each number represented by ? or n.

(a) $19 + 25 = 25 + ?$ (c) $n + 543 = 543 + 975$ (e) $24 + 82 + ? = 13 + 24 + 82$

(b) $43 + 57 = ? + 43$ (d) $65 + n = 14 + 65$ (f) $30 + 27 + 85 = n + 27 + 85$

Illustrative Solution   (c) Interchanging addends does not change their sum. Since $975 + 543 = 543 + 975$, $n = 975$.

Ans. (a) 19 (b) 57 (d) 14 (e) 13 (f) 30

2.2   APPLYING THE ADDITIVE IDENTITY PROPERTY

State each number represented by ? or n.

(a) $? + 0 = 59$ (c) $589 = n + 0$ (e) $5 + ? + 0 = 69 + 5$

(b) $0 + ? = 150$ (d) $1{,}234 = 0 + n$ (f) $77 + n = 8 + 0 + 77$

Illustrative Solution   (d) Adding zero to any number results in the same number. Since $1{,}234 = 0 + 1{,}234$, then $n = 1{,}234$.

Ans. (a) 59 (b) 150 (c) 589 (e) 69 (f) 8

2.3   USING THE NUMBER LINE TO CHECK ADDITION

Check each addition using the same number line.

(a) $3 + 5 = 8$ (c) $10 + 1 = 11$ (e) $14 + 0 = 14$ (g) $3 + 3 + 3 = 9$

(b) $5 + 3 = 8$ (d) $9 + 5 = 14$ (f) $3 + 4 + 5 = 12$ (h) $1 + 3 + 10 = 14$

Illustrative Solutions   (*d*) To check $9 + 5 = 14$, begin at 9 on the number line, then go right 5 units. The point reached is 14 (*correct*). (*e*) To check $14 + 0 = 14$, begin at 14 on the number line and go right 0 units. Since no progress is made, the point reached is 14 (*correct*). The remaining solutions are left to the student.

2.4   **CONSTRUCTING AND USING AN ADDITION TABLE**

(1) Construct an addition table for addends from 10 through 14. Shade the boxes that contain the doubles of the addends. (2) Check each using the addition table:

(*a*) $10 + 10 = 20$	(*c*) $11 + 12 = 23$	(*e*) $12 + 11 = 23$
(*b*) $10 + 13 = 23$	(*d*) $12 + 12 = 24$	(*f*) $14 + 12 = 26$

Solutions

(1) See Table 2-2.

Table 2-2

Addends	10	11	12	13	14
10	20	21	22	23	24
11	21	22	23	24	25
12	22	23	24	25	26
13	23	24	25	26	27
14	24	25	26	27	28

(2) *Illustrative Solutions*   (*c*) Check $11 + 12 = 23$ by going horizontally along the 11-row as far as the 12-column. The entry reached is 23 (*correct*). (*e*) Check $12 + 11 = 23$ by going horizontally along the 12-row as far as the 11-column. The entry reached is 23 (*correct*). Each of the remaining checks is left for the student.

2.5   **APPLYING THE ASSOCIATIVE LAW OF ADDITION**

Using two groupings, find each sum without interchanging addends.

(*a*) $1 + 2 + 3$	(*c*) $3 + 7 + 10$	(*e*) $1 + 10 + 20$
(*b*) $1 + 3 + 2$	(*d*) $7 + 3 + 10$	(*f*) $1 + 20 + 10$

Solutions

(*a*) $(1 + 2) + 3 = 1 + (2 + 3)$
$3 + 3 = 1 + 5$
$6 = 6$

(*c*) $(3 + 7) + 10 = 3 + (7 + 10)$
$10 + 10 = 3 + 17$
$20 = 20$

(*e*) $(1 + 10) + 20 = 1 + (10 + 20)$
$11 + 20 = 1 + 30$
$31 = 31$

(*b*) $(1 + 3) + 2 = 1 + (3 + 2)$
$4 + 2 = 1 + 5$
$6 = 6$

(*d*) $(7 + 3) + 10 = 7 + (3 + 10)$
$10 + 10 = 7 + 13$
$20 = 20$

(*f*) $(1 + 20) + 10 = 1 + (20 + 10)$
$21 + 10 = 1 + 30$
$31 = 31$

3.   ADDING NUMBERS

Adding Numbers: No Carry Needed

Whole numbers are added by arranging them vertically so that digits having the same place value are aligned in a vertical column; that is, units digits are aligned with units digits, tens digits with tens digits, and so on. In this way, digits having the same place value can be added together.

To Add Whole Numbers: No Carry Needed

Find the sum: $123 + 456$.

PROCEDURE

1. Arrange the addends so that digits having the same place value are aligned in the same vertical column:

2. Add units with units, tens with tens, hundreds with hundreds, etc.:

SOLUTION

1. Arrange addends: *THINK*

$$123 \rightarrow 1 \text{ hundred } + 2 \text{ tens} + 3 \text{ units}$$
$$\underline{+\ 456} \rightarrow \underline{4 \text{ hundreds} + 5 \text{ tens} + 6 \text{ units}}$$

2. Add: $579 \leftarrow 5 \text{ hundreds} + 7 \text{ tens} + 9 \text{ units}$

Ans. 579

Note. The work shown under *THINK* should be done mentally.

Added Numbers: Carry Needed

In the following model problem, it is necessary to carry 10 from the units place to 1 ten in the tens place. Note under *THINK* how 5 tens + 12 units becomes 6 tens + 2 units.

To Add Whole Numbers: Carry Needed

Find the sum: $123 + 439$.

PROCEDURE

1. Arrange the addends vertically so that digits having the same place value are aligned in the same vertical column:

2. Add units with units, tens with tens, hundreds with hundreds, etc.:

3. If a sum in any column has two digits, bring down the units digit and *carry* the tens digit to the preceding column on the left, as shown:

SOLUTION

1. Arrange addends: *THINK*

$$123 \rightarrow 1 \text{ hundred } + 2 \text{ tens} + 3 \text{ units}$$
$$\underline{+\ 439} \rightarrow \underline{4 \text{ hundreds} + 3 \text{ tens} + 9 \text{ units}}$$

2. Add: $5 \text{ hundreds} + 5 \text{ tens} + 12 \text{ units}$

3. $\boxed{1}$ Carry $5 \text{ hundreds} + (5 \text{ tens} + \overline{1 \text{ ten}}) + 2 \text{ units}$

$$\begin{array}{ccc} 1 & 2 & 3 \\ +4 & 3 & 9 \\ \hline 5\ 6 & \boxed{1} & 2 \end{array} \leftarrow 5 \text{ hundreds} + \quad 6 \text{ tens} \quad +2 \text{ units}$$

Important Note: The carry of 1 is written in the above solution. However, once the process is understood, a carry should be done mentally instead.

Illustrative Solution   Add: $173 + 439$.

$$\begin{array}{ccc} \boxed{1} & \boxed{1} & \text{Carry} \\ 1 & 7 & 3 \\ +4 & 3 & 9 \\ \hline 6\boxed{1} & 1 & \boxed{1}2 \\ (c)(b) & & (a) \end{array}$$

Ans. 612

How to Think When Adding

(a) $3 + 9 = 12$ Write 2 in units column.
Carry 1 to tens column.

(b) $7 + 3 = 10$ 10 plus carry of 1 equals 11.
Write 1 in tens column. Carry 1 to hundreds column.

(c) $1 + 4 = 5$ 5 plus carry of 1 equals 6.

Extending the Commutative and Associative Laws of Addition

The Commutative Law of Addition enables us to interchange addends, and the Associative Law of Addition enables us to change the pairing of addends. By means of both laws, addends can be rearranged in any way we may choose.

Rule for Rearranging Addends: Rearranging addends does not change their sum.

Thus, $99 + 78 + 1 + 2$ can be rearranged as $99 + 1 + 78 + 2$ to simplify the addition; then, $(99 + 1) + (78 + 2) = 100 + 80 = 180$.

Checking Sums

A sum of three or more numbers may be found by adding upwards as well as downwards. Adding numbers in the opposite direction after a sum has been found is justified by the Rule for Rearranging Addends.

Thus, the sum of 12, 23, and 44 can be found by adding down and checked by adding up, as follows:

```
                              79 ↑
                              12
         Add DOWN │   23   Check UP
                  │   44
                  ↓   ──
                      79
```

Adding down leads to the sum for $12 + 23 + 44$. Checking up leads to the sum for $44 + 23 + 12$, a rearrangement of addends.

3.1   ADDING NUMBERS: NO CARRY NEEDED

Add:
(a) $36 + 41$ (d) 1,722   (e)   423   (f) 6,390,415
(b) $367 + 130$ + 173   3,240   205,344
(c) $572 + 2,217 + 45,100$ + 15   +1,003,020

Illustrative Solutions   In (a), (b), and (c), align the addends vertically.

```
(a)   36        (b)   367        (c)      572
     +41             +130             2,217
     ──              ───            +45,100
      77  Ans.        497  Ans.      ───────
                                     47,889  Ans.
```

Ans.   (d) 1,895   (e) 3,678   (f) 7,598,779

3.2   ADDING NUMBERS: CARRY NEEDED

Add and show each carry:
(a) $25 + 59$ (c) $618 + 2,793$ (d)   27
(b) $146 + 394 + 572$ + 59

Solutions   In (a), (b), and (c), align the addends vertically.

```
(a)   1      (b)   21     (c)   1 11    (d)   1
     25          146           618          27
    +59          394         +2,793        +59
    ──           ───         ──────        ──
     84        +  572         3,411         86
               ─────
               1,112
```

3.3   FINDING AND CHECKING SUMS

Find each sum and check your answer.
(a) $23 + 15 + 41$ (c) $142 + 106 + 126 + 245$ (e) $58,073 + 69,154 + 9,860$
(b) $204 + 135 + 120$ (d) $8,142 + 7,106 + 3,162 + 8,254$ (f) $1,481,592 + 673,017 + 50,001$

Solutions   Add down and check up, as symbolized by the arrows.

```
(a)   23    (b)   204   (c)    11    (d)    11    (e)   211    (f)  12 11
      15          135        142         8,142        58,073      1,481,592
    + 41        + 120        106         7,106        69,154        673,017
    ──          ───          126         3,162      +  9,860     +  50,001
      79          459      + 245       + 8,254       ───────      ─────────
                           ───         ──────        137,087      2,204,610
                           619         26,664
```

3.4   FINDING DISTANCES USING A MILEAGE CHART

Referring to Fig. 2-2, find the distance in kilometers of the shortest routes in each case.

(*a*) between *A* and *C*   (*d*) between *C* and *E*
(*b*) between *B* and *D*   (*e*) between *C* and *F*
(*c*) between *B* and *E*   (*f*) between *E* and *F*

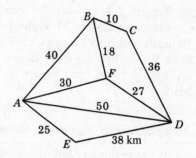

Solutions

(*a*) 50 kilometers by way of *B*   (*d*) 74 kilometers by way of *D*
(*b*) 45 kilometers by way of *F*   (*e*) 28 kilometers by way of *B*
(*c*) 65 kilometers by way of *A*   (*f*) 55 kilometers by way of *A*

Fig. 2-2

3.5   PROBLEM SOLVING INVOLVING ADDITION

(*a*) How many miles were traveled in a three-day trip if the separate distances covered in kilometers were (1) 345, 365, and 410; (2) 268, 192, and 548; (3) 352, 423, and 396?

(*b*) A comet appeared in 1905. In what years will it make its next three visits if it reappears every 75 years?

(*c*) Find the total amount of three purchases if the amounts spent are (1) $450, $560, and $1,625; (2) $360, $275, and $975; (3) $2,430; $4,585, and $6,820.

(*d*) In a match of three games, the following scores were made by three bowlers of a team.

	1st Game	2nd Game	3rd Game
Player A	162	237	174
Player B	178	165	188
Player C	168	186	229

(1) Find the total score of each player. (2) Find the team score for each game. (3) Show that the total of the players' scores equals the total of the team scores.

Illustrative Solution

(*b*) Year of 1st visit $= 1905 + 75 = 1980$ Year of 2nd visit $= 1980 + 75 = 2055$
Year of 3rd visit $= 2055 + 75 = 2130$

Ans.   (*a*) (1) 1,120   (2) 1,008   (3) 1,171
(*c*) (1) $2,635   (2) $1,610   (3) $13,835
(*d*) (1) A, 573; B, 531; C, 583
(2) 1st game, 508; 2nd game, 588; 3rd game, 591   (3) Both totals are 1,687.

4. FUNDAMENTAL PROPERTIES OF MULTIPLICATION

Terms: Factors, Product, Multiplicand, Multiplier

Factors are numbers being multiplied. Their *product* is the answer obtained. Thus, in $2 \times 3 = 6$, the factors are 2 and 3. Their product is 6.

In any product of two factors, the first factor is the *multiplicand* and the second factor is the *multiplier*. Thus, in $2 \times 3 = 6$, the multiplicand is 2 and the multiplier is 3.

Commutative Law of Multiplication

> **Commutative Law of Multiplication**
> Interchanging factors does not change their product.

Thus, $2 \times 3 = 3 \times 2$. In general, for any two numbers *a* and *b*, $ab = ba$.

Figure 2-3 provides a concrete illustration of the Commutative Law of Multiplication. In (*a*) a framework of 20 beads consists of 5 rows, each having 4 beads. In (*b*) the same framework has been turned on its side so that there are now 4 rows, each having 5 beads. There are still 20 beads.

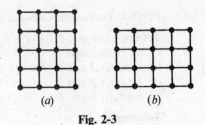

(*a*)   (*b*)

Fig. 2-3

Multiplication as Repeated Addition of the Same Number

We may think of multiplication as the repeated addition of the same number. Using the framework in Fig. 2-3, we see on the left that $4 + 4 + 4 + 4 + 4 = 20$; that is, five 4s or 5×4 equals 20. On the right, we see that $5 + 5 + 5 + 5 = 20$; that is, four 5s or 4×5 equals 20.

Methods of Showing Multiplication

Multiplication may be shown in the following ways:

1. Using the multiplication sign:	6×2	$6 \times a$	$a \times b$
2. Using a raised dot:	$6 \cdot 2$	$6 \cdot a$	$a \cdot b$
3. Using no sign:	(See Note.)	$6a$	ab
4. Using parentheses and no sign:	$6(2)$	$6(a)$	$a(b)$

Note: Of course, 6×2 cannot be written as 62. The multiplication sign may not be omitted between numbers being multiplied.

Multiplicative Identity Property: Multiplicative Identity

> **Multiplicative Identity Property**
> Multiplying any number by 1 results in the same number.

Thus, $549 \times 1 = 549$. In general, for any number n, $n \times 1 = n$ and $1 \times n = n$.

One (1) is the *multiplicative identity*, since multiplying any number by 1 results in identically the same number.

Multiplicative Property of Zero

> **Multiplicative Property of Zero**
> The product of any number and zero is 0.

Thus, $549 \times 0 = 0$. In general, for any number n, $n \times 0 = 0$ and $0 \times n = 0$.

Using a Number Line to Perform Multiplication

To Multiply Two Numbers Using a Number Line

Using a number line, multiply: (*a*) 3×4 (*b*) 4×3

PROCEDURE   SOLUTIONS

1. Begin at 0:
2. Make equal jumps to the right, the number of jumps being the first factor, and the length of each jump being the second factor:
3. The product is the coordinate of the point reached:

1. Begin at 0, Fig. 2-4.
2. To the right, make 3 equal jumps of 4 units each.

3. Reach 12, the product.
   Ans. 12

Begin at 0, Fig. 2-5
To the right, make 4 equal jumps of 3 units each.

Reach 12, the product.
Ans. 12

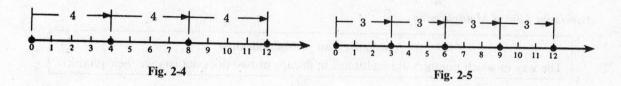

Fig. 2-4   Fig. 2-5

Note that Figs. 2-4 and 2-5 illustrate the Commutative Law of Multiplication by showing that $3 \times 4 = 4 \times 3$.

Constructing a Multiplication Table of Basic Products

Table 2-3.   Multiplication Table for Digits from 2 to 9

Factors	2	3	4	5	6	7	8	9
2	4	6	8	10	12	14	16	18
3	6	9	12	15	18	21	24	27
4	8	12	16	20	24	28	32	36
5	10	15	20	25	30	35	40	45
6	12	18	24	30	36	42	48	54
7	14	21	28	35	42	49	56	63
8	16	24	32	40	48	56	64	72
9	18	27	36	45	54	63	72	81

The factors in the multiplication table are the digits 2, 3, 4, 5, 6, 7, 8, and 9. The digit 0 is not included as a factor since multiplying any number by 0 results in 0. The digit 1 is not used as a factor since multiplying any number by 1 results in the same number.

A *basic product* is the product of two digits. The multiplication table contains 64 basic products. Thus, multiplying 9 and 7, or 7 and 9, we obtain the basic product 63.

Basic Products in Table 2-3

1. The basic products in the shaded boxes are the *squares* of the digits. Thus, the shaded 9 equals 3×3; that is, 9 is the square of 3.

2. Going from left to right in any row, or from top to bottom in any column, each new number is obtained by *adding* the factor involved to the previous number. Thus, after 4, by adding 2 each time, we obtain 6, 8, 10, 12, 14, 16, and 18.

3. For each basic product to the left of a shaded box, there is the same basic product above the shaded box. Thus, the basic product 6, one box to the left of 9, is the same as the basic product 6 which is one box above 9. This illustrates the Commutative Law of Multiplication since one of the 6s is 2×3 and the other 6 is 3×2.

Memorizing the Basic Products

Each basic product should be memorized. To this end, each of the 8 shaded squares must be memorized and also each of the 28 basic products either to the left of the shaded boxes or to the right of them. When you have memorized one group of basic products, you can recall the other group of 28 by means of the Commutative Law of Multiplication.

Associative Law of Multiplication

Associative Law of Multiplication
The way in which numbers are multiplied in groups of two does not change their product.

Thus, the product $3 \times 4 \times 10$ may be found by using two different methods. In each of these methods, a different pair of factors is used. As shown in the following illustrative problem, parentheses are used to pair factors.

To Multiply Three Numbers

Find the product: $3 \times 4 \times 10$.

METHOD 1	METHOD 2
1. Pair the first two numbers.	1. Pair the last two numbers
$(3 \times 4) \times 10$	$3 \times (4 \times 10)$
2. Multiply the pair in parentheses.	2. Multiply the pair in parentheses.
12×10	3×40
3. Multiply the resulting two numbers.	3. Multiply the resulting two numbers.
120 *Ans.*	120 *Ans.*

Since the results are the same, $(3 \times 4) \times 10 = 3 \times (4 \times 10)$.

In general, $abc = (ab)c = a(bc)$ for any three numbers a, b, and c.

As in the following example, a product may be simplified by using both the Commutative and Associative Laws of Multiplication:

$$25 \times (37 \times 4) = 25 \times (4 \times 37) \text{ Commutative Law of Multiplication}$$
$$= (25 \times 4) \times 37 \text{ Associative Law of Multiplication}$$
$$= 100 \times 37 = 3,700 \quad Ans.$$

4.1   APPLYING THE COMMUTATIVE LAW OF MULTIPLICATION

State each number represented by ? or n.

(a) $14 \times 25 = 25n$ (c) $? \times 35 = 35(50)$ (e) $7(9)(11) = 11(7)n$
(b) $12 \times 15 = ? \times 12$ (d) $50n = 75(50)$ (f) $25 \cdot 40n = 20 \cdot 40 \cdot 25$

Illustrative Solution   (b) Interchanging factors does not change their product. Since $12 \times 15 = 15 \times 12$, then ? = 15.

Ans.   (a) 14   (c) 50 (multiplication indicated by parentheses and no sign) (d) 75   (e) 9   (f) 20
(multiplication indicated by a raised dot)

4.2   APPLYING THE MULTIPLICATIVE IDENTITY PROPERTY

State each number represented by ? or n.

(a) $1 \cdot n = 145$ (c) $15(25) = 1(25)n$ (e) $5 \cdot ? \cdot 7 = 7 \cdot 5$
(b) $258 = n \cdot 1$ (d) $50 \cdot 9 \cdot ? = 9 \cdot 50$ (f) $75n = 8 \cdot 1 \cdot 75$

Illustrative Solution   (c) Multiplying any number by 1 results in the same number. Since $1(25)n = 25n$, then $15(25) = 25n$ and $n = 15$.

Ans.   (a) 145   (b) 258   (d) 1   (e) 1   (f) 8

4.3   APPLYING THE MULTIPLICATIVE PROPERTY OF ZERO

State each number represented by ? or n.

(a) $75 \times 0 = ?$ (c) $5(6)(0) = 7(?)$ (e) $n - 5 = 25(50)(75)0$
(b) $25 \times 0 = ? - 1$ (d) $25n = 25(9)0$ (f) $50 \cdot 0 = n \cdot 0$

Illustrative Solution   (*d*) The product of any number and zero is 0. Since $25(9)0 = 0$, then $25n = 0$ and $n = 0$.

Ans.   (*a*) 0   (*b*) 1   (*c*) 0   (*e*) 5   (*f*) *n* may be any number

4.4   USING THE NUMBER LINE TO PERFORM MULTIPLICATION

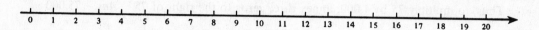

Check each multiplication using the number line.
(*a*) $2 \times 5 = 10$ (*c*) $8 \times 0 = 0$ (*e*) $10 \times 2 = 20$ (*g*) $2 \times 2 \times 3 = 12$
(*b*) $6 \times 1 = 6$ (*d*) $2 \times 10 = 20$ (*f*) $3 \times 6 = 18$ (*h*) $3 \times 3 \times 2 = 18$

Illustrative Solution   (*d*) To check $2 \times 10 = 20$, begin at 0 on the number line, then to the right make 2 jumps of 10 units each. The point reached is 20 (*correct*).   (*c*) To check $8 \times 0 = 0$, begin at 0 on the number line and to the right make 8 jumps of 0 units each. Since no progress is made, the point reached is 0 (*correct*). The remaining solutions are left to the student.

4.5   CONSTRUCTING AND USING A MULTIPLICATION TABLE

(1) Construct a multiplication table for factors from 10 to 14. Shade the boxes that contain the numbers that are the squares of a factor.   (2) Check each using the multiplication table:
(*a*) $10 \times 10 = 100$ (*c*) $11 \times 13 = 143$ (*e*) $13 \times 11 = 143$
(*b*) $10 \times 12 = 120$ (*d*) $12 \times 12 = 144$ (*f*) $14 \times 11 = 154$

Solutions

(1) See Table 2-4. Entries in shaded boxes are the squares of the factors.

Table 2-4

Factors	10	11	12	13	14
10	100	110	120	130	140
11	110	121	132	143	154
12	120	132	144	156	168
13	130	143	156	169	182
14	140	154	168	182	196

(2) *Illustrative Solutions*   (*e*) Check $13 \times 11 = 143$ by going horizontally along the 13-row as far as the 11-column. The entry reached is 143 (*correct*). Each of the remaining checks is left for the student.

4.6   APPLYING THE ASSOCIATIVE LAW OF MULTIPLICATION

Using two groupings, find each product without interchanging factors.
(*a*) $1 \times 2 \times 4$ (*c*) $3 \times 7 \times 10$ (*e*) $3 \times 5 \times 5$
(*b*) $4 \times 1 \times 2$ (*d*) $3 \times 2 \times 5$ (*f*) $10 \times 4 \times 2$

Solutions

(*a*) $(1 \cdot 2)4 = 1 \cdot (2 \cdot 4)$ (*c*) $(3 \cdot 7)10 = 3(7 \cdot 10)$ (*e*) $(3 \cdot 5) \cdot 5 = 3(5 \cdot 5)$
$\qquad 2 \cdot 4 = 1 \cdot 8$ $\qquad 21 \cdot 10 = 3(70)$ $\qquad 15 \cdot 5 = 3(25)$
$\qquad\quad 8 = 8$ $\qquad\quad 210 = 210$ $\qquad\quad 75 = 75$
(*b*) $(4 \cdot 1) \cdot 2 = 4(1 \cdot 2)$ (*d*) $(3 \cdot 2) \cdot 5 = 3(2 \cdot 5)$ (*f*) $(10 \cdot 4) \cdot 2 = 10(4 \cdot 2)$
$\qquad 4 \cdot 2 = 4(2)$ $\qquad 6 \cdot 5 = 3(10)$ $\qquad 40 \cdot 2 = 310(8)$
$\qquad\quad 8 = 8$ $\qquad\quad 30 = 30$ $\qquad\quad 80 = 80$

5. MULTIPLICATION BY A PLACE VALUE OR BY A MULTIPLE OF A PLACE VALUE

Multiplying by 10, 100, 1,000, or Any Other Place Value

Rule: To multiply a whole number by 10, 100, 1,000, or any other place value, annex as many zeros to the right of the number as there are zeros in the place value.

Thus, to multiply 75 by 1,000, annex three zeros to the right of 75.   *Ans.*   75,000

Multiplying by a Multiple of Any Place Value

A *multiple of 10* is a whole number having 10 as one of its factors. Thus, 30, 90, and 140 are multiples of 10.

A *multiple of 100* has a factor of 100, a multiple of 1,000 has a factor of 1,000, and so on for multiples of any place value. Thus, 300 is a multiple of 100. Also, 7,000 is a multiple of 1,000.

To Multiply a Number by a Multiple of a Place Value

Multiply 5 by 7,000.

PROCEDURE

1. Omit the zeros at the right end of the multiple, then multiply the number that remains by the other factor:

2. Annex the omitted zeros to the right of the resulting product:

SOLUTION

1. Omit the 3 zeros at the right end of 7,000; then multiply 7, the number that remains, by 5. $7 \times 5 = 35$

2. To the resulting product, 35, annex 3 zeros. *Ans.* 35,000

5.1   MULTIPLYING BY 10, 100, 1,000, OR ANY PLACE VALUE

Express each in digits:
(*a*) 12 tens   (*c*) 25 thousands   (*e*) 415 hundred thousands   (*g*) 400 billions
(*b*) 12 hundreds   (*d*) 6 ten thouands   (*f*) 30 millions   (*h*) 555 trillions (1 trillion equals 1,000 billions)

Find each product:
(*i*) 16×10 (*k*) 153×100 (*m*) $2,450 \times 1,000$ (*o*) $40,000 \times 1,000,000$
(*j*) 16×100 (*l*) $153 \times 1,000$ (*n*) $40,000 \times 100$ (*p*) $12,345 \times 1,000,000,000$

Illustrative Solutions (*e*) Since one hundred thousand is 100,000 (that is, 1 followed by 5 zeros), annex 5 zeros to 415.   *Ans.* 41,500,000

Ans. (*a*) 120   (*f*) 30,000,000   (*j*) 1,600   (*n*) 4,000,000
       (*b*) 1,200   (*g*) 400,000,000,000   (*k*) 15,300   (*o*) 40,000,000,000
       (*c*) 25,000   (*h*) 555,000,000,000,000 (*l*) 153,000   (*p*) 12,345,000,000,000
       (*d*) 60,000   (*i*) 160   (*m*) 2,450,000

5.2   MULTIPLYING BY A MULTIPLE OF ANY PLACE VALUE

Find each product.
(*a*) 12×30 (*b*) 12×300 (*c*) $8 \times 4,000$ (*d*) $15 \times 2,000$ (*e*) $800 \times 4,000$

Solutions

(*a*) $12 \times 3 = 36$. Annex 1 zero.   *Ans.* 360   (*d*) $15 \times 2 = 30$. Annex 3 zeros.   *Ans.* 30,000
(*b*) $12 \times 3 = 36$. Annex 2 zeros.   *Ans.* 3,600   (*e*) $800 \times 4 = 3,200$. Annex 3 zeros.   *Ans.* 3,200,000
(*c*) $8 \times 4 = 32$. Annex 3 zeros.   *Ans.* 32,000

6. DISTRIBUTIVE LAW

> **Distributive Law**
>
> To multiply a number by the sum of two addends, multiply the number by each addend and then add the resulting products.

Thus, $5(6 + 4) = 5 \times 6 + 5 \times 4$, which can be written $5(6 + 4) = 5 \cdot 6 + 5 \cdot 4$. To show that $5(6 + 4) = 5 \cdot 6 + 5 \cdot 4$, let us find the total earnings of Mr. Able, who is paid \$5 an hour for his work. If he works for 6 hours the first day and then 4 hours the second day, his earnings can be calculated in two ways:

(1) He worked for $6 + 4$, or 10, hours. Hence, he has earned $\$5(6 + 4)$, or \$50.
(2) The first day, he earned $\$5 \cdot 6$, or \$30. The second day, he learned $\$5 \cdot 4$, or \$20. Hence, he earned $\$5 \cdot 6 + \$5 \cdot 4$, or \$50.

Since either method gives the same result, $5(6 + 4) = 5 \cdot 6 + 5 \cdot 4$.

In general, for any three numbers a, b, and c: $a(b + c) = ab + ac$.

Extending the Distributive Law

The Distributive Law may be extended to cover the multiplication of a number by the sum of three or more addends. Thus,

$$5(2 + 3 + 4) = 5 \cdot 2 + 5 \cdot 3 + 5 \cdot 4 \qquad 10(1 + 2 + 3 + 4) = 10 \cdot 1 + 10 \cdot 2 + 10 \cdot 3 + 10 \cdot 4$$
$$5(9) = 10 + 15 + 20 \qquad\qquad 10(10) = 10 + 20 + 30 + 40$$
$$45 = 45 \qquad\qquad\qquad 100 = 100$$

6.1   APPLYING THE DISTRIBUTIVE LAW

Apply the Distributive Law to find each product.

(a) $2 \times (3 + 6)$ (b) $3 \times (4 + 8)$ (c) $10 \times (4 + 6)$ (d) $5(10 + 2)$ (e) $5(100 + 2)$

Solutions   Apply $a(b + c) = ab + ac$ to find each answer in two ways.

(a) $2(3 + 6) = 2 \cdot 3 + 2 \cdot 6$ (c) $10(4 + 6) = 10 \cdot 4 + 10 \cdot 6$ (e) $5(100 + 2) = 5 \cdot 100 + 5 \cdot 2$
$\quad 2 \cdot 9 = 6 + 12$ $\quad 10 \cdot 10 = 40 + 60$ $\quad 5 \cdot 102 = 500 + 10$
$\quad\quad 18 = 18$ $\quad\quad 100 = 100$ $\quad\quad 510 = 510$

(b) $3(4 + 8) = 3 \cdot 4 + 3 \cdot 8$ (d) $5(10 + 2) = 5 \cdot 10 + 5 \cdot 2$
$\quad 3 \cdot 12 = 12 + 24$ $\quad 5 \cdot 12 = 50 + 10$
$\quad\quad 36 = 36$ $\quad\quad 60 = 60$

6.2   EXTENDING THE DISTRIBUTIVE LAW

Apply the Extended Distributive Law to find each product.

(a) $10(3 + 4 + 5)$ (b) $2(1 + 3 + 5 + 7)$ (c) $5(2 + 4 + 6 + 7 + 9)$

Solutions   In each, find the answer in two ways.

(a) $10(3 + 4 + 5) = 10 \cdot 3 + 10 \cdot 4 + 10 \cdot 5$ (b) $2(1 + 3 + 5 + 7) = 2 \cdot 1 + 2 \cdot 3 + 2 \cdot 5 + 2 \cdot 7$
$\quad\quad 10 \cdot 12 = 30 + 40 + 50$ $\quad\quad 2 \cdot 16 = 2 + 6 + 10 + 14$
$\quad\quad\quad 120 = 120$ $\quad\quad\quad 32 = 32$

(c) $5(2 + 4 + 6 + 7 + 9) = 5 \cdot 2 + 5 \cdot 4 + 5 \cdot 6 + 5 \cdot 7 + 5 \cdot 9$
$\quad\quad\quad 5 \cdot 28 = 10 + 20 + 30 + 35 + 45$
$\quad\quad\quad\quad 140 = 140$

7. MULTIPLYING BY A ONE-DIGIT NUMBER

Multiplying by a One-Digit Number: No Carry Needed

In the following illustrative solution, the product for 512×3 is found. Note in the procedure how each of the digits of the multiplicand, 512, is multiplied by the one digit of the multiplier. In this case, no carry is needed.

To Multiply by a One-Digit Number: No Carry Needed

Multiply: 512×3.

PROCEDURE	SOLUTION	THINK
1. Arrange the work vertically with units digits aligned:	1. 5 1 2 × 3	Align the units digits 2 and 3.
2. Beginning at the right and going in order to the left, multiply each digit of the multiplicand by the multiplier:	2. 15 3 6 (c) (b)(a) 1,536 *Ans.*	**Multiplication Order** (a) $2 \times 3 = 6$ Align 6 in units column. (b) $1 \times 3 = 3$ Write 3 to left of 6. (c) $5 \times 3 = 15$ Write 15 to left of 3.

Understanding the Multiplication Procedure. The procedure is based on:
  (1) The use of the expanded form of the multiplicand.

$$\text{Thus, } 512 = 500 + 10 + 2 = 5 \text{ hundreds} + 1 \text{ tens} + 2 \text{ units}$$

  (2) The application of the Distributive Law.

$$\text{Thus, } 3(500 + 10 + 2) = 3 \cdot 5 \text{ hundreds} + 3 \cdot 1 \text{ tens} + 3 \cdot 2 \text{ units}$$
$$= 15 \text{ hundreds} + 3 \text{ tens} + 6 \text{ units}$$

Partial Products and Their Basic Products

Each of the products obtained when one of the digits of the multiplicand is multiplied by the multiplier is a *basic product* having no more than two digits. Each basic product combined with its associated place value results in a *partial product*.

Thus, in the illustrative solution above, the resulting product, 1,536, or 15 hundreds + 3 tens + 6 units, is the sum of the partial products 1,500, 30, and 6. The basic products 15, 3, and 6 resulted from the multiplying of one digit by one digit.

Multiplying by a One-Digit Number: Carry Needed

In the following illustrative solution, it is necessary to carry 10 from the units place to the tens place. Note under *THINK* how 3 tens + 1 ten becomes 4 tens.

To Multiply by a One-Digit Number: Carry Needed

Multiply: 514×3.

PROCEDURE	SOLUTION	THINK
1. Arrange the work horizontally with units digits aligned:	1. 5 1 4 × 3	Align the units digits 4 and 3.
2. Beginning at the right and going in order to the left, multiply each digit of the multiplicand by the multiplier. If a product has two digits, bring down the units digit and *carry* the tens digit to the preceding column on the left:	2. [1] Carry 5 1 4 × 3 15 4[1]2 (c) (b)(a) 1,542 *Ans.*	**Multiplication Order** (a) $4 \times 3 = 12$ Since $12 = 1$ ten + 2 units, carry 1 to tens column and write 2 in units column. (b) $1 \times 3 = 3$ 3 tens + 1 ten = 4 tens. Write 4 to the left of 2. (c) $5 \times 3 = 15$ Write 15 to the left of 4.

Important Note: The carry of 1 is written in the above solution. However, once the process is understood, a carry should be done mentally instead.

Illustrative Solution   Multiply: 957×3.

How to Think When Multiplying

```
  1  2  Carry
  9  5  7
×        3
28 7  2 1
(c)(b)  (a)
2,871  Ans.
```

(a) $7 \times 3 = 21$ Write 1 in units column.
Carry 2 to tens column.

(b) $5 \times 3 = 15$ 15 plus carry of 2 equals 17.   Write 7 in tens column.
Carry 1 to hundreds column.

(c) $9 \times 3 = 27$ 27 plus carry of 1 equals 28.   Write 28 to left of 7.

7.1   **MULTIPLYING BY A ONE-DIGIT NUMBER: NO CARRY NEEDED**

Multiply:

(a) 43×3 (b) 423×2 (c) $50,221 \times 4$ (d) 310×9 (e) $900,303 \times 3$

Solutions

(a)	(b)	(c)	(d)	(e)
43	423	50,221	310	900,303
× 3	× 2	× 4	× 9	× 3
129	846	200,884	2,790	2,700,909

7.2   **MULTIPLYING BY A ONE-DIGIT NUMBER: CARRY NEEDED**

Multiply and show each carry:

(a) 45×7 (b) 56×7 (c) 145×2 (d) 415×6 (e) 609×4 (f) $1,008 \times 7$

Illustrative Solution

```
(b)    4
     5  6
   ×    7
    39 4 2    Ans. 392
```

(i) $6 \times 7 = 42$ Bring down 2 and carry 4.
(ii) $5 \times 7 = 35$ 35 plus carry of 4 equals 39.

Ans. (a)	(c)	(d)	(e)	(f)
3	1	3	3	5
45	145	415	609	1,008
× 7	× 2	× 6	× 4	× 7
315	290	2,490	2,436	7,056

7.3   **MULTIPLYING BY A ONE-DIGIT NUMBER USING DISTRIBUTIVE LAW**

Multiply using expanded forms and the Distributive Law:

(a) 45×7 (b) 56×7 (c) 145×2 (d) 415×6

Solutions

(a) $(40 + 5)7$
$40 \cdot 7 + 5 \cdot 7$
$280 + 35$
315 *Ans.*

(b) $(50 + 6)7$
$50 \cdot 7 + 6 \cdot 7$
$350 + 42$
392 *Ans.*

(c) $(100 + 40 + 5)2$
$100 \cdot 2 + 40 \cdot 2 + 5 \cdot 2$
$200 + 80 + 10$
290 *Ans.*

(d) $(400 + 10 + 5)6$
$400 \cdot 6 + 10 \cdot 6 + 5 \cdot 6$
$2400 + 60 + 30$
$2,490$ *Ans.*

7.4   **MULTIPLYING BY A ONE-DIGIT MULTIPLIER: VARIED TYPES**

Multiply:

(a) 22×3 (b) 13×3 (c) 19×5 (d) 64×7 (e) 82×6 (f) 92×9

(g)	(h)	(i)	(j)	(k)	(l)	(m)
103	123	214	301	425	1,312	5,679
× 2	× 3	× 4	× 9	× 8	× 4	× 9

Illustrative Solution (*f*)   92   Note that the carry of 1 is not written over 9.

$$\frac{\times\ 9}{828}\ Ans.$$

When understood, carrying should be a mental process.

Ans. (*a*) 66   (*b*) 39   (*c*) 95   (*d*) 448   (*e*) 492   (*g*) 206   (*h*) 369   (*i*) 856   (*j*) 2,709
        (*k*) 3,400   (*l*) 5,248   (*m*) 51,111

7.5   PROBLEM SOLVING INVOLVING A ONE-DIGIT MULTIPLIER

(*a*) At a rate of 8 kilometers per minute, how far will a plane fly in (1) 20 minutes, (2) 45 minutes, (3) 1 hour?

(*b*) If the price of each shirt is $5, how much is the cost of (1) 10 shirts, (2) 1 dozen shirts, (3) 1 gross of shirts? (1 gross = 144)

(*c*) If a carton contains 6 cans of soup, how many cans of soup are in (1) 100 cartons, (2) 300 cartons, (3) 4,000 cartons?

(*d*) Find the total score of each of the following football players, counting 6 points for each touchdown and 3 points for each field goal:

	Touchdowns	Field Goals	Points after Touchdown
Player A	12	8	7
Player B	15	4	2
Player C	13	6	4

Illustrative Solutions

(*a*) (3) Since 1 hour = 60 minutes, the plane will fly 60×8 or 480 km.

(*d*)   A: $12(6) + 8(3) + 7 = 72 + 24 + 7 = 103$ *Ans.*   103 points
           B: $15(6) + 4(3) + 2 = 90 + 12 + 2 = 104$ *Ans.*   104 points
           C: $13(6) + 6(3) + 4 = 78 + 18 + 4 = 100$ *Ans.*   100 points

Ans.   (*a*) (1) 160 km   (2) 360 km
         (*b*) (1) $50   (2) $60   (3) $720
         (*c*) (1) 600 cans   (2) 1,800 cans   (3) 24,000 cans

8.   MULTIPLYING NUMBERS OF ANY SIZE

In this section, illustrative solutions are shown of problems in which both multiplicand and multiplier have more than one digit.

To Multiply by a Two-Digit Number

Multiply: 54×32.

PROCEDURE

1. Arrange the work vertically with units digits aligned:

2. Beginning at the right and going in order to the left, multiply the multiplicand by each digit of the multiplier:

3. Add the resulting products:

SOLUTION

1.   54
      $\times\ 32$

Align the units digits 4 and 2.

Multiplication Order

2. (*a*)   108

(*b*)   162

(*a*) $54 \times 2 = 108$

Align 8 in units column.

(*b*) 54×3 tens $= 162$ tens

To allow for tens, write 162 tens *by shifting one place to left.*

3.   1,728   *Ans.*

Checking Multiplication Results by Commutative Law

To check multiplication results, interchange numbers. Thus, check 54×32 by multiplying 32 by 54:

$$
\begin{array}{r}
32 \\
\times\ 54 \\
\hline
128 \\
160 \\
\hline
1728\ \textit{Check}
\end{array}
$$

To Multiply by a Three-Digit Number

Illustrative Solution   Multiply: 742×416.

	742		Align the units digits 2 and 6.
	$\times\ 416$		**Multiplication Order**
(a)	4452	(a) $742 \times 6 = 4,452$	Align 2 in units column.
(b)	742	(b) 742×1 ten $= 742$ tens	Write 742 tens by shifting 742 one place to the left.
(c)	2968	(c) 742×4 hundreds $= 2,968$ hundreds.	Write 2,968 hundreds by shifting 2,968 two places to left.
	308672 *Ans.*	Add the resulting products.	

To Multiply by a Four-Digit Number

Illustrative Solution   Multiply: $342 \times 4,016$.

	342		Align the units digits 2 and 6.
	$\times\ 4016$		**Multiplication Order**
(a)	2052	(a) $342 \times 6 = 2,052$	Align units digit 2 in units column.
(b)	342	(b) 342×1 ten $= 342$ tens.	Write 342 tens by shifting 342 one place to the left.
(c)	13680	(c) 342×40 hundreds $= 13,680$ hundreds.	Write 13,680 hundreds by shifting 13,680 two places to left.
	1373472 *Ans.*	Add the resulting products.	

8.1   MULTIPLYING BY A TWO-DIGIT NUMBER

Multiply:

(a) 95×34 (b) 126×23 (c) 458×49 (d) 407×73 (e) 802×82 (f) $1,412 \times 75$

$$
\begin{array}{llllll}
(g)\ 39 & (h)\ 76 & (i)\ 108 & (j)\ 312 & (k)\ 801 & (l)\ 3572 & (m)\ 3002 \\
\times\ 68 & \times\ 61 & \times\ 55 & \times\ 45 & \times\ 39 & \times\ 28 & \times\ 84
\end{array}
$$

Illustrative Solution

$$
\begin{array}{rl}
(d) \quad 407 & \\
\times\ 73 & \\
\hline
1221 & (407 \times 3 = 1,221) \\
2849 & (407 \times 7 \text{ tens} = 2,849 \text{ tens. } \textit{Shift left} \text{ one place.}) \\
\hline
29711 & \textit{Ans.}
\end{array}
$$

Ans.   (a) 3,230   (b) 2,898   (c) 22,442   (e) 65,764   (f) 105,900   (g) 2,652   (h) 4,636
(i) 5,940   (j) 14,040   (k) 31,239   (l) 100,016   (m) 252,168

8.2   MULTIPLYING BY A THREE-DIGIT NUMBER

Multiply, and check the result by interchanging numbers:
(a) 124×415 (b) 408×812 (c) $8,503 \times 540$

Solutions

(a)
```
      124        Check:    415
    × 415                × 124
      620                 1660
      124                  830
      496                  415
    51460                51460
```
Ans.   51,460

(b)
```
      408        Check:    812
    × 812                × 408
      816                 6496
      408                 3248
     3264               331296
   331296
```
Ans.   331,296

(c)
```
     8503        Check:    540
    × 540                × 8503
    34012                 1620
    42515                27000
  4591620                 4320
                       4591620
```
Ans.   4,591,620

8.3   MULTIPLYING BY A NUMBER HAVING FOUR OR MORE DIGITS

Multiply and check the result in each
(a) 431 × 2,015   (b) 504 × 8,305   (c) 1,357 × 2,468   (d) 12,345 × 20,067

Illustrative Solution   (b)
```
       504      Check:    8305
     × 8305              × 504
      2520               33220
      1512              415250
      4032             4185720
    4185720
```

Ans.   (a) 868,465   (c) 3,349,076   (d) 247,727,115

8.4   PROBLEM SOLVING INVOLVING NUMBERS OF ANY SIZE

(a) At a rate of 400 km per hour, how many km will a plane fly in (1) 11 hours, (2) one-half day, (3) 15 hours?

(b) If a newpaper has 48 pages, how many pages are in (1) 30 newspapers; (2) 45 newspapers; (3) 10 packages, each containing 500 newspapers?

(c) If a car averages 15 miles per gallon, how many miles will a car travel on (1) 20 gallons, (2) 32 gallons, (3) 150 gallons?

(d) If sound travels 1,100 feet per second, how many feet will it travel in (1) 30 seconds, (2) 2 minutes, (3) 1 hour?

(e) Complete the following purchase lists by finding the cost of each group of articles purchased and the total cost:

(1) Merchant's Purchase List

Articles	Number	($) Price	($) Cost
Shirts	18	25	
Ties	30	15	
Jackets	14	75	
Hats	15	30	

Total Cost:

(2) Consumer's Purchase List

Articles	Number	($) Price	($) Cost
Shirts	3	18	
Ties	5	12	
Shoes	2	75	
Suits	3	175	

Total Cost:

Illustrative Solution   (d) (2) Since 2 minutes = 120 seconds, sound will travel 1,100 × 120 or 132,000 feet.

Ans.   (a) (1) 4,400   (2) 4,800   (3) 6,000
(b) (1) 1,440   (2) 2,160   (3) 240,000
(c) (1) 300   (2) 480   (3) 2,250
(d) (1) 33,000   (3) 3,960,000
(e) (1) shirts, $450; ties, $450; jackets, $1,050; hats, $450; total cost, $2,400
    (2) shirts, $54; ties, $60; shoes, $150; suits, $525; total cost, $789

9. RELATING SUBTRACTION AND ADDITION

Terms Used in Subtraction: Minuend, Subtrahend, Difference

If 7 is subtracted from 10, then 10 is the *minuend*, 7 is the *subtrahend*, and 3, the result, is the *difference*. The result in subtraction may be called the *remainder*. Thus,

Minuend	10	(the number being diminished)
Subtrahend	−7	(the number being subtracted)
Difference	3	(the result)

Rule 1:   MINUEND − SUBTRAHEND = DIFFERENCE

Relating Addition and Subtraction Facts

Subtraction can be thought of in terms of addition. For example, the statement "3 is left if 7 is subtracted from 10" is related to the statement "3 added to 7 equals 10." Hence, the subtraction fact $10 - 7 = 3$ is related to the addition fact $3 + 7 = 10$. By interchanging, we obtain the subtraction fact, $10 - 3 = 7$, and the related addition fact, $7 + 3 = 10$.

Set of Four Related Addition and Subtraction Facts

ADDEND + ADDEND = SUM					MINUEND − ONE ADDEND = OTHER ADDEND				
3	+	7	=	10	10	−	7	=	3
7	+	3	=	10	10	−	3	=	7

Important Note: The minuend in a subtraction fact is the sum in each related addition fact. This truth is the basis for Rule 2.

Rule 2:   MINUEND = SUBTRAHEND + DIFFERENCE

Thus, in $6 - 4 = 2$, the minuend, 6, is the sum in $2 + 4 = 6$.

Checking Subtraction Using Addition (Rule 2)

Subtraction fact $55 - 22 = 33$ can be checked by the related addition fact $33 + 22 = 55$. Note below how we may subtract down and check by adding up.

Subtract Down		Add Up
$(55 - 22 = 33)$	$\begin{array}{r} 55 \\ -22 \\ \hline 33 \end{array}$	$(33 + 22 = 55)$

Using an Addition Table to Obtain Subtraction Facts

The interrelation between addition and subtraction enables us to gather sets of four related addition and subtraction facts from an addition table, such as Table 2-5.

Note in the table that the shaded 17-box is the intersection of the 8-row and 9-column, telling us that $8 + 9 = 17$. If we go across the 8-row to 17, we can go up the column to find 9 at the top. In this way, we learn that $17 - 8 = 9$. Now find $9 + 8 = 17$ and also $17 - 9 = 8$.

Table 2-5

+	6	7	8	9	10
6	12	13	14	15	16
7	13	14	15	16	17
8	14	15	16	17	18
9	15	16	17	18	19
10	16	17	18	19	20

9.1   OBTAINING SETS OF RELATED ADDITION AND SUBTRACTION FACTS

For each addition fact, obtain the three related addition and subtraction facts:
(a) $8 + 7 = 15$ (b) $10 + 8 = 18$ (c) $33 + 44 = 77$ (d) $138 + 125 = 263$

Solutions

(a)	(b)	(c)	(d)
$7 + 8 = 15$	$8 + 10 = 18$	$44 + 33 = 77$	$125 + 138 = 263$
$15 - 7 = 8$	$18 - 8 = 10$	$77 - 44 = 33$	$263 - 125 = 138$
$15 - 8 = 7$	$18 - 10 = 8$	$77 - 33 = 44$	$263 - 138 = 125$

9.2   CHECKING SUBTRACTION RESULTS USING ADDITION (Rule 2)

Check each subtraction result by using a related addition fact:

(a)	(b)	(c)	(d)	(e)
77	138	254	$7,310$	$5,895,720$
-33	-35	-121	$-5,300$	$-2,444,310$
44	103	133	$2,010$	$3,451,410$

Solutions

33	35	121	$5,300$	$2,444,310$
$+44$	$+103$	$+133$	$+2,010$	$+3,451,410$
77 ✓	138 ✓	254 ✓	$7,310$ ✓	$5,895,720$ ✓

9.3   FINDING UNKNOWNS IN A SUBTRACTION

Find each unknown using a related addition fact (Rule 2):
(a) $14 - ? = 9$ (c) $9 - ? = 4$ (e) $20 - n = 5$ (g) $n - 5 = 20$
(b) $? - 14 = 9$ (d) $? - 9 = 4$ (f) $n - 20 = 5$ (h) $15 = 20 - n$

Illustrative Solutions

(a) Apply Rule 2; MINUEND = SUBTRAHEND + DIFFERENCE. Since $14 - ? = 9$, then also $14 = ? + 9$. The related addition fact is $14 = 5 + 9$. Hence, $? = 5$.
(b) Since $? - 14 = 9$, then $? = 14 + 9$. Hence, $? = 23$.

Ans.   (c) $? = 5$ (d) $? = 13$ (e) $n = 15$ (f) $n = 25$ (g) $n = 25$ (h) $n = 5$

10.   SUBTRACTING WHOLE NUMBERS

To Subtract: No Replacement Needed

No replacement is needed when each digit of the subtrahend is less than the corresponding digit of the minuend.

Illustrative Problem

Subtract and check: (a) $59 - 23$ (b) $594 - 231$

PROCEDURE   SOLUTIONS

1. Arrange the numbers vertically with units under units, tens under tens, etc.:

59	594
-23	-231

2. Subtract in the units column first, then in the tens column, then in the hundreds column, etc.:

36 *Ans.* 363 *Ans.*

3. Check the result by addition:

36	363
$+23$	$+231$
59 ✓	594 ✓

Understanding Subtraction   To understand subtraction, use the expanded form of the numbers, as follows:

$$
\begin{array}{ll}
(a) & 59 = -50 + 9 \\
& -23 = -20 - 3 \\
& \overline{ 30 + 6} \\
& 36 \leftarrow \quad 36
\end{array}
\qquad
\begin{array}{ll}
(b) & 594 = 500 + 90 + 4 \\
& -231 = -200 - 30 - 1 \\
& \overline{ 300 + 60 + 3} \\
& 363 \leftarrow \quad 363
\end{array}
$$

To Subtract: Replacement Needed

Replacement is needed when any digit of the subtrahend is greater than the corresponding digit of the minuend.

Using Method 1

Subtract and check: (a) $53 - 28$ (b) $531 - 287$

PROCEDURE   SOLUTIONS

1. Arrange the numbers vertically with units under units, tens under tens, etc.:

$$
\begin{array}{r} 53 \\ -28 \end{array}
\qquad
\begin{array}{r} 531 \\ -287 \end{array}
$$

2. Subtract in the units column first, then in the tens column, then in the hundreds column, etc.:

(8 cannot be subtracted from 3.)   (7 cannot be subtracted from 1.)

3. If, in any column, subtraction is not possible, increase the smaller digit by 10 and decrease the digit to its immediate left by 1:

$$
\begin{array}{r} 4 \\ 5^{1}3 \\ -2\ 8 \\ \hline 2\ 5 \end{array} \ \textit{Ans.}
\qquad
\begin{array}{r} 4^{1}2 \\ 5\ 3^{1}1 \\ -2\ 8\ 7 \\ \hline 2\ 4\ 4 \end{array} \ \textit{Ans.}
$$

4. Check the result by addition:

$$
\begin{array}{r} 25 \\ +28 \\ \hline 53 \end{array}
\qquad
\begin{array}{r} 244 \\ +287 \\ \hline 531 \end{array}
$$

Using Method 2

Instead of decreasing a digit of the minuend by 1, *the digit of the subtrahend in the same column is increased 1.* The result is still the same, as shown:

Add 1 to the subtrahend digit. (a)

$$
\begin{array}{r} 5^{1}3 \\ -2\ 8 \\ \hline 1 \\ 2\ 5 \end{array} \ \textit{Ans.}
$$

(b)

$$
\begin{array}{r} 5^{1}3^{1}1 \\ -2\ 8\ 7 \\ \hline 1\ 1\ 4 \\ 2\ 4\ 4 \end{array} \ \textit{Ans.}
$$

Think: $2 + 1 = 3$
or $8 + 1 = 9$.

10.1 To Subtract: No Replacement Needed

Subtract and check each. In (e), use the expanded form to subtract.

$$
(a)\ \begin{array}{r}48 \\ -35\end{array}
\quad (b)\ \begin{array}{r}92 \\ -50\end{array}
\quad (c)\ \begin{array}{r}138 \\ -26\end{array}
\quad (d)\ \begin{array}{r}407 \\ -206\end{array}
\quad (e)\ \begin{array}{r}417 \\ -206\end{array}
$$

Solutions.   Answers and check are shown in each.

$$
(a)\ \begin{array}{r}13\ \textit{Ans.} \\ +35 \\ \hline 48\ \checkmark\end{array}
\quad (b)\ \begin{array}{r}42\ \textit{Ans.} \\ +50 \\ \hline 92\ \checkmark\end{array}
\quad (c)\ \begin{array}{r}112\ \textit{Ans.} \\ +26 \\ \hline 138\ \checkmark\end{array}
\quad (d)\ \begin{array}{r}201\ \textit{Ans.} \\ +206 \\ \hline 407\ \checkmark\end{array}
$$

(e)
$$
\begin{array}{r} 400 + 10 + 7 \\ -200 - 6 \\ \hline 200 + 10 + 1 \\ = 211 \end{array}
\qquad
\begin{array}{r} 211\ \textit{Ans.} \\ +206 \\ \hline 417\ \checkmark \end{array}
$$

10.2 To Subtract: Replacement Needed

Subtract:

$$
(a)\ \begin{array}{r}75 \\ -48\end{array}
\quad (b)\ \begin{array}{r}324 \\ -167\end{array}
\quad (c)\ \begin{array}{r}4,521 \\ -1,807\end{array}
\quad (d)\ \begin{array}{r}24,571 \\ -\ 7,148\end{array}
$$

Solutions

In the following, Method 1 is used. When replacement is needed, the upper digit (minuend) is decreased 1.

(a)
```
       6
    7 ¹5
   -4  8
   -----
    2  7
```

(b)
```
    2 ¹1
    3 2 ¹4
   -1 6  7
   --------
    1 5  7
```

(c)
```
    3    1
    4 ¹5 2 ¹1
   -1 8  0  7
   -----------
    2 7  1  4
```

(d)
```
    1     6
    2 ¹4 5 7 ¹1
   -   7 1 4 8
   -------------
    1 7 4 2 3
```

In the following, Method 2 is used. When replacement is needed, the lower digit (subtrahend) is increased 1.

(a)
```
    7 ¹5
   -4  8
    1
   -----
    2  7
```

(b)
```
    3 ¹2 ¹4
   -1 6  7
    1  1
   --------
    1 5  7
```

(c)
```
    4 ¹5 2 ¹1
   -1 8  0  7
    1     1
   -----------
    2 7  1  4
```

(d)
```
    2 ¹4 5 7 ¹1
   -    7 1 4 8
    1      1
   -------------
    1 7 4 2 3
```

10.3   PROBLEM SOLVING INVOLVING SUBTRACTION

(a) How much higher is the taller hill if the heights of two hills are (1) 567 meters and 203 meters? (2) 567 meters and 283 meters?

(b) How many more pages of a book remain to be read if the book has 305 pages and (1) 101 pages have been read? (2) 147 pages have been read?

(c) How many girls are there in a school if there are (1) 450 boys in an enrollment of 875 students? (2) 450 boys in an enrollment of 826 students?

(d) Find the difference in scores if in a basketball game the scores are (1) 95 points and 72 points, (2) 95 points and 77 points, (3) 102 points and 73 points.

(e) How much heavier is one weight than another if the weights are (1) 604 pounds and 501 pounds? (2) 604 pounds and 512 pounds? (3) 604 pounds and 568 pounds?

(f) How much faster does the fast plane travel if the speeds of two planes are (1) 575 miles per hour and 421 miles per hour? (2) 683 miles per hour and 458 miles per hour? (3) 1,000 feet per second and 572 feet per second?

Illustrative Solutions For (a),

(1)
```
    567            364
   -203           +203
   -----          -----
    374   Ans.     567 ✓
```

(2)
```
     4
    5 ¹6 7         5 ¹6 7          284
   -2  8 3    or  -2  8 3         +283
   --------       --------        -----
    2  8 4         1              567 ✓
                   --------
                   2  8 4  Ans.
```

Ans. (b) (1) 204 pages   (2) 158 pages
   (c) (1) 425 girls   (2) 376 girls
   (d) (1) 23 points   (2) 18 points   (3) 29 points
   (e) (1) 103 pounds   (2) 92 pounds   (3) 36 pounds
   (f) (1) 154 miles per hour   (2) 225 miles per hour   (3) 428 miles per hour

11. RELATING DIVISION AND MULTIPLICATION

Terms Used in Division: Dividend, Divisor, Quotient

If 20 is divided by 4, then 20 is the *dividend*, 4 is the *divisor*, and 5, the result, is the *quotient*.

In $20 \div 4 = 5$ $\begin{cases} \text{20 is the dividend, the number being divided.} \\ \text{4 is the divisor, the number by which we divide.} \\ \text{5 is the quotient, the result in division.} \end{cases}$

Rule 1: DIVIDEND ÷ DIVISOR = QUOTIENT

Relating Multiplication and Division Facts

Division can be thought of in terms of multiplication. For example, the statement "5 results if 20 is divided by 4" is related to the statement "5 multiplied by 4 equals 20." Hence, the division fact $20 \div 4 = 5$ is related to the multiplication fact $5 \times 4 = 20$. By interchanging, we obtain the division fact $20 \div 5 = 4$ and the multiplication fact $4 \times 5 = 20$.

Set of Four Related Multiplication and Division Facts

FACTOR	×	FACTOR	=	PRODUCT		DIVIDEND	÷	ONE FACTOR	=	OTHER FACTOR
4	×	5	=	20		20	÷	5	=	4
5	×	4	=	20		20	÷	4	=	5

Important Note: The dividend of a division fact is the product in each related multiplication fact. This truth is the basis for Rule 2.

Rule 1:   DIVIDEND = DIVISOR × QUOTIENT

Thus, in $20 \div 4 = 5$, the dividend 20 is the product in $5 \times 4 = 20$.

Checking Division by Applying Rule 2

Division fact $24 \div 3 = 8$ can be checked by the related multiplication fact $8 \times 3 = 24$.

Using a Multiplication Table to Obtain Division Facts

The interrelation between multiplication and division enables us to gather sets of four related multiplication and division facts from a multiplication table, such as Table 2-6.

Table 2-6

×	6	7	8	9	10
6	36	42	48	54	60
7	42	49	56	63	70
8	48	56	64	72	80
9	54	63	72	81	90
10	60	70	80	90	100

Note in the multiplication table that the shaded 72-box is the intersection of the 8-row and the 9-column, telling us that $8 \times 9 = 72$. If we go across the 8-row to 72, we can go up the column to find 9 at the top. In this way, we learn that $72 \div 8 = 9$. Now find $9 \times 8 = 72$ and also $72 \div 9 = 8$.

Division by 1

Rule 3:   Dividing any number by 1 results in the same number.

Thus, $57 \div 1 = 57$. In general, if n represents any number, then $n \div 1 = n$.

Rule 4:   If zero is divided by any nonzero number, the quotient is 0.

Thus, $0 \div 99 = 0$. In general, if n is any nonzero number, then $0 \div n = 0$.

Division by Zero Is Impossible

Since division by zero would lead to results that are either valueless or of any value, *division by zero must be considered to be impossible.*

11.1   SETS OF RELATED MULTIPLICATION AND DIVISION FACTS

For each multiplication fact, obtain the three related multiplication and division facts.
(a) $8 \times 7 = 56$ (b) $25 \times 4 = 100$ (c) $33 \times 44 = 1,452$ (d) $138 \times 125 = 17,250$

Solutions

(a) $7 \times 8 = 56$ (b) $4 \times 25 = 100$ (c) $44 \times 33 = 1,452$ (d) $125 \times 138 = 17,250$
     $56 \div 7 = 8$ $100 \div 4 = 25$ $1,452 \div 44 = 33$ $17,250 \div 125 = 138$
     $56 \div 8 = 7$ $100 \div 25 = 4$ $1,452 \div 33 = 44$ $17,250 \div 138 = 125$

11.2   CHECKING DIVISION RESULTS USING MULTIPLICATION (Rule 2)

Check each division result by using a related multiplication fact:
(a) $56 \div 14 = 4$ (b) $425 \div 17 = 25$ (c) $41,162 \div 3,742 = 11$ (d) $67,228 \div 1,372 = 49$

Solutions Below, read "$\underline{?}$" as "should equal."

(a) $4 \times 14 \underline{?} 56$ (b) $25 \times 17 \underline{?} 425$ (c) $11 \times 3,742 \underline{?} 41,162$ (d) $49 \times 1,372 \underline{?} 67,228$
     $56 = 56$ ✓ $425 = 425$ ✓ $41,162 = 41,162$ ✓ $67,228 = 67,228$ ✓

11.3   FINDING UNKNOWNS IN A DIVISION

Find each unknown using a related multiplication fact (Rule 2):
(a) $20 \div ? = 4$ (c) $? \div 20 = 4$ (e) $n \div 30 = 5$ (g) $30 \div n = 10$ (i) $? \div 1 = 0$
(b) $? \div 4 = 20$ (d) $30 \div n = 5$ (f) $n \div 5 = 30$ (h) $n \div 30 = 10$ (j) $0 \div ? = 0$

Illustrative Solutions

(a) Apply Rule 2: DIVIDEND $=$ DIVISOR $\times$ QUOTIENT
    Since $20 \div ? = 4$, then $20 = ? \times 4$.
    Related multiplication fact: $20 = 5 \times 4$. Hence, $? = 5$ *Ans.*
(b) Since $? \div 4 = 20$, then $? = 4 \times 20$. Hence, $? = 80$ *Ans.*

Ans. (c) $? = 80$ (e) $n = 150$ (g) $n = 3$ (i) $? = 0$
        (d) $n = 6$ (f) $n = 150$ (h) $n = 300$ (j) $? =$ any number except 0

11.4   DIVISION BY 1, DIVISION BY 0, DIVISION OF 0

Divide:
(a) $12 \div 0$ (b) $0 \div 2$ (c) $1 \div 1$ (d) $0 \div 1$ (e) $1 \div 0$ (f) $12 \div 1$ (g) $0 \div 0$

Solutions

"Division by zero is impossible" applies to (a), (e), and (g); (b) 0 (Rule 4); (c) 1 (Rule 3); (d) 0 (Rule 3); (f) 12 (Rule 3).

11.5   PROBLEM SOLVING INVOLVING DIVISION

(a) How many dots will each row contain if 20 dots are to be divided equally among (1) 4 rows, (2) 5 rows, (3) 6 rows?
(b) How many rows will there be if 20 dots are divided into rows, each having (1) 4 dots, (2) 5 dots, (3) 3 dots?

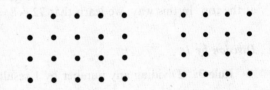

Fig. 2-6   Fig. 2-7

Illustrative Solutions

(a) (1) In Fig. 2-6, the array of 20 dots divided equally among 4 rows shows that each row will contain 5 dots. *Ans.*
(b) (1) In Fig. 2-7, the array of 20 dots divided into rows, each having 4 dots, shows that there will be 5 rows. *Ans.*

Ans. (a) (2) 4 dots   (3) 3 dots and 2 dots remaining
        (b) (2) 4 rows   (3) 6 rows and 2 dots remaining

12. DIVISION USING REPEATED SUBTRACTION

Division as Repeated Subtraction

Just as multiplication may be understood to be repeated addition, division may be understood to be *repeated subtraction*. The process of repeated subtraction may be used to discover that an array of 20 dots arranged in rows of 4 dots each, has 5 such rows (see Fig. 2-7).

To Find a Quotient by Using Repeated Subtraction

How many rows does an array of 20 dots have if each row contains 4 dots?

PROCEDURE

(Each subtraction of 4 dots leaves 1 row less.)
1. Subtract 4 from 20, leaving a remainder of 16:

2. Subtract 4 from 16, leaving a remainder of 12:

3. Subtract 4 from 12, leaving a remainder of 8:

4. Subtract 4 from 8, leaving a remainder of 4:

5. Subtract 4 from the remainder of 4,

    leaving no remainder

SOLUTION

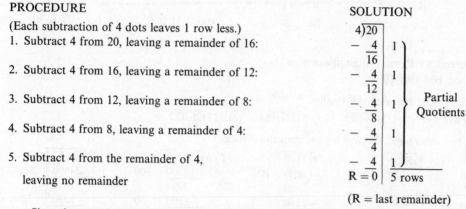

(R = last remainder)

Since there are 5 subtractions and each subtraction leaves 1 row less, there are 5 rows. Since the quotient 5 is the sum of five 1s, each 1 is a *partial quotient*. The division may be checked by using multiplication:

$$5 \times 4 \stackrel{?}{=} 20$$

$$20 = 20 \checkmark$$

Finding a Quotient by Repeated Subtraction
of Multiples of 10 and 100

The process of repeated subtraction becomes very time-consuming if the number of subtractions is a large one. For example, to determine how many tennis cans are needed to hold 729 balls if each can holds 3 balls, the number of repeated subtractions would be 243 if only 3 is subtracted each time!

Instead of subtracting 3 each time, it is better to subtract 300 as many times as possible. Each 300 that is subtracted takes care of the filling of 100 cans. When the remainder is less than 300, subtract 30 as many times as possible. Each 30 that is subtracted takes care of the filling of 10 cans. When the remainder is less than 30, subtract 3 as many times as possible. Note how all this is done in the following solution:

Quotient by Repeated Subtraction of Multiples of 10 and 100

How many tennis cans, each holding 3 balls, can be filled with 729 balls?

PROCEDURE

1. Subtract 300 twice from 729 leaving a remainder of 129.
   (Subtracting 600 takes care of the filling of 200 cans.)
2. Subtract 30 four times from the remainder of 129, leaving a new remainder of 9:
   (Subtracting 120 takes care of the filling of 40 cans.)
3. Subtract 3 three times from the remainder of 9, leaving no remainder:
   (Subtracting 9 takes care of the filling of 3 cans.)

SOLUTION

```
 3)729
  -600 | 200 ⎫
   129 |
  -120 |  40 ⎬ Partial
     9 |      Quotients
  -  9 |   3 ⎭
R = 0 | 243   Ans.   243
```

Since we have taken care of the filling of 200, 40, and 3 cans, the total of cans filled is 243. Since the quotient 243 is the sum of 200, 40, and 3, then 200, 40, and 3 are *partial quotients*.

12.1   FINDING QUOTIENTS BY REPEATED SUBTRACTION OF MULTIPLES OF 10

Find each quotient using repeated subtraction.

(a) 5)65   (b) 7)385   (c) 12)288   (d) 14)1,000   (e) 45)3,495

Solutions   The last remainder, R, must be less than the divisor.

(a) 5)65   (b) 7)385   (c) 12)288   (d) 14)1000   (e) 45)3495
    −50 | 10   −350 | 50   −240 | 20   −980 | 70   −3150 | 70
    ─── | ─── | ─── | ─── | ──── |
     15 | 35 | 48 | 20 | 345 |
    −15 | 3   −35 | 5   −48 | 4   −14 | 1   −315 | 7
    ─── | ─── | ─── | ─── | ──── |
    R = 0 | 13   R = 0 | 55   R = 0 | 24   R = 6 | 71   R = 30 | 77

    Ans. 13   *Ans.* 55   *Ans.* 24   *Ans.* 71 R 6   *Ans.* 77 R 30

12.2   FINDING QUOTIENTS BY REPEATED SUBTRACTION OF MULTIPLES OF 100 AND 10

Find each quotient using repeated subtraction.

(a) 4)568   (b) 9)5,859   (c) 28)5,180   (d) 63)21,184   (e) 123)26,422

Solutions   The last remainder, R, must be less than the divisor.

(a) 4)568   (b) 9)5859   (c) 28)5180   (d) 63)21184   (e) 123)26422
    −400 | 100   −5400 | 600   −2800 | 100   −18900 | 300   −24600 | 200
    ──── | ──── | ──── | ───── | ───── |
     168 | 459 | 2380 | 2284 | 1822 |
    −160 | 40   −450 | 50   −2240 | 80   −1890 | 30   −1230 | 10
    ──── | ──── | ──── | ───── | ───── |
       8 | 9 | 140 | 394 | 592 |
     −8 | 2   −9 | 1   −140 | 5   −378 | 6   −492 | 4
    ──── | ──── | ──── | ───── | ───── |
    R = 0 | 142   R = 0 | 651   R = 0 | 185   R = 16 | 336   R = 100 | 214

    Ans. 142   *Ans.* 651   *Ans.* 185   *Ans.* 336 R 16   *Ans.* 214 R 100

13. LONG DIVISION AND SHORT DIVISION

Long Division

The process of division using repeated subtraction can be simplified into a process of "long division." In the process of long division, partial quotients, subtraction signs, and unnecessary final zeros are omitted.

Comparing Long Division and Division by Repeated Subtraction

Divide 651 ÷ 14.

SOLUTIONS   **By Long Division**   **By Repeated Subtraction**

            46   14)651
        14)651   −560 | 40 ⎱
           56   Final 0 omitted ◄──   91 | ⎰ Partial
           ──   −84 | 6   Quotients
           91   ──── |
           84   R = 7 | 46   *Ans.* 46 R 7
           ──
        R = 7   *Ans.* 46 R 7

Long Division Sequences of Four Steps

In each sequence of a long division process, there are the following four steps:

(1) *Divide*   (2) *Multiply*   (3) *Subtract*   (4) *Bring down the next dividend digit*

As long as there is another digit to be brought down, a new sequence is needed.

The first sequence of four steps used above in dividing 651 by 14 is set forth step by step as follows:

First Sequence Used in Dividing 651 by 14 by Long Division

In the first sequence, 65 is the partial dividend.

Step 1   **Divide** 65 by 14 to obtain 4, the first digit of the quotient.
Place the result 4 over 5, the last digit of 65.

Think:
$$\frac{4 \text{ tens} + \ldots\ldots}{14)\overline{65 \text{ tens} + 1 \text{ unit}}}$$

Write: $14)\overline{651}$ with 4 above

Step 2   **Multiply** 4 by 14 to obtain 56.
Place the result 56 under 65.

Write:
$$\begin{array}{r} 4 \\ 14)\overline{651} \\ 56 \end{array}$$

Step 3   **Subtract** 56 from 65 to obtain 9.
The remainder 9 must be less than the divisor, 14.

Write:
$$\begin{array}{r} 4 \\ 14)\overline{651} \\ 56 \\ \hline 9 \end{array}$$

Step 4   **Bring down 1, the next digit** of 651.
Place the new digit 1 after the remainder 9 obtained in step 3.
(91 becomes the new partial dividend.)

Write:
$$\begin{array}{r} 4 \\ 14)\overline{651} \\ 56 \\ \hline 91 \end{array}$$

You are now ready to learn the generalized long division process. Note how the four-step sequence is applied twice to solve the division of 651 by 14.

To Find a Quotient by Long Division

Divide: $651 \div 14$.

PROCEDURE FOR EACH SEQUENCE:	1st SEQUENCE	2nd SEQUENCE
1. **Divide** a partial dividend by the divisor to obtain a digit of the quotient: (Place the result over the last digit of the partial dividend.)	$14)\overline{651}$ with 4 above (65 is the first partial dividend)	$\begin{array}{r}46\\14)\overline{651}\\56\\\hline 91\end{array}$ (91 is the second partial dividend)
2. **Multiply** the quotient digit obtained in step 1 by the divisor and place the product under the partial dividend: (If the product exceeds the partial dividend, decrease the quotient digit by 1.)	$\begin{array}{r}4\\14)\overline{651}\\56\end{array}$ $(56 = 14 \times 4)$	$\begin{array}{r}46\\14)\overline{651}\\56\\\hline 91\\84\end{array}$ $(84 = 14 \times 6)$
3. **Subtract** the product obtained in step 3 from the partial dividend: (The remainder must be less than the divisor.)	$\begin{array}{r}4\\14)\overline{651}\\56\\\hline 9\end{array}$	$\begin{array}{r}46\\14)\overline{651}\\56\\\hline 91\\84\\\hline\end{array}$ Completed Solution $R = 7$
4. **Bring down the next digit** of the dividend and annex it to the remainder obtained in step 3: (The result becomes the partial dividend for the next sequence.)	$\begin{array}{r}4\\14)\overline{651}\\56\\\hline 91\end{array}$	(The 4th step is not needed in the 2nd sequence since there is no digit to bring down.) *Ans.* 46 R 7

Checking Quotients in Division Problems

Recall Rule 2, the rule used to check quotients in division problems.

Rule 2:   DIVIDEND = DIVISOR × QUOTIENT   (Use Rule 2 when remainder is 0.)

Rule 3:   DIVIDEND = DIVISOR × QUOTIENT + REMAINDER

Thus, to check the result 46 R 7 in the division problem $651 \div 14$, multiply 46×14, then add 7:

$$46 \times 14 + 7 \overset{?}{=} 651$$
$$644 + 7 \overset{?}{=} 651$$
$$651 = 651 \checkmark$$

In the following problem, three sequences are needed for the long division process. In each sequence, keep the four basic steps in mind:

(1) *Divide*   (2) *Multiply*   (3) *Subtract*   (4) *Bring down the next dividend digit*

To Find a Quotient Using Long Division

Divide: $5,901 \div 24$.

1st SEQUENCE	2nd SEQUENCE	3rd SEQUENCE	COMPLETE SOLUTION
$\begin{array}{r} 2 \\ 24\overline{)5901} \\ 48 \\ \hline 110 \end{array}$	$\begin{array}{r} 24 \\ 24\overline{)5901} \\ 48 \\ \hline 110 \\ 96 \\ \hline 141 \end{array}$	$\begin{array}{r} 245 \\ 24\overline{)5901} \\ 48 \\ \hline 110 \\ 96 \\ \hline 141 \\ 120 \\ \hline 21 \end{array}$	$\begin{array}{r} 245 \\ 24\overline{)5901} \\ 48 \\ \hline 110 \\ 96 \\ \hline 141 \\ 120 \\ \hline 21 \end{array}$
(59 is the first partial dividend)	(110 is the second partial dividend)	(141 is the third partial dividend)	*Ans.* 245 R 21

Short Division

Short division can be used when a number is divided by a divisor of one digit. The work is done mentally except for the writing of the digits of the quotient. If the remainders are written, each remainder is written to the left and slightly above the next digit of the dividend.

To Find a Quotient Using Short Division

Divide 2,860 by   (a) 5,   (b) 6,   (c) 7,   (d) 9.

Solutions

(a) $\dfrac{5\ 7\ 2}{5\overline{)28^36^10}}$ *Ans.*   (b) $\dfrac{4\ 7\ 6}{6\overline{)28^46^40}}$ R 4 *Ans.*   (c) $\dfrac{4\ 0\ 8}{7\overline{)28^06^60}}$ R 4 *Ans.*   (d) $\dfrac{3\ 1\ 7}{9\overline{)28^16^70}}$ R 7   *Ans.*

You will find it helpful to compare the short division solutions with the long division solutions of the same problems.

(a) $\begin{array}{r} 572 \\ 5\overline{)2860} \\ 25 \\ \hline 36 \\ 35 \\ \hline 10 \\ 10 \\ \hline R = 0 \end{array}$ (b) $\begin{array}{r} 476 \\ 6\overline{)2860} \\ 24 \\ \hline 46 \\ 42 \\ \hline 40 \\ 36 \\ \hline R = 4 \end{array}$ (c) $\begin{array}{r} 408 \\ 7\overline{)2860} \\ 28 \\ \hline 60 \\ 56 \\ \hline R = 4 \end{array}$ (d) $\begin{array}{r} 317 \\ 9\overline{)2860} \\ 27 \\ \hline 16 \\ 9 \\ \hline 70 \\ 63 \\ \hline R = 7 \end{array}$

572   *Ans.*   476 R 4   *Ans.*   408 R 4   *Ans.*   317 R 7   *Ans.*

Averages

To find the average of two or more numbers, first add the numbers, then divide the sum obtained by the number of the given numbers. Thus, the average of two numbers is their sum divided by 2; the average of three numbers is their sum divided by 3, etc.

For example, the average of 20 and 24 is

$$\frac{20 + 24}{2} = \frac{44}{2} = 22$$

and the average of 20, 24, and 25 is

$$\frac{20 + 24 + 25}{2} = \frac{69}{3} = 23$$

13.1   DIVIDING BY A ONE-DIGIT DIVISOR: LONG DIVISION

Divide:

(a) $7\overline{)567}$ (b) $3\overline{)1,096}$ (c) $6\overline{)244}$ (d) $9\overline{)15,372}$ (e) $8\overline{)163,269}$

Solutions   In each, carefully place the first digit of the quotient over the last digit of the first partial dividend.

(a)
```
      81
7)567
   56
    7
    7
```

(b)
```
      365
3)1096
   9
   19
   18
    16
    15
   R = 1
```

(c)
```
      40
6)244
   24
  R = 4
```

(d)
```
      1708
9)15372
   9
   63
   63
    72
    72
```

(e)
```
      20408
8)163269
   16
    32
    32
     69
     64
    R = 5
```

Ans. 81   Ans. 365 R 1   Ans. 40 R 4   Ans. 1,708   Ans. 20,408 R 5

(Do short division solutions, then compare these with the long division solutions shown.)

13.2   DIVIDING BY A ONE-DIGIT DIVISOR: SHORT DIVISION

Divide 1,890 by each and check:

(a) 2   (b) 3   (c) 5   (d) 7   (e) 8

Solutions C

Ans.   Check using Rule 2 in (a) through (d) and Rule 3 in (e).

(a)
```
    9 4 5
2)1 8⁰9¹0
```
Check:
945 × 2 ≟ 1890
1890 = 1890

(b)
```
    6 3 0
3)1 8⁰9⁰0
```
Check:
630 × 3 ≟ 1890
1890 = 1890

(c)
```
    3 7 8
5)1 8³9⁴0
```
Check:
378 × 5 ≟ 1890
1890 = 1890

(d)
```
    2 7 0
7)1 8⁴9⁰0
```
Check:
270 × 7 ≟ 1890
1890 = 1890

(e)
```
    2 3 6 R 2
8)1 8²9⁵0
```
Check:
236 × 8 + 2 ≟ 1890
1888 + 2 ≟ 1890
1890 = 1890

(Do long division solutions, then compare these with the short division solutions shown.)

13.3   DIVIDING BY A TWO-DIGIT DIVISOR

Divide:

(a) $12\overline{)72}$ (d) $25\overline{)1,425}$ (g) $57\overline{)5,814}$ (j) $89\overline{)41,385}$

(b) $10\overline{)520}$ (e) $34\overline{)2,006}$ (h) $60\overline{)12,300}$ (k) $92\overline{)92,644}$

(c) $15\overline{)1,170}$ (f) $46\overline{)5,658}$ (i) $71\overline{)22,081}$ (l) $98\overline{)1,211,084}$

Illustrative Solutions

(f)
```
          123  Ans.
   46)5,658
      4 6
      1 05
        92
        138
        138
```

(k)
```
        1,007  Ans.
  92)92,644
     92
        644
        644
```

(l)
```
          12,358  Ans.
  98)1,211,084
     98
        231
        196
          35 0
          29 4
           5 68
           4 90
             784
             784
```

Ans. (a) 6   (b) 52   (c) 78   (d) 57   (e) 59   (g) 102   (h) 205   (i) 311   (j) 465

13.4 MORE DIFFICULT DIVISIONS

Divide:

(a) 120)840   (f) 619)60,043   (k) 1,001)627,627   (p) 9,506)47,615,554
(b) 206)4,120   (g) 780)66,300   (l) 1,725)603,750   (q) 10,370)85,511,050
(c) 317)7,925   (h) 853)90,418   (m) 3,640)2,893,800   (r) 24,652)229,682,884
(d) 472)17,936   (i) 909)245,430   (n) 5,922)7,935,480   (s) 53,988)1,091,745,886
(e) 581)35,441   (j) 990)476,190   (o) 7,008)14,506,560

Illustrative Solutions

(k)
```
           627
  1,001)627,627
        600 6
         27 02
         20 02
          7 007
          7 007
```

(q)
```
              8,246
  10,370)85,511,050
         82 960
          2 551 0
          2 074 0
            477 05
            414 80
             62 250
             62 220
                 30
```

(r)
```
                 9,317
  24,652)229,682,884
         221 868
           7 814 8
           7 395 6
             419 28
             246 52
             172 764
             172 564
                 200
```

Ans. 627   *Ans.* 8,246 R 30   *Ans.* 9,317 R 200

Ans. (a) 7   (b) 20   (c) 25   (d) 38   (e) 61   (f) 97   (g) 85   (h) 106   (i) 270   (j) 481
(l) 350   (m) 795   (n) 1,340   (o) 2,070   (p) 5,009   (s) 20,222 R 550

13.5 PROBLEM SOLVING INVOLVING DIVISION

(a) A car averages 15 miles per gallon of gasoline. How many gallons are needed to drive a distance of (1) 480 miles, (2) 525 miles, (3) 1,155 miles.

(b) If each egg box holds 12 eggs, how many boxes are needed for (1) 540 eggs, (2) 1,500 eggs, (3) 3,168 eggs?

(c) A coal truck delivers a total of 21,000 pounds of coal. How many pounds on the average were carried per trip if the number of trips was (1) 5, (2), 6, (3) 7?

(d) At a bond sale, $225,000 worth of bonds were sold. How many bonds were sold if each bond was sold for (1) $25, (2) $50, (3) $150?

(e) Find the average of the following marks: (1) 72 and 94; (2) 72, 94, and 65; (3) 72, 94, 65, and 81.

(f) On a football team, the weights of the seven linemen in pounds are 180, 190, 190, 188, 174, 215, and 172 and the weights of the four backfield men in pounds are 188, 172, 165, and 179. Find the average weight per man (1) of the seven men in the line, (2) of the four men in the backfield, and (3) of the eleven men on the team.

Illustrative Solutions

(a) (1)
$$
\begin{array}{r}
32 \quad Ans. \\
15\overline{)480} \\
45 \\
\overline{30} \\
30 \\
\overline{}
\end{array}
$$

(c) (1)
$$
\begin{array}{r}
4,\ 200 \quad Ans. \\
5\overline{)21,\ ^{1}000}
\end{array}
$$

Ans. (a) (2) 35   (2) 77
     (b) (1) 45   (2) 125   (3) 264
     (c) (2) 3,500   (3) 3,000
     (d) (1) 9,000   (2) 4,500   (3) 1,500
     (e) (1) 83   (2) 77   (3) 78
     (f) (1) 187   (2) 176   (3) 183

14. FACTORS, MULTIPLES, DIVISORS, AND PRIMES

Factor, Multiple, and Divisor

The terms *factor*, *multiple*, and *divisor* are used only with natural numbers. In the multiplication fact $3 \times 4 = 12$, 3 and 4 are each a *factor* of 12. On the other hand, 12 is a *multiple* of 3 and, also, 12 is a *multiple* of 4.

If a number is a factor of another number, then it is a *divisor* of that number. The statements "3 is a factor of 12" and "3 is a divisor of 12" mean the same thing. However, factor and divisor mean the same thing only in *exact division*; that is, in division, where there is no remainder.

Prime and Composite

The terms *prime* and *composite* are used only with natural numbers. A *prime* is a natural number greater than 1 that has no factors other than 1 and itself. Thus, among the natural numbers from 1 to 10, only 2, 3, 5, and 7 are primes. For example, 5 is a prime because its only factors are 1 and 5.

All natural numbers other than 1 and the primes are *composites* or composite numbers. Thus, 6 is a composite, since 2 and 3 are factors of 6. Also, all even numbers except 2 are composites.

Fundamental Theorem of Arithmetic: Prime Factorization

The Fundamental Theorem of Arithmetic states that *every composite number is the unique product of prime factors*.

Thus,

$$12 = 2 \cdot 2 \cdot 3$$

shows that the composite number 12 is the product of the prime factors 2, 2, and 3. (Notice that a particular prime number may be repeated.)

It is also true that

$$12 = 2 \cdot 3 \cdot 2 \quad \text{and} \quad 12 = 3 \cdot 2 \cdot 2$$

since reordering the factors does not change the product. According to the Fundamental Theorem, there are *no further ways* of expressing 12 as the product of prime numbers.

The *prime factorization* of a composite number is the process of finding its unique prime factors. One such process is the "peeling" process, which is applied as follows:

To Find the Prime Factors of a Number

Find the prime factors of: (a) 120   (b) 2,002

PROCEDURE   SOLUTIONS

1. Divide the number by the smallest prime divisor as
   many times as this can be done:

$$
\begin{array}{r} 60 \\ 2\overline{)120} \end{array}
\qquad
\begin{array}{r} 1001 \\ 2\overline{)2002} \end{array}
$$

$$
\begin{array}{r} 30 \\ 2\overline{)60} \end{array}
$$

$$
\begin{array}{r} 15 \\ 2\overline{)30} \end{array}
$$

2. Divide the resulting quotient by *its* smallest prime
   divisor as many times as this can be done, and so on
   until a quotient is reached which is not a composite
   number:

$$
\begin{array}{r} 5 \\ 3\overline{)15} \end{array}
\qquad
\begin{array}{r} 143 \\ 7\overline{)1001} \end{array}
$$

$$
\begin{array}{r} 13 \\ 11\overline{)143} \end{array}
$$

3. Combine all the prime divisors into the prime factors $120 = 2 \cdot 2 \cdot 2 \cdot 3 \cdot 5$ $2,002 = 2 \cdot 7 \cdot 11 \cdot 13$
   of the given number: *Ans.* 2, 2, 2, 3, 5   *Ans.* 2, 7, 11, 13

Highest Common Factor (H.C.F.) or Greatest Common Divisor (G.C.D.)

The *highest common factor (H.C.F.)* or the *greatest common divisor (G.C.D.)* of two or more given numbers is the product of their common prime factors.

Thus, since $36 = 2 \cdot 2 \cdot 3 \cdot 3$ and $90 = 2 \cdot 3 \cdot 3 \cdot 5$, the highest common factor of 36 and 90 is 18, the product of the common factors 2, 3, and 3.

Least Common Multiple (L.C.M.)

A number is a *common multiple* of two or more given numbers if it is a multiple of each of the given numbers.

Thus, the common multiples of 9 and 12 are 36, 72, 108, ... This can be seen if the multiples of 9 and 12 are listed separately and the common multiples are underlined:

Multiples of 12:   12, 24, 36, 48, 60, 72, 84, ...
Multiples of 9:   9, 18, 27, 36, 45, 54, 63, 72, ...

The *least common multiple (L.C.M.)* of two or more given numbers is the least of their common multiples. Thus the L.C.M. of 9 and 12 is 36.

Simple Tests of Divisibility of Natural Numbers

1. A number is divisible by 2 only if its last digit is 0, 2, 4, 6, or 8; that is, its last digit is an even number.
2. A number is divisible by 4 only if the number named by its last two digits is divisible by 4.
   Thus, 912, 2,512, and 10,312 are divisible by 4 since 12 is divisible by 4.
3. A number is divisible by 8 only if the number named by its last three digits is divisible by 8.
   Thus, 1,136, 23,136, and 6,785,136 are divisible by 8 since 136 is divisible by 8.
4. A number is divisible by 3 only if the sum of its digits is divisible by 3.
   Thus, 1,245 is divisible by 3 since $1 + 2 + 4 + 5$, or 12, is divisible by 3.
5. A number is divisible by 9 only if the sum of its digits is divisible by 9.
   Thus, 1,746 is divisible by 9 since $1 + 7 + 4 + 6$, or 18, is divisible by 9.
6. A number is divisible by 5 only if its last digit is 0 or 5.
   Thus, 12,345 and 12,340 are divisible by 5.
7. A number is divisible by 10 only if its last digit is 0.
   Thus, 123,450 is divisible by 10.
8. A number is divisible by 25 only if the number named by its last two digits is divisible by 25.
   Thus, 1,225, 1,250, 1,275, and 1,200 are divisible by 25.

14.1 FACTORS, DIVISORS, AND MULTIPLES

State the set of: (a) factors of (1) 8, (2) 12, (3) 30;   (b) divisors of (1) 18, (2) 42, (3) 105;   (c) two-digit multiples of (1) 10, (2) 12, (3) 15.

Illustrative Solutions

(b) (1) A divisor of a number must be a factor of the number; that is, divisible into the number without remainder. A divisor of a number must be a natural or counting number. The factors of 18 are 1 and the number itself, 18, as well as 2, 3, 6, and 9.

Ans. {1, 2, 3, 6, 9, 18}

(c) (2) A multiple of a number is the product of the number and a natural number. Hence, the set of two-digit multiples of 12 is {12, 24, 36, 48, 60, 72, 84, 96}. *Ans.*

Ans. (a) (1) {1, 2, 4, 8}   (2) {1, 2, 3, 4, 6, 12}   (3) {1, 2, 3, 5, 6, 10, 15, 30}
   (b) (2) {1, 2, 3, 6, 7, 14, 21, 42}   (3) {1, 3, 5, 7, 15, 21, 35, 105}
   (c) (1) {1, 20, 30, 40, 50, 60, 70, 80, 90}   (3) {15, 30, 45, 60, 75, 90}

14.2 PRIMES AND PRIME FACTORS

(a) State the set of prime factors of:
(1) 6   (2) 125   (3) 128   (4) 1,000   (5) 1,001
(b) State each as a product of prime factors:
(1) 70   (2) 100   (3) 120   (4) 279   (5) 256

Illustrative Solution   (a) (4) The prime factors of 1,000 are 2, 2, 2, 5, 5, and 5. However, in a set, members are not repeated. *Ans.* {2, 5}

Ans. (a) (1) {2, 3}   (2) {5}   (3) {2}   (5) {7, 11, 13}
   (b) (1) $2 \cdot 5 \cdot 7$ (2) $2 \cdot 2 \cdot 5 \cdot 5$ (3) $2 \cdot 2 \cdot 2 \cdot 3 \cdot 5$ (4) $3 \cdot 3 \cdot 31$
   (5) $2 \cdot 2 \cdot 2 \cdot 2 \cdot 2 \cdot 2 \cdot 2 \cdot 2$

14.3 H.C.F. AND L.C.M.

Given: $12 = 2 \cdot 2 \cdot 3$ $36 = 2 \cdot 2 \cdot 3 \cdot 3$ $54 = 2 \cdot 3 \cdot 3 \cdot 3$ $60 = 2 \cdot 2 \cdot 3 \cdot 5$
Find the H.C.F. and L.C.M. of the given numbers in each:
(a) 12 and 36   (b) 12 and 54   (c) 36 and 54   (d) 54 and 60   (e) 12, 36, and 54

Illustrative Solution   (c) The H.C.F. of 36 and 54 is the product of their common prime factors. Since these are 2, 3, and 3, H.C.F. = $2 \cdot 3 \cdot 3 = 18$. The L.C.M. of 36 and 54 is the least of their common multiples. By obtaining the multiples of the greater number, 54, note that 108, a multiple of 54, is also a multiple of 36. Hence, the L.C.M. of 36 and 54 is 108.

Ans. (a) H.C.F. = 12,   L.C.M. = 36   (d) H.C.F. = 6,   L.C.M. = 540
   (b) H.C.F. = 6,   L.C.M. = 108   (e) H.C.F. = 6,   L.C.M. = 108

14.4 TESTING DIVISIBILITY BY 2, 4, AND 8

Given:

$h = 5,482$ $m = 5,112$ $q = 71,236$ $s = 1,357,552$ $u = 1,137,542$
$k = 5,412$ $p = 71,136$ $r = 71,246$ $t = 1,137,532$

Which of the given numbers are divisible   (a) by 2 but not by 4 or 8,   (b) by 2 and 4 but not by 8,   (c) by 2, 4, and 8?

Solutions

(a) *Ans.* h, r, u (The last digit of these numbers is divisible by 2. These numbers, however, do not pass the test for divisibility by 4 or by 8.)
(b) *Ans.* k, q, t (The number named by the last two digits of these numbers is divisible by 4. These numbers, however, do not pass the test for divisibility by 8.)
(c) *Ans.* m, p, s (The number named by the last three digits of these numbers is divisible by 8.)

14.5   TESTING DIVISIBILITY BY 3 AND 9

Given:

$h = 5,482$ $m = 5,472$ $q = 71,138$ $s = 1,137,554$ $u = 1,137,552$
$k = 5,412$ $p = 71,136$ $r = 71,241$ $t = 1,137,555$

Which of the given numbers is divisible   (*a*) by 3 but not by 9,   (*b*) by 3 and 9,   (*c*) neither by 3 nor by 9?

Solutions

(*a*) *Ans.*   *k, r, u*   (The sum of the digits is divisible by 3 but not by 9.)
(*b*) *Ans.*   *m, p, t*   (The sum of the digits is divisible by both 3 and 9.)
(*c*) *Ans.*   *h, q, s*   (The sum of the digits is not divisible by 3 or by 9.)

14.6   TESTING DIVISIBILITY BY 5, 10, AND 25

Given:

$h = 5,485$ $m = 5,470$ $q = 71,125$ $s = 1,137,550$ $u = 1,137,540$
$k = 5,475$ $p = 71,130$ $r = 71,115$ $t = 1,137,555$

Which of the given numbers is divisible   (*a*) by 5 but not by 10 or 25,   (*b*) by 5 and 10 but not by 25,
(*c*) by 5 and 25 but not by 10?

Solutions

(*a*) *Ans.*   *h, r, t*   (The last digit is 5 but not 0, and the number named by the last two digits is not divisible by 25.)
(*b*) *Ans.*   *m, p, u*   (The last digit is 0 and the number named by the last two digits is not divisible by 25.)
(*c*) *Ans.*   *k, p, s*   (The number named by the last two digits is divisible by 25 and the last digit is not 0.)

15.   USING A CALCULATOR TO PERFORM THE FUNDAMENTAL OPERATIONS OF ARITHMETIC

The calculator can be used to perform arithmetic calculations with extreme ease. We assume, for this text, that the student will use a basic calculator which accommodates the four basic operations, as well as signed numbers, squares, and square roots. Figure 2-8 illustrates such a calculator.

15.1   SIMPLE CALCULATIONS

Evaluate each using a calculator:
(*a*) $430 + 920$ (*b*) 671×284 (*c*) $29\overline{)5063}$

Solutions

(*a*) Press: | 4 | | 3 | | 0 | | + | | 9 | | 2 | | 0 | | = |

Answer on screen 1,350

(*b*) Press: | 6 | | 7 | | 1 | | * | | 2 | | 8 | | 4 | | = |

Answer on screen

(*c*) Press: | 5 | | 0 | | 6 | | 3 | | ÷ | | 2 | | 9 | | = |

Answer on screen

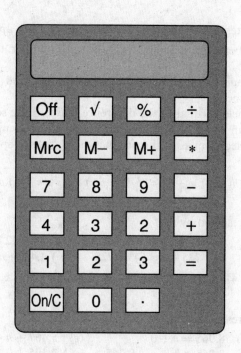

Fig. 2-8

15.2 DIVISIBILITY TESTS

Use a calculator to determine whether: (a) 8754 is divisible by 8   (b) 1234789 is divisible by 7.

Solutions

(a) Press: 8   7   5   4   ÷   8   =

Answer (a) on screen shows the division yields a remainder

(b) Press: 1   2   3   4   7   8   9   ÷   7   =

Answer (b) on screen proves the division yields a remainder

Supplementary Problems

2.1.   State each product without multiplication signs.   **(1.1)**

    (a) $100 \times a$ (c) $10 \times c \times w$ (e) $10 \times 2 \times e \times z$

    (b) $20 \times 4 \times b$ (d) $c \times d \times y$ (f) $3 \times 5 \times f \times w \times z$

    Ans.   (a) $100a$ (b) $80b$ (c) $10cw$ (d) cdy (e) $20ez$ (f) $15fwz$

2.2. Change each verbal statement to an equivalent statement using appropriate letters   **(1.2)**

(*a*) If five times a number is decreased by twice the same number, the result is equivalent to three times the number.

(*b*) If sum of twice a number, three times the same number, and 75 is equal to 75 increased by five times the number.

(*c*) The cost of an article is equal to the difference between the selling price and the profit.

(*d*) The perimeter of an equilateral triangle is three times the length of a side.

(*e*) The distance traveled in miles is the product of the rate of speed in miles per hour and the time in hours.

Ans. (*a*) $5n - 2n = 3n$ (*b*) $2n + 3n + 75 = 75 + 5n$ (*c*) $c = s - p$ (*d*) $p = 3s$ (*e*) $d = rt$

2.3. State each number represented by *n*.   **(2.1)**

(*a*) $20 + 50 = 50 + n$ (*c*) $n + 729 = 729 + 975$ (*e*) $45 + 72 + n = 75 + 45 + 72$

(*b*) $58 + 42 = n + 58$ (*d*) $725 + n = 643 + 725$ (*f*) $1,456 + n + 1,654 = 1,654 + 1,456 + 1,546$

Ans. (*a*) 20 (*b*) 42 (*c*) 975 (*d*) 643 (*e*) 75 (*f*) 1,546

2.4. State each number represented by *n*.   **(2.2)**

(*a*) $n + 0 = 597$ (*c*) $8,965 = n + 0$ (*e*) $5,467 + n + 0 = 6,537 + 5,467$

(*b*) $0 + n = 7,689$ (*d*) $8,965 + 0 = n + 8,965$ (*f*) $2,345 + 0 + n = 0 + 15,000 + 2,345$

Ans. (*a*) 597 (*b*) 7,689 (*c*) 8,965 (*d*) 0 (*e*) 6,537 (*f*) 15,000

2.5. Check each addition using the number line.   **(2.3)**

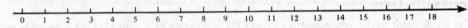

(*a*) $1 + 4 = 5$ (*c*) $10 + 4 = 14$ (*e*) $15 + 5 = 20$ (*g*) $2 + 3 + 5 = 10$

(*b*) $4 + 1 = 5$ (*d*) $15 + 0 = 15$ (*f*) $12 + 6 = 18$ (*h*) $0 + 4 + 13 = 17$

Ans. Each check is left for the student.

2.6. (1) Construct an addition table for even-number addends from 2 to 12. Shade the   **(2.4)** boxes that contain the numbers that are the doubles of an addend. (2) Check each using the addition table:

(*a*) $6 + 6 = 12$ (*c*) $8 + 10 = 18$ (*e*) $10 + 8 = 18$

(*b*) $4 + 8 = 12$ (*d*) $12 + 4 = 16$ (*f*) $10 + 12 = 22$

Ans. (1) See Table 2-7. Entries in shaded boxes are doubles of addends.
(2) Each check is left for the student.

Table 2-7

Addends	2	4	6	8	10	12
2	4	6	8	10	12	14
4	6	8	10	12	14	16
6	8	10	12	14	16	18
8	10	12	14	16	18	20
10	12	14	16	18	20	22
12	14	16	18	20	22	24

2.7. Using two groupings, find each sum without interchanging addends.

(a) $2 + 3 + 7$ (c) $10 + 1 + 9$ (e) $10 + 5 + 5$
(b) $12 + 4 + 3$ (d) $8 + 2 + 6$ (f) $3 + 20 + 7$

Ans. (a) $(2 + 3) + 7 = 2 + (3 + 7)$
$5 + 7 = 2 + 10$
$12 = 12$

(b) $(12 + 4) + 3 = 12 + (4 + 3)$
$16 + 3 = 12 + 7$
$19 = 19$

(c) $(10 + 1) + 9 = 10 + (1 + 9)$
$11 + 9 = 10 + 10$
$20 = 20$

(d) $(8 + 2) + 6 = 8 + (2 + 6)$
$10 + 6 = 8 + 8$
$16 = 16$

(e) $(10 + 5) + 5 = 10 + (5 + 5)$
$15 + 5 = 10 + 10$
$20 = 20$

(f) $(3 + 20) + 7 = 3 + (20 + 7)$
$23 + 7 = 3 + 27$
$30 = 30$

2.8. Add: (3.1)

(a) $21 + 45$ (c) $92 + 35$ (e) $3,407 + 8,391$ (g) $28,514 + 81,364$
(b) $83 + 16$ (d) $218 + 311$ (f) $5,237 + 4,631$ (h) $176,706 + 912,173$

(i)   681   (j)   4,368   (k)   34,713   (l)   618,903   (m)   1,245,713
    + 715   + 7,430   + 44,135   + 761,074   + 9,413,265

Ans. (a) 66 (b) 99 (c) 127 (d) 529 (e) 11,798 (f) 9,868 (g) 109,878 (h) 1,088,879
(i) 1,396 (j) 11,798 (k) 78,848 (l) 1,379,977 (m) 10,658,978

2.9. Add and show each carry: (3.2)

(a)   34   (b)   63   (c)   756   (d)   7,836   (e) 947,394
    + 78   75   + 3,485   8,995   509
             + 67   + 46,304   86,088
                                                    + 3,752

Ans. (a) *1*   (b) *1*   (c) *1 11*   (d) *22 11*   (e) *111 22*
         34   63   756   7,836   947,394
       + 78   75   + 3,485   8,995   509
         112   + 67   4,241   + 46,304   86,088
                  205   63,135   + 3,752
                                                      1,037,743

2.10. Add: **(3.3)**

(a) $3,407 + 5,092 + 1,649 + 9,157$ (b) $63,702 + 7,039 + 85,007 + 123,497$

(c)   8,125,903   (d)   3,615,782   (e)   621,317,468
      4,207,715   3,071   203,725
      5,378,520   72,930   12,015,161
    + 6,401,191   + 15,314,715   4,708
                                                      + 724,971,284

In (a) and (b), align the addends vertically and indicate the check.

Ans.   (a)   19,305   (b)   279,245
            3,407   63,702
            5,092   7,039
            1,649   85,007
          + 9,157   Check   + 123,497   Check
           19,305   279,245

      (c) 24,113,329   (d) 19,006,498   (e) 1,358,512,346

2.11. Figure 2-9 gives distances in kilometers between towns. Find the length of the   **(3.4)**
shortest route between   (a) B and E,   (b) A and D,   (c) B and D,   (d) C and E,   (e) D and
F,   (f) E and A.

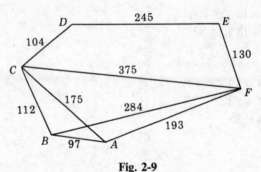

Fig. 2-9

Ans.   (a) 414   (b) 279   (c) 216   (d) 349   (e) 375   (f) 323

2.12. (a) How many employees does a company have if, in its three separate departments,   **(3.5)**
the numbers of employees are: (1) 215, 274, and 539; (2) 3,410, 4,572, and 2,914; (3) 20,450,
48,765, and 62,317?

(b) How many students are there in a school if the numbers of students in each of its four sections
are: (1) 666, 458, 312, and 305; (2) 1,950, 1,764, 1,315, and 960; (3) 1,508, 1,278, 1,114, and
1,093?

(c) What is the total of the sales in a department store if the sales of its three separate departments are: (1) $14,500, $16,780, and $42,535; (2) $135,745, $312,900, and $458,960;
(3) $1,100,000, $2,450,000, and $3,585,000?

(d) What is the total weight of four cargoes if the numbers of kilograms of their separate weights
are: (1) 450, 235, 350, and 750; (2) 1,250, 3,550, 4,750, and 2,150; (3) 13,700, 15,680, 24,593,
and 49,870?

Ans.   (a) (1) 1,028   (2) 10,896   (3) 131,532
       (b) (1) 1,741   (2) 5,989   (3) 4,993
       (c) (1) $73,815   (2) $907,605   (3) $7,135,000
       (d) (1) 1,785 kg   (2) 11,700 kg   (3) 103,843 kg

2.13. State each number represented by n.   **(4.1)**

    (a) $35 \cdot 47 = 47n$ (c) $n \cdot 145 = 145 \cdot 541$ (e) $6 \cdot 16 \cdot 61 = 61 \cdot 6 \cdot n$

    (b) $79 \cdot 97 = n \cdot 97$ (d) $521n = 512 \cdot 521$ (f) $374 \cdot 437 \cdot n = 437 \cdot 734 \cdot 374$

    Ans.   (a) 35   (b) 79   (c) 541   (d) 512   (e) 16   (f) 734

2.14. State each number represented by n.   **(4.2)**

    (a) $1 \cdot n = 517$ (b) $729 = n \cdot 1$ (c) $27 \cdot 72 = 1 \cdot 72n$ (d) $58 \cdot n \cdot 1 = 85 \cdot 58$

    Ans.   (a) 517   (b) 729   (c) 27   (d) 85

2.15. State each number represented by n.   **(4.3)**

    (a) $83 \cdot 0 = n$ (b) $164 \cdot 0 = 1 - n$ (c) $43n = 0$ (d) $n - 4 = 34 \cdot 0$ (e) $0 \cdot n = 0 \cdot 34$

    Ans.   (a) 0   (b) 1   (c) 0   (d) 4   (e) n may represent any number

2.16. Check each multiplication using the number line.   **(4.4)**

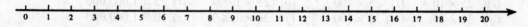

    (a) $3 \times 5 = 15$ (c) $8 \times 1 = 8$ (e) $3 \times 6 = 18$ (g) $2 \times 3 \times 3 = 18$

    (b) $5 \times 3 = 15$ (d) $8 \times 2 = 16$ (f) $4 \times 4 = 16$ (h) $2 \times 2 \times 5 = 20$

    Ans.   Each check is left for the student.

2.17. (1) Construct a multiplication table for odd-number factors from 1 to 11. Shade the   **(4.5)**
boxes that contain the numbers that are the squares of a factor.   (2) Check each using the
multiplication table:

    (a) $3 \times 9 = 27$ (c) $9 \times 11 = 99$ (e) $7 \times 7 = 49$

    (b) $5 \times 9 = 45$ (d) $11 \times 7 = 77$ (f) $9 \times 5 = 45$

    Ans.   (1) See Table 2-8. Shaded boxes contain squares of factors.

Table 2-8

Factors	1	3	5	7	9	11
1	1	3	5	7	9	11
3	3	9	15	21	27	33
5	5	15	25	35	45	55
7	7	21	35	49	63	77
9	9	27	45	63	81	99
11	11	33	55	77	99	121

    (2) Each check is left for the student.

2.18. Using two groupings, find each product without interchanging factors.   **(4.6)**

    (a) $2 \times 3 \times 6$ (c) $11 \times 4 \times 10$ (e) $3 \times 16 \times 2$

    (b) $5 \times 8 \times 7$ (d) $1 \times 20 \times 10$ (f) $16 \times 2 \times 3$

Ans. (a) $(2 \times 3) \times 6 = 2 \times (3 \times 6)$ 　　(d) $(1 \times 20) \times 10 = 1 \times (20 \times 10)$
　　　　　$6 \times 6 = 2 \times 18$ 　　　　　　　　$20 \times 10 = 1 \times 200$
　　　　　$36 = 36$ 　　　　　　　　　　　$200 = 200$

　　(b) $(5 \times 8) \times 7 = 5 \times (8 \times 7)$ 　　(e) $(3 \times 16) \times 2 = 3 \times (16 \times 2)$
　　　　　$40 \times 7 = 5 \times 56$ 　　　　　　　　$48 \times 2 = 3 \times 32$
　　　　　$280 = 280$ 　　　　　　　　　　$96 = 96$

　　(c) $(11 \times 4) \times 10 = 11 \times (4 \times 10)$ 　(f) $(16 \times 2) \times 3 = 16 \times (2 \times 3)$
　　　　　$44 \times 10 = 11 \times 40$ 　　　　　　　$32 \times 3 = 16 \times 6$
　　　　　$440 = 440$ 　　　　　　　　　　$96 = 96$

2.19. Express each in digits:　　　　　　　　　　　　　　　　　　　**(5.1)**

　(a) 15 tens　　(c) 32 thousands　　(e) 42 hundred thousands　(g) 40 billions
　(b) 15 hundreds　(d) 8 ten thousands　(f) 7 millions　　　　(h) 7,892 trillions

Find each product:

　(i) 25×10　(l) 25×100　(o) $25 \times 1,000$　(r) $25 \times 1,000,000$
　(j) 30×10　(m) 30×100　(p) $30 \times 1,000$　(s) $30 \times 1,000,000$
　(k) 425×10　(n) 425×100　(q) $425 \times 1,000$　(t) $425 \times 1,000,000$

Ans. (a) 150　(e) 4,200,000　　(i) 250　(m) 3,000　(q) 425,000
　　(b) 1,500　(f) 7,000,000　　(j) 300　(n) 42,500　(r) 25,000,000
　　(c) 32,000　(g) 40,000,000,000　(k) 4,250　(o) 25,000　(s) 30,000,000
　　(d) 80,000　(h) 7,892,000,000,000,000　(l) 2,500　(p) 30,000　(t) 425,000,000

2.20. Find each product.　　　　　　　　　　　　　　　　　　　**(5.2)**

　(a) 25×20　(e) 11×30　(i) 20×30　(m) 15×40
　(b) 25×40　(f) 11×400　(j) 30×400　(n) $150 \times 2,000$
　(c) 25×800　(g) $11 \times 5,000$　(k) $50 \times 5,000$　(o) $1,500 \times 30,000$
　(d) $25 \times 30,000$　(h) $11 \times 60,000$　(l) $70 \times 60,000$　(p) $15,000 \times 500,000$

Ans. (a) 500　(e) 330　(i) 600　(m) 600
　　(b) 1,000　(f) 4,400　(j) 12,000　(n) 300,000
　　(c) 20,000　(g) 55,000　(k) 250,000　(o) 45,000,000
　　(d) 750,000　(h) 660,000　(l) 4,200,000　(p) 7,500,000,000

2.21. Apply the Distributive Law to find each product.　　　　　　　**(6.1)**

　(a) $3 \times (4 + 6)$　(c) $7 \times (9 + 11)$　(e) $9 \times (100 + 2)$
　(b) $4 \times (8 + 12)$　(d) $8 \times (10 + 3)$　(f) $11 \times (1,000 + 8)$

Ans. (a) $3(4+6) = 3 \cdot 4 + 3 \cdot 6$　　(d) $8(10+3) = 8 \cdot 10 + 8 \cdot 3$
　　　　$3(10) = 12 + 18$ 　　　　　　$8(13) = 80 + 24$
　　　　$30 = 30$ 　　　　　　　　$104 = 104$

　　(b) $4(8+12) = 4 \cdot 8 + 4 \cdot 12$　　$9(100+2) = 9 \cdot 100 + 9 \cdot 2$
　　　　$4(20) = 32 + 48$ 　　(e)　$9(102) = 900 + 18$
　　　　$80 = 80$ 　　　　　　　$918 = 918$

　　(c) $7(9+11) = 7 \cdot 9 + 7 \cdot 11$　(f) $11(1,000+8) = 11 \cdot 1,000 + 11 \cdot 8$
　　　　$7(20) = 63 + 77$ 　　　　$11(1,008) = 11,000 + 88$
　　　　$140 = 140$ 　　　　　　$11,088 = 11,088$

2.22. Apply the Extended Distributive Law to find each product.   **(6.2)**

(a) $3(100 + 60 + 3)$ (b) $11(1,000 + 300 + 70 + 2)$

Ans.   (a) $3(100 + 60 + 3) = 3 \cdot 100 + 3 \cdot 60 + 3 \cdot 3$

$3(163) = 300 + 180 + 9$

$489 = 489$

(b) $11(1,000 + 300 + 70 + 2) = 11 \cdot 1,000 + 11 \cdot 300 + 11 \cdot 70 + 11 \cdot 2$

$11(1,372) = 11,000 + 3,300 + 770 + 22$

$15,092 = 15,092$

2.23. Multiply: **(7.1)**

(a) 34×2 (b) 132×3 (c) 401×5 (d) $7,201 \times 4$ (e) $802,312 \times 3$ (f) $9,201,012 \times 4$

Ans.   (a) 68   (b) 396   (c) 2,005   (d) 28,804   (e) 2,406,936   (f) 36,804,048

2.24. Multiply and show each carry: **(7.2)**

(a) 57×6 (b) 178×9 (c) $2,438 \times 7$ (d) $24,649 \times 5$ (e) $471,693 \times 8$

Ans.

(a)	(b)	(c)	(d)	(e)
4	77	3 25	23 24	515 72
57	178	2,438	24,649	471,693
× 6	× 9	× 7	× 5	× 8
342	1,602	17,066	123,245	3,773,544

2.25. Multiply using expanded forms and the Distributive Law: **(7.3)**

(a) 234×2 (b) 175×8 (c) $2,345 \times 9$

Ans.   (a) $(200 + 30 + 4) \times 2$ (b) $(100 + 70 + 5) \times 8$

$200 \cdot 2 + 30 \cdot 2 + 4 \cdot 2$ $100 \cdot 8 + 70 \cdot 8 + 5 \cdot 8$

$400 + 60 + 8$ $800 + 560 + 40$

468 *Ans.* $1,400$ *Ans.*

(c) $(2,000 + 300 + 40 + 5) \times 9$

$2,000 \cdot 9 + 300 \cdot 9 + 40 \cdot 9 + 5 \cdot 9$

$18,000 + 2,700 + 360 + 45$

$21,105$ *Ans.*

2.26. Multiply: **(7.4)**

(a) 23×3 (d) 58×6 (g) 92×8 (j) 315×4 (m) 839×7

(b) 21×4 (e) 94×5 (h) 123×2 (k) 406×8 (n) 560×9

(c) 18×7 (f) 88×8 (i) 203×3 (l) 725×5

(o)	(p)	(q)	(r)	(s)	(t)
267	2,428	9,004	31,607	80,415	314,117
× 5	× 7	× 6	× 8	× 4	× 9

Ans.   (a) 69   (e) 470   (i) 609   (m) 5,873   (q) 54,024

(b) 84   (f) 792   (j) 1,260   (n) 5,040   (r) 252,856

(c) 126   (g) 736   (k) 3,248   (o) 1,335   (s) 321,660

(d) 348   (h) 246   (l) 3,625   (p) 16,996   (t) 2,827,053

2.27. (a) At a rate of 5 miles per second, how far will a satellite travel in (1) 20 seconds,   **(7.5)**
  (2) 48 seconds, (3) 1 minute?
 (b) If the cost of a tie is $20, how much will the cost be for (1) 20 ties, (2) 2 dozen ties?
 (c) If each package contains 240 bars of soap, how many bars are contained in (1) 20 packages,
  (2) 125 packages, (3) 3,472 packages?

Ans. (a) (1) 100 miles   (2) 240 miles   (3) 300 miles
  (b) (1) $400   (2) $480
  (c) (1) 4,800   (2) 30,000   (3) 833,280

2.28. Multiply: **(8.1)**
 (a) 93×25 (b) 116×32 (c) 304×41 (d) 592×18 (e) $1,304 \times 55$

Ans. (a) 2,325   (b) 3,712   (c) 12,464   (d) 10,656   (e) 71,720

2.29. Multiply and check the result in each: **(8.2)**
 (a) 214×356 (b) 305×704 (c) $8,214 \times 450$ (d) $9,105 \times 20,035$

Ans. (a) 76,184   (b) 214,720   (c) 3,696,300   (d) 182,418,675

2.30. Multiply and check the result in each: **(8.1, 8.2, 8.3)**

 (a) 42×71 (e) 321×145 (i) 903×703
 (b) 54×67 (f) 301×145 (j) $1,204 \times 2,145$
 (c) 125×34 (g) 413×129 (k) $23,450 \times 4,708$
 (d) 241×83 (h) 851×402 (l) 74×35

(m) 81	(n) 152	(o) 302	(p) 710	(q) 824	(r) 2,814	(s) 3,702	(t) 80,314
$\times 64$	$\times 77$	$\times 54$	$\times 88$	$\times 145$	$\times 260$	$\times 1,042$	$\times 2,546$

Ans. (a) 2,982   (e) 46,545   (i) 634,809   (m) 5,184   (q) 119,480
  (b) 3,618   (f) 43,645   (j) 2,582,580   (n) 11,704   (r) 731,640
  (c) 4,250   (g) 53,277   (k) 110,402,600   (o) 16,308   (s) 3,857,484
  (d) 20,003   (h) 342,102   (l) 2,590   (p) 62,480   (t) 204,479,444

2.31. (a) At a rate of 320 miles per hour, how many miles will a plane fly in (1) 10 hours,   **(8.4)**
  (2) 15 hours, (3) one-half day?
 (b) If the cost of a radio set is $50, how much is the cost of (1) 14 sets, (2) 45 sets, (3) 215 sets?
 (c) If each sheet of stamps contains 80 stamps, how many stamps are there in (1) 25 sheets,
  (2) 75 sheets, (3) 150 sheets?
 (d) If each box of oranges contains 64 oranges, how many oranges are there in (1) 35 boxes, (2)
  115 boxes, (3) 3,450 boxes?
 (e) Complete the following purchase lists by finding the cost of each group of articles purchased
  and the total cost:

(1) Merchant's Purchase List

Articles	Number	($) Price	($) Cost
Shoes	20	50	
Suits	12	150	
Coats	8	130	
Ties	35	24	

Total Cost:

(2) Consumer's Purchase List

Articles	Number	($) Price	($) Cost
Ties	4	10	
Gloves	3	36	
Shirts	5	18	
Hats	2	50	

Total Cost:

Ans.   (*a*) (1) 3,200   (2) 4,800   (3) 3,840
　　　(*b*) (1) $700   (2) $2,250   (3) $10,750
　　　(*c*) (1) 2,000   (2) 6,000   (3) 12,000
　　　(*d*) (1) 2,240   (2) 7,360   (3) 220,800
　　　(*e*) (1) shoes, $1,000; suits, $1,800; coats, $1,040; ties, $840; total cost, $4,680
　　　　　(2) ties, $40; gloves, $108; shirts, $90; hats, $100; total cost, $338

2.32.   For each addition fact, obtain the three related addition and subtraction facts: **(9.1)**

(*a*) $9 + 6 = 15$ (*b*) $10 + 22 = 32$ (*c*) $33 + 55 = 88$ (*d*) $107 + 132 = 239$ (*e*) $372 + 425 = 797$

Ans.   (*a*)　$6 + 9 = 15$　　　(*b*)　$22 + 10 = 32$　　(*c*)　$55 + 33 = 88$
　　　　　　$15 - 6 = 9$　　　　　　$32 - 22 = 10$　　　　　$88 - 55 = 33$
　　　　　　$15 - 9 = 6$　　　　　　$32 - 10 = 22$　　　　　$88 - 33 = 55$

　　　　(*d*)　$132 + 107 = 239$　　(*e*)　$425 + 372 = 797$
　　　　　　$239 - 132 = 107$　　　　$797 - 425 = 372$
　　　　　　$239 - 107 = 132$　　　　$797 - 372 = 425$

2.33.   Check each subtraction result by using a related addition fact: **(9.2)**

(*a*)	(*b*)	(*c*)	(*d*)	(*e*)	(*f*)	(*g*)	(*h*)
15	32	88	239	797	8,576	8,341	45,892
−9	−22	−55	−132	−372	−3,241	−4,579	−31,285
6	10	33	107	425	5,335	3,762	14,607

Ans.   (*a*)　6　　　(*c*)　33　　　(*e*)　425　　(*g*)　3,762
　　　　　+9　　　　　+55　　　　+372　　　　+4,579
　　　　　15 ✓　　　　88 ✓　　　　797 ✓　　　　8,341 ✓

　　　　(*b*)　10　　　(*d*)　107　　(*f*)　5,335　(*h*)　14,607
　　　　　+22　　　　+132　　　　+3,241　　　+31,285
　　　　　32 ✓　　　　239 ✓　　　8,576 ✓　　　45,892 ✓

2.34.   Find the missing number by using a related addition fact: **(9.3)**

(*a*) $30 - ? = 10$　　(*c*) $9 - ? = 5$　　(*e*) $25 - n = 15$　　(*g*) $15 = 20 - n$　　(*i*) $20 = n - 15$
(*b*) $? - 30 = 10$　　(*d*) $? - 9 = 5$　　(*f*) $n - 15 = 25$　　(*h*) $15 = n - 20$

Ans.   (*a*) $30 = 20 + 10$ *Ans.* 20　(*d*) $14 = 9 + 5$ *Ans.* 14　　(*g*) $20 = 5 + 15$ *Ans.* 5
　　　(*b*) $40 = 30 + 10$ *Ans.* 40　(*e*) $25 = 10 + 15$ *Ans.* 10　(*h*) $35 = 20 + 15$ *Ans.* 35
　　　(*c*) $9 = 4 + 5$ *Ans.* 4　　(*f*) $40 = 15 + 25$ *Ans.* 40　(*i*) $35 = 15 + 20$ *Ans.* 35

2.35.   Subtract and check each: **(10.1)**

(*a*)	(*b*)	(*c*)	(*d*)	(*e*)
73	98	285	609	4,780
−21	−60	−31	−105	−2,150

Ans.   (*a*)　52 *Ans.*　(*b*)　38 *Ans.*　(*c*)　254 *Ans.*　(*d*)　504 *Ans.*　(*e*)　2,630 *Ans.*
　　　　　+21　　　　　+60　　　　　+31　　　　　+105　　　　　+2,150
　　　　　73 ✓　　　　98 ✓　　　　285 ✓　　　　609 ✓　　　　4,780 ✓

2.36.   Subtract: **(10.2)**

(*a*)	(*b*)	(*c*)	(*d*)
71	281	4,750	47,684
−23	−35	−2,180	−9,325

Ans.   (*a*) 48   (*b*) 246   (*c*) 2,570   (*d*) 38,359

2.37. Subtract: (10.1, 10.2)

$$
\begin{array}{llllll}
(a) & \begin{array}{r} 46 \\ -25 \end{array} &
(d) & \begin{array}{r} 589 \\ -214 \end{array} &
(g) & \begin{array}{r} 500 \\ -289 \end{array} \\
\end{array}
$$

(a)	46 −25	(d)	589 −214	(g)	500 −289	(j)	9,540 −1,268	(m)	24,700 −11,005
(b)	45 −26	(e)	584 −219	(h)	9,568 −1,240	(k)	24,750 −11,040	(n)	24,000 −11,050
(c)	40 −26	(f)	514 −289	(i)	9,560 −1,248	(l)	24,740 −11,050	(o)	21,030 − 4,750

Ans. (a) 21   (d) 375   (g) 211   (j) 8,272   (m) 13,695
      (b) 19   (e) 365   (h) 8,328   (k) 13,710   (n) 12,950
      (c) 14   (f) 225   (i) 8,312   (l) 13,690   (o) 16,280

2.38. (a) How much higher is the higher of two hills if the heights of the hills in meters are   (10.3)
(1) 684 and 451, (2) 681 and 454, (3) 601 and 454?

(b) How many boys are there in a school if out of an enrollment of 3,000 students, the number of girls is (1) 1,400, (2) 2,049, (3) 2,409?

(c) How much faster does the faster of two planes travel if the speeds of the two planes are (1) 692 miles per hour and 371 miles per hour, (2) 671 miles per hour and 392 miles per hour, (3) 1,000 feet per second and 736 feet per second?

(d) Find the difference in basketball scores if the scores are (21) 96 points and 73 points, (2) 93 points and 76 points, (3) 90 points and 76 points?

(e) How many tables remain unsold if out of a total of 1,500 tables the number sold is (1) 400, (2) 800, (3) 840, (4) 1,284?

Ans. (a) (1) 233 m   (2) 227 m   (3) 147 m
      (b) (1) 1,600   (2) 951   (3) 591
      (c) (1) 321 mph   (2) 279 mph   (3) 264 ft/sec
      (d) (1) 23 points   (2) 17 points   (3) 14 points
      (e) (1) 1,100   (2) 700   (3) 660   (4) 216

2.39. For each multiplication fact, obtain the three related multiplication and divsion facts.   (11.1)

(a) $9 \times 6 = 54$ (b) $11 \times 22 = 242$ (c) $38 \times 55 = 1,815$ (d) $250 \times 175 = 43,750$

Ans. (a) $6 \times 9 = 54$ (b) $22 \times 11 = 242$ (c) $55 \times 33 = 1,815$ (d) $175 \times 250 = 43,750$
          $54 \div 6 = 9$ $242 \div 22 = 11$ $1,815 \div 55 = 33$ $43,750 \div 175 = 250$
          $54 \div 9 = 6$ $242 \div 11 = 22$ $1,815 \div 33 = 55$ $43,750 \div 250 = 175$

2.40. Check each division result by using a related multiplication fact: (11.2)

(a) $54 \div 9 = 6$ (b) $242 \div 22 = 11$ (c) $1,815 \div 33 = 55$ (d) $43,750 \div 250 = 175$

Ans. (a) $9 \times 6 \underset{?}{=} 54$ (b) $22 \times 11 \underset{?}{=} 242$ (c) $33 \times 55 \underset{?}{=} 1,815$ (d) $250 \times 175 \underset{?}{=} 43,750$
          $54 = 54$ ✓ $242 = 242$ ✓ $1,815 = 1,815$ ✓ $43,750 = 43,750$ ✓

2.41. Find each unknown using a related multiplication fact: (11.3)

(a) $50 \div ? = 10$ (d) $40 \div n = 5$ (g) $n \div 12 = 6$
(b) $? \div 10 = 55$ (e) $n \div 5 = 40$ (h) $12 \div n = 6$
(c) $? \div 55 = 10$ (f) $5 = n \div 40$ (i) $n \div 6 = 12$

Ans. (a) $50 = ? \times 10$ Ans. 5   (d) $40 = n \times 5$ Ans. 8   (g) $n = 12 \times 6$ Ans. 72
      (b) $? = 10 \times 55$ Ans. 550   (e) $n = 40 \times 5$ Ans. 200   (h) $12 = n \times 6$ Ans. 2
      (c) $? = 55 \times 10$ Ans. 550   (f) $5 \times 40 = n$ Ans. 200   (i) $n = 6 \times 12$ Ans. 72

2.42. Divide: **(11.4)**

    (a) $18 \div 1$ (b) $18 \div 0$ (c) $0 \div 18$ (d) $1 \div 0$ (e) $0 \div 1$ (f) $0 \div 0$ (g) $0 \div 123{,}456$

    Ans.   (a) 18   (b) no value   (c) 0   (d) no value   (e) 0   (f) no value   (g) 0

2.43. (a) How many dimes will each row contain if 24 dimes are to be divided equally   **(11.5)**
       among (1) 3 rows, (2) 6 rows, (3) 8 rows, (4) 9 rows?

    (b) How many rows will there be if 24 dimes are divided into rows, each having (1) 3 dimes, (2) 6
       dimes, (3) 5 dimes, (4) 7 dimes?

    Ans.   (a) (1) 8 dimes   (2) 4 dimes   (3) 3 dimes   (4) 2 dimes and remainder 6 dimes
          (b) (1) 8 rows   (2) 4 rows   (3) 4 rows and remainder 4 dimes   (4) 3 rows and remainder 3 dimes

2.44. Find each quotient using repeated subtraction.   **(12.1)**

    (a) $4\overline{)92}$ (b) $6\overline{)81}$ (c) $11\overline{)495}$ (d) $21\overline{)1{,}448}$ (e) $35\overline{)2{,}265}$

```
Ans.  (a)  4)92              (b)  6)81             (c)  11)495           (d)  21)1448          (e)  35)2265
          −80 | 20               −60 | 10              −440 | 40             −1260 | 60            −2100 | 60
           12                     21                    55                    188                   165
          −12 |  3               −18 |  3              − 55 |  5             − 168 |  8            − 140 |  4
            0 | 23            R = 3 | 13                  0 | 45           R = 20 | 68           R = 25 | 64
```

        Ans. 23   *Ans.* 13 R 3   *Ans.* 45   *Ans.* 71 R 20   *Ans.* 64 R 25

2.45. Find each quotient using repeated subtraction.   **(12.2)**

    (a) $5\overline{)565}$ (b) $8\overline{)1{,}725}$ (c) $25\overline{)3{,}100}$ (d) $42\overline{)13{,}482}$ (e) $132\overline{)28{,}400}$

```
Ans.  (a)  5)565            (c)  25)3100             (e)  132)28400
          −500 | 100            −2500 | 100             −26400 | 200
            65                    600                     2000
          − 50 |  10           − 500 |  20             − 1320 |  10
            15                    100                      680
          − 15 |   3           − 100 |   4             −  660 |   5
             0 | 113               0 | 124           R = 20 | 215
```

      Ans. 113   *Ans.* 124   *Ans.* 215 R 20

```
      (b)  8)1725            (d)  42)13482
          −1600 | 200            −12600 | 300
            125                    882
          − 80 |  10            − 840 |  20
            45                     42
          − 40 |   5            − 42 |   1
        R = 5 | 215               0 | 321
```

      Ans. 215 R 5   *Ans.* 321

2.46. Divide: **(13.1)**

    (a) $8\overline{)592}$ (b) $5\overline{)2{,}107}$ (c) $6\overline{)3{,}170}$ (d) $9\overline{)31{,}540}$ (e) $8\overline{)60{,}416}$ (f) $7\overline{)3{,}516{,}899}$

    Ans.   (a) 74   (b) 421 R 2   (c) 528 R 2   (d) 3,504 R 4   (e) 7,552   (f) 502,414 R 1

2.47. Divide 2,548 by each and check: **(13.2)**

    (a) 3 (b) 4 (c) 5 (d) 7 (e) 8 (f) 9

Ans. (a) $\dfrac{8\ 4\ 9\,R\,1}{3)25^12^28}$ (c) $\dfrac{5\ 0\ 9\,R\,3}{5)25^04^48}$ (e) $\dfrac{3\ 1\ 8\,R\,4}{8)25^14^68}$

        (b) $\dfrac{6\ 3\ 7}{4)25^12^28}$ (d) $\dfrac{3\ 6\ 4}{7)25^44^28}$ (f) $\dfrac{2\ 8\ 3\,R\,1}{9)25^74^28}$

(Each check is left to the student.)

2.48. Divide: **(13.3)**

    (a) $12)\overline{432}$ (e) $32)\overline{2,688}$ (i) $74)\overline{7,326}$ (m) $91)\overline{31,850}$

    (b) $15)\overline{765}$ (f) $41)\overline{3,608}$ (j) $78)\overline{8,200}$ (n) $95)\overline{45,800}$

    (c) $18)\overline{1,116}$ (g) $55)\overline{5,005}$ (k) $82)\overline{10,004}$ (o) $97)\overline{50,000}$

    (d) $25)\overline{1,875}$ (h) $67)\overline{6,231}$ (l) $89)\overline{21,100}$ (p) $99)\overline{75,000}$

Ans. (a) 36   (e) 84   (i) 99   (m) 350

        (b) 51   (f) 88   (j) 105 R 10   (n) 482 R 10

        (c) 62   (g) 91   (k) 122   (o) 515 R 45

        (d) 75   (h) 93   (l) 237 R 7   (p) 757 R 57

2.49. Divide: **(13.4)**

    (a) $105)\overline{630}$ (g) $295)\overline{11,505}$ (m) $475)\overline{40,000}$ (s) $34,725)\overline{16,042,950}$

    (b) $110)\overline{1,320}$ (h) $305)\overline{13,725}$ (n) $531)\overline{49,400}$ (t) $58,314)\overline{40,994,742}$

    (c) $162)\overline{2,916}$ (i) $365)\overline{17,520}$ (o) $670)\overline{68,340}$ (u) $136,703)\overline{12,166,600}$

    (d) $213)\overline{4,473}$ (j) $391)\overline{19,550}$ (p) $792)\overline{99,000}$ (v) $200,035)\overline{18,203,185}$

    (e) $258)\overline{6,450}$ (k) $413)\overline{24,000}$ (q) $4,275)\overline{782,325}$ (w) $312,007)\overline{312,631,014}$

    (f) $280)\overline{9,520}$ (l) $420)\overline{28,980}$ (r) $5,075)\overline{1,608,775}$

Ans. (a) 6   (d) 21   (g) 39   (j) 50   (m) 84 R 100 (p) 125   (s) 462   (v) 91

        (b) 12   (e) 25   (h) 45   (k) 58 R 46 (n) 93 R 17 (q) 183   (t) 703   (w) 1,002

        (c) 18   (f) 34   (i) 48   (l) 69   (o) 102   (r) 317   (u) 89 R 33

2.50. (a) At an average speed of 45 miles per hour, how many hours does it take to travel a   **(13.5)**
      distance of (1) 945 miles, (2) 1,620 miles, (3) 2,340 miles?

    (b) How much does a worker earn each week if his yearly salary is (1) $16,480, (2) $65,000,
      (3) $71,500?

    (c) Profits for the year of $306,000 are to be divided equally among stockholders. How much will
      each stockholder receive if the number of them is (1) 68, (2) 150, (3) 225?

    (d) How many pieces of lumber can be cut from a length of lumber that is 336 inches if the
      number of cm in each piece is (1) 6, (2) 8, (3) 14, (4) 21?

Ans. (a) (1) 21 (2) 36 (3) 52

        (b) (1) $315 (2) $1,250 (3) $1,375

        (c) (1) $4,500 (2) $2,040 (3) $1,360

        (d) (1) 56 (2) 42 (3) 24 (4) 16

2.51. State the set of   (*a*) factors of (1) 10, (2) 18, (3) 32;   (*b*) divisors of (1) 21, (2), 36,   **(14.1)**
(3) 117;   (*c*) two-digit multiples of (1) 11, (2) 17, (3) 25.

> *Ans.*   (*a*) (1) {1, 2, 5, 10}   (2) {1, 2, 3, 6, 9, 18}   (3) {1, 2, 4, 8, 16, 32}
> (*b*) (1) {1, 3, 7, 21}   (2) {1, 2, 3, 4, 6, 9, 12, 18, 36}   (3) {1, 3, 9, 13, 39, 117}
> (*c*) (1) {11, 22, 33, 44, 55, 66, 77, 88, 99}   (2) {17, 34, 51, 68, 85}   (3) {25, 50, 75}

2.52. (*a*) State the set of prime factors of (1) 12, (2) 105, (3) 660, (4) 9,009,   **(14.2)**
(5) 1,000,000.   (*b*) State each as the product of prime factors: (1) 12, (2) 105, (3) 660, (4) 9,009,
(5) 1,000,000.

> *Ans.*   (*a*) (1) {2, 3}   (2) {3, 5, 7}   (3) {2, 3, 5, 11}   (4) {3, 7, 11, 13}   (5) {2, 5}
> (*b*) (1) $2 \cdot 2 \cdot 3$ (2) $3 \cdot 5 \cdot 7$ (3) $2 \cdot 2 \cdot 3 \cdot 5 \cdot 11$ (4) $3 \cdot 3 \cdot 7 \cdot 11 \cdot 13$
> (5) $2 \cdot 2 \cdot 2 \cdot 2 \cdot 2 \cdot 2 \cdot 5 \cdot 5 \cdot 5 \cdot 5 \cdot 5 \cdot 5$

2.53. Given: $12 = 2 \cdot 2 \cdot 3$, $27 = 3 \cdot 3 \cdot 3$, $30 = 2 \cdot 3 \cdot 5$, and $42 = 2 \cdot 3 \cdot 7$. Find the H.C.F.   **(14.3)**
and L.C.M. of each given pair of numbers:

(*a*) 12 and 27   (*b*) 12 and 30   (*c*) 27 and 30   (*d*) 27 and 42   (*e*) 30 and 42

> *Ans.*   (*a*) H.C.F. = 3, L.C.M. = 108   (*c*) H.C.F. = 3, L.C.M. = 270   (*e*) H.C.F. = 6, L.C.M. = 210
> (*b*) H.C.F. = 6, L.C.M. = 60   (*d*) H.C.F. = 3, L.C.M. = 378

2.54. Given: $a = 86,364$, $b = 86,634$, $c = 86,432$, $d = 125,482$, $e = 125,824$, $f = 125,428$.   **(14.4)**
Which of the given numbers are divisible   (*a*) by 2 but not by 4 or 8,   (*b*) by 2 and 4 but not
by 8,   (*c*) by 2, 4, and 8?

> *Ans.*   (*a*) *b*, *d*   (*b*) *a*, *f*   (*c*) *c*, *e*

2.55. Given: $h = 7,812$, $i = 46,032$, $j = 40,632$, $k = 111,303$, $l = 131,108$, $m = 2,012,022$,   **(14.5)**
$n = 2,012,111$. Which of the given numbers are divisible   (*a*) by 3 but not by 9,   (*b*) by 3 and
9,   (*c*) neither by 3 nor by 9?

> *Ans.*   (*a*) *i*, *j*   (*b*) *h*, *k*, *m*   (*c*) *l*, *n*

2.56. Given: $p = 7,215$, $q = 5,120$, $r = 25,255$, $s = 75,075$, $t = 300,010$, $u = 300,125$. Which   **(14.6)**
of the given numbers are divisible   (*a*) by 5 but not by 10 or 25,   (*b*) by 5 and 10 but not by
25,   (*c*) by 5 and 25 but not by 10?

> *Ans.*   (*a*) *p*, *r*   (*b*) *q*, *t*   (*c*) *s*, *u*

2.57. Use a calculator to find: **(15.1)**

(*a*) $5,032 + 41,786$ (*b*) $71,408 - 16,309$
(*c*) $1,060 \times 5,004$ (*d*) $41\overline{)82,041}$

> *Ans.*   (*a*) 46,818   (*b*) 55,099   (*c*) 5,304,240   (*d*) 2,001

2.58. Use a calculator to determine if the following are True or False: **(15.2)**

(*a*)   5,723 is divisible by 7
(*b*)   4,984 is divisible by 3
(*c*)   51,087 is divisible by 3

> *Ans.*   (*a*) F   (*b*) F   (*c*) T

Chapter 3

Fractions

1. UNDERSTANDING FRACTIONS

The word "fraction" comes from a Latin word meaning *broken*. A "broken" whole may be a part of a stick, a part of a loaf of bread, or a part of an hour. For example, we may talk about 3/4 of a stick, 1/2 of a load of bread, or 2/3 of a circle.

Terms of a Fraction: Numerator and Denominator

The terms of a fraction are its *numerator*, the number above the fraction line, and its *denominator*, the number below the fraction line. The denominator may tell us into how many equal parts the object is divided, in which case the numerator tells us the number of equal parts that are being taken.

Thus, 3/4 of a strip of paper means that the strip is divided into four equal parts and three of these parts have been taken. Each of the equal parts is one-quarter, 1/4 of the strip of paper, Fig. 3-1.

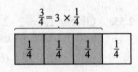

Fig. 3-1

Meanings of a Fraction

1. A fraction may mean *a part of a whole thing* or *a part of a group of things*. Thus, 2/3 may refer to two-thirds of a circle (Fig. 3-2) or to two out of three circles (Fig. 3-3).

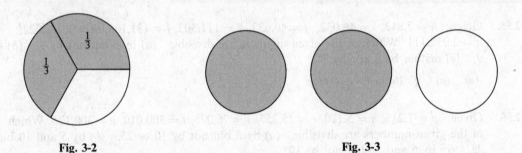

Fig. 3-2　　　　　　　**Fig. 3-3**

2. A fraction may mean *division*. Thus the fraction 3/4 may mean $3 \div 4$ or $4\overline{)3}$. If a division is written as a fraction, the dividend becomes the numerator of the fraction and the divisor becomes its denominator.

3. A fraction may mean *a ratio* of two numbers. To obtain the ratio of two measurements, both measurements must be in the same unit of measure. Thus, the ratio of 1 inch to 1 foot is the ratio of 1 inch to 12 inches. In fraction form, the ratio of 1 inch to 1 foot is 1/12, and in ratio form it is 1:12. (Ratios will be fully treated in Chapter 11.)

Common Fractions, Proper Fractions, Improper Fractions, Unit Fractions

A *common fraction* is a fraction whose numerator is a whole number and whose denominator is a *nonzero* whole number. Thus, 3/4, 4/4, and 5/4 are common fractions.

A *proper fraction* is a common fraction whose numerator is less than its denominator. Thus, 3/4 is a proper fraction. The value of a proper fraction is less than 1.

An *improper fraction* is a common fraction whose numerator is equal to or greater than its denominator. Thus, 5/4 and 4/4 are improper fractions.

$$\frac{5}{4} = 1 + \frac{1}{4} \qquad \frac{4}{4} = 1$$

The value of an improper fraction is 1 or more than 1. A *unit fraction* is a common fraction whose numerator is 1. Any common fraction may be expressed as the product of its numerator and a unit fraction whose denominator is the same as the denominator of the common fraction. Thus, 1/2 and 1/9 are unit fractions. Note in Fig. 3-1 that

$$\frac{3}{4} = 3 \times \frac{1}{4} = 3 \text{ quarters}$$

Mixed Numbers

A *mixed number* is a number having both a whole number part and a fractional part. An improper fraction such as 5/4 may be expressed as a mixed number. Thus, 5/4 is expressible as $1\frac{1}{4}$. The mixed number $4\frac{1}{3}$ equals $4 + \frac{1}{3}$, read, "four and one-third." Hence, a mixed number is the sum of its whole number and its fraction.

Fraction Rules Involving Zero

Since a fraction may mean division, the division rules involving zero apply to fractions.

Rule 1 (Zero Numerator): The value of a fraction is 0 if the numerator is 0 and the denominator is not zero. Thus, $0/10 = 0$. (If zero is divided by a nonzero number, the quotient is 0.)

Rule 2 (Zero Denominator): A fraction has no value if the denominator is 0. (Division by 0 is impossible.)

Rule 3 (Zero Fraction): If the value of a fraction is 0 and the denominator is not 0, the numerator is 0. Thus, if $n/10 = 0$, then $n = 0$.

1.1   EXPRESSING QUANTITIES AS FRACTIONS OR MIXED NUMBERS

Express each of the following as a fraction or a mixed number: (*a*) one-sixth,   (*b*) five-twentieths,   (*c*) four and one-half,   (*d*) $4 \div 5$,   (*e*) 5:9,   (*f*) 7 out of 8,

$$(g) \; 4 \times \frac{1}{7} \qquad (h) \; 10 + \frac{3}{11} \qquad (i) \; 10 + \left(5 \times \frac{1}{7}\right)$$

Illustrative Solutions

(*e*) Convert ratio form 5:9 to fraction form 5/9.   *Ans.*   5/9

(*h*) Combine 10 as the whole number and 3/11 as the fraction of the mixed number.   *Ans.* $10\frac{3}{11}$

Ans.   (*a*) $\dfrac{1}{6}$ (*b*) $\dfrac{5}{20}$ (*c*) $4\frac{1}{2}$ (*d*) $\dfrac{4}{5}$ (*f*) $\dfrac{7}{8}$ (*g*) $\dfrac{4}{7}$ (*i*) $10\frac{5}{7}$

1.2   EXPRESSING PART OF A WHOLE FIGURE AS A FRACTION

State the fractions that indicate the shaded and unshaded parts of the various sections of Fig. 3-4.

Illustrative Solution   (*c*) 9/18 indicates the shading of 9 of the 18 boxes, while 1/2 indicates the shading of 1 of every 2 boxes. 9/18 and 1/2 may also indicate the unshaded part.

Ans.

	Shaded Part	Unshaded Part		Shaded Part	Unshaded Part
(*a*)	3/5	2/5	(*d*)	8/21	13/21
(*b*)	5/6	1/6	(*e*)	5/12	7/12

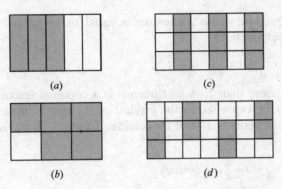

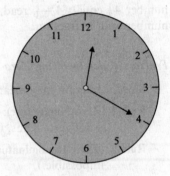

Fig. 3-4

1.3   EXPRESSING A PART OF A GROUP OF THINGS AS A FRACTION

(a) In a class of 25 students, 12 are boys. What part of the class is boys? girls?

(b) In a box of 40 cookies, 25 were eaten. What part of the box was eaten? uneaten?

(c) If 3 of a dozen eggs are broken, what part of the dozen is broken? unbroken?

(d) If 30¢ of a dollar has been spent, what part of the dollar has been spent? unspent?

(e) If 7 ounces of a pound of butter is used for baking, what part of the pound is used? unused?

(f) The time on the clock is 12:20 o'clock, Fig. 3-5. Since 12 o'clock, through what part of a complete turn has the minute hand moved? Through what part of a complete turn will the minute hand move during the rest of the hour?

(g) At 3 P.M., what part of the day has elapsed? remains?

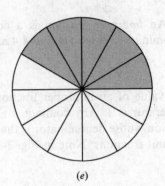

Fig. 3-5

Illustrative Solution   (b) 25/40 is the part of the box that was eaten. We may also use 5/8, since 5 out of every 8 cookies were eaten.   *Ans.*   25/40 or 5/8

Use 15/40 to indicate that 15 out of the 40 cookies were not eaten. You may also use 3/8, since 3 out of every 8 cookies were not eaten.   *Ans.*   15/40 or 3/8

Ans.   (a) 12/25, 13/25   (d) 30/100 or 3/10, 70/100 or 7/10   (f) 20/60 or 1/3, 40/60 or 2/3
            (c) 3/12 or 1/4, 9/12 or 3/4   (e) 7/16, 9/16   (g) 15/24 or 5/8, 9/24 or 3/8

1.4   EXPRESSING DIVISION AS A FRACTION OR VICE VERSA

(a) Express each of the following as a fraction:

(1) 11 divided by 15   (2) 15 ÷ 20   (3) 2)3̄   (4) 3)2̄   (5) 5)1̄

(b) Express each fraction as a division using the division symbol,)‾ :

(1) 7/5   (2) 1/3   (3) 19/25   (4) 25/19

Illustrative Solution

(a)   (2) Make the dividend, 15, the numerator and the divisor, 20, the denominator.   *Ans.*   15/20
(b)   (4) Make the numerator, 25, the dividend and the denominator, 19, the divisor.   *Ans.*   19)25̄

Ans.   (a) (1) 11/15   (3) 3/2   (4) 2/3   (5) 1/5   (b) (1) 5)7̄   (2) 3)1̄   (3) 25)19̄

1.5   EXPRESSING A RATIO AS A FRACTION

(a) An inch is what part of a foot? yard?   (b) The value of a penny is what part of the value of a nickel? a dollar?   (c) The value of a dime is what part of the value of quarter? a half-dollar?   (d) A year is what part of a decade? a century?   (e) Four weeks is what part of a year of 365 days?   (f) 2 meters is what part of 3 kilometers?

Illustrative Solution

(a) $1 \text{ inch}:1 \text{ foot} = \dfrac{1 \text{ inch}}{12 \text{ inches}} = \dfrac{1}{12}$ $1 \text{ inch}:1 \text{ yard} = \dfrac{1 \text{ inch}}{36 \text{ inches}} = \dfrac{1}{36}$

Ans.   (b) 1/5 of a nickel, 1/100 of a dollar   (d) 1/10 of a decade, 1/100 of a century
         (c) 10/25 or 2/5 of a quarter, 10/50 or 1/5 of a half-dollar   (e) 28/365 of a year
         (f) $\dfrac{2}{3,000} = \dfrac{1}{1,500}$ of a kilometer

1.6   SELECTING PROPER FRACTIONS, IMPROPER FRACTIONS, MIXED NUMBERS

Which of the following are   (a) proper fractions,   (b) improper fractions,   (c) mixed numbers?

$$k = \frac{7}{10} \qquad m = \frac{10}{7} \qquad n = 1\tfrac{5}{8} \qquad p = 2\tfrac{3}{5} \qquad q = \frac{23}{5} \qquad r = \frac{5}{23}$$

Solutions

(a) In 7/10 and 5/23, the numerator is less than the denominator. *Ans.* k, r
(b) In 10/7 and 23/5, the numerator is greater than the denominator. *Ans.* m, q
(c) $1\tfrac{5}{8}$ and $2\tfrac{3}{5}$ consist of a whole number and a fraction. *Ans.* n, p

1.7   ZERO NUMERATORS, ZERO DENOMINATORS, FRACTIONS WHOSE VALUE IS ZERO

In (a) to (e), find each value.

$$(a)\ \frac{0}{1} \qquad (b)\ \frac{1}{0} \qquad (c)\ \frac{0}{0} \qquad (d)\ \frac{0}{50} \qquad (e)\ \frac{50}{0}$$

In (f) to (h), find the value of n if the value of the fraction is 0.

$$(f)\ \frac{n}{5} \qquad (g)\ \frac{8-n}{5} \qquad (h)\ \frac{8 \times n}{5}$$

Solutions   Apply fraction rules involving zero.

(a) 0 (Rule 1)   (c) no value (Rule 2)   (e) no value (Rule 2)   (g) 8 (Rule 3)
(b) no value (Rule 2) (d) 0 (Rule 1)   (f) 0 (Rule 3)   (h) 0 (Rule 3)

2. EQUIVALENT FRACTIONS: REDUCTION TO LOWEST TERMS

Equivalent Fractions

Equivalent fractions are fractions having the same value although they have different terms. Thus, since 2/3 = 20/30, then 2/3 and 20/30 are equivalent fractions.

Equivalence Rules

To obtain equivalent fractions, use one of the following rules:

Equivalence Rule 1: The value of a fraction is not changed if its numerator and denominator are both **multiplied by the same number**, excluding zero.

Thus, $3/4 = 30/40$. Here both 3 and 4 are multiplied by 10.
Note: Rule 1 makes use of the multiplicative identity property of 1, as follows:

$$\frac{3}{4} = \frac{3}{4} \cdot 1 = \frac{3}{4} \cdot \frac{10}{10} = \frac{30}{40}$$

Equivalence Rule 2: The value of a fraction is not changed if its numerator and denominator are both **divided by the same number,** excluding zero.

Thus, $30/40 = 3/4$. Here both 30 and 40 are divided by 10.

Reducing a Fraction to Lowest Terms

A fraction is *reduced to lowest terms* when its numerator and denominator have no common factor except 1. A fraction not in lowest terms is equivalent to just one fraction that is in lowest terms. Thus, 40/80 is equivalent to 1/2 when reduced to lowest terms.

To Reduce a Fraction to Lowest Terms

	(*a*) 40/60	(*b*) 63/150
PROCEDURE	**SOLUTIONS**	
1. Factor its numerator and denominator:	$\dfrac{1}{\dfrac{20 \times 2}{20 \times 3}}$	$\dfrac{1}{\dfrac{3 \times 3 \times 7}{3 \times 5 \times 10}}$
2. Divide both terms by every common factor:	$\dfrac{1}{1}$	$\dfrac{1}{1}$
3. Multiply the remaining factors:	$\dfrac{2}{3}$ *Ans.*	$\dfrac{21}{50}$ *Ans.*

2.1   APPLYING EQUIVALENCE RULE 1

Change each of the following to equivalent fractions by multiplying its numerator and denominator by 2, 3, 5, 10, 20, 25, and 100:   (*a*) 2/3,   (*b*) 1/5,   (*c*) 7/10.

Solutions

(*a*) $\dfrac{2}{3} = \dfrac{4}{6} = \dfrac{6}{9} = \dfrac{10}{15} = \dfrac{20}{30} = \dfrac{40}{60} = \dfrac{50}{75} = \dfrac{200}{300}$

(*b*) $\dfrac{1}{5} = \dfrac{2}{10} = \dfrac{3}{15} = \dfrac{5}{25} = \dfrac{10}{50} = \dfrac{20}{100} = \dfrac{25}{125} = \dfrac{100}{500}$

(*c*) $\dfrac{7}{10} = \dfrac{14}{20} = \dfrac{21}{30} = \dfrac{35}{50} = \dfrac{70}{100} = \dfrac{140}{200} = \dfrac{175}{250} = \dfrac{700}{1,000}$

2.2   APPLYING EQUIVALENCE RULE 2

Change each of the following to equivalent fractions by dividing its numerator and denominator by 2, 3, 4, 6, 10, and 12:   (*a*) 60/480,   (*b*) 240/360,   (*c*) 120/900.

Solutions

(*a*) $\dfrac{60}{480} = \dfrac{30}{240} = \dfrac{20}{160} = \dfrac{15}{120} = \dfrac{10}{80} = \dfrac{6}{48} = \dfrac{5}{40}$

(*b*) $\dfrac{240}{360} = \dfrac{120}{180} = \dfrac{80}{120} = \dfrac{60}{90} = \dfrac{40}{60} = \dfrac{24}{36} = \dfrac{20}{30}$

(*c*) $\dfrac{120}{900} = \dfrac{60}{450} = \dfrac{40}{300} = \dfrac{30}{225} = \dfrac{20}{150} = \dfrac{12}{90} = \dfrac{10}{75}$

2.3 CHANGING FRACTIONS TO EQUIVALENT FRACTIONS

In each equation, show how the first fraction can be changed into the equivalent second fraction.

(a) $\dfrac{6}{8} = \dfrac{3}{4}$ (c) $\dfrac{125}{150} = \dfrac{5}{6}$ (e) $\dfrac{70}{98} = \dfrac{5}{7}$

(b) $\dfrac{6}{8} = \dfrac{720}{960}$ (d) $\dfrac{125}{150} = \dfrac{3,750}{4,500}$ (f) $\dfrac{4}{7} = \dfrac{4,400}{7,700}$

Solutions   Either multiply both terms by the same number, Equivalence Rule 1, or divide both terms by the same number, Equivalence Rule 2.

(a) $\dfrac{6 \div 2}{8 \div 2}$ (c) $\dfrac{125 \div 25}{150 \div 25}$ (e) $\dfrac{70 \div 14}{98 \div 14}$

(b) $\dfrac{6 \times 120}{8 \times 120}$ (d) $\dfrac{120 \times 30}{150 \times 30}$ (f) $\dfrac{4 \times 1,100}{7 \times 1,100}$

2.4 REDUCING FRACTIONS TO LOWEST TERMS

Reduce each fraction to lowest terms:

(a) $\dfrac{5}{10}$ (c) $\dfrac{100}{700}$ (e) $\dfrac{1,200}{14,400}$ (g) $\dfrac{1,450}{15,950}$ (i) $\dfrac{90}{60}$ (k) $\dfrac{250}{350}$ (m) $\dfrac{700}{210}$

(b) $\dfrac{6}{36}$ (d) $\dfrac{105}{735}$ (f) $\dfrac{2,345}{7,035}$ (h) $\dfrac{6}{9}$ (j) $\dfrac{15}{25}$ (l) $\dfrac{56}{64}$ (n) $\dfrac{54}{90}$

Illustrative Solution   In (c), divide out the common factor 100:

$$\frac{100}{700} = \frac{\overset{1}{\cancel{100}} \times 1}{\underset{1}{\cancel{100}} \times 7} = \frac{1}{7} \quad Ans.$$

Or, use Equivalence Rule 2:

$$\frac{100 \div 100}{700 \div 100} = \frac{1}{7} \quad Ans.$$

Ans.   (a) $\dfrac{1}{2}$ (d) $\dfrac{1}{7}$ (f) $\dfrac{1}{3}$ (h) $\dfrac{2}{3}$ (j) $\dfrac{3}{5}$ (l) $\dfrac{7}{8}$ (n) $\dfrac{3}{5}$

(b) $\dfrac{1}{6}$ (e) $\dfrac{1}{12}$ (g) $\dfrac{1}{11}$ (i) $\dfrac{3}{2}$ (k) $\dfrac{5}{7}$ (m) $\dfrac{10}{3}$

3. MULTIPLYING FRACTIONS

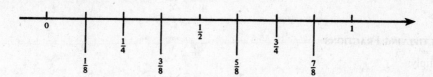

Fig. 3-6   Number Line in 8ths from 0 to 1

Using a Number Line, Fig. 3-6 to Multiply Two Fractions

1. Note that $\dfrac{1}{8}$ is one of two equal parts of $\dfrac{1}{4}$, or $\dfrac{1}{2} \times \dfrac{1}{4}$. Hence, $\dfrac{1}{2} \times \dfrac{1}{4} = \dfrac{1}{8}$.

2. Note that $\dfrac{1}{4}$ is one of three equal parts of $\dfrac{3}{4}$, or $\dfrac{1}{3} \times \dfrac{3}{4}$. Hence, $\dfrac{1}{3} \times \dfrac{3}{4} = \dfrac{1}{4}$.

Fundamental Principle for Multiplication of Fractions

> **To Multiply Fractions**
> (1) Multiply their numerators to obtain the numerator of the product.
> (2) Multiply their denominators to obtain the denominator of the product.

Thus,
$$\frac{1}{2} \times \frac{1}{4} = \frac{1 \times 1}{2 \times 4} = \frac{1}{8}$$

The fundamental multiplication principle can be extended to multiply any number of fractions. Thus,

$$\frac{1}{3} \times \frac{2}{5} \times \frac{7}{11} = \frac{1 \times 2 \times 7}{3 \times 5 \times 11} = \frac{14}{165}$$

Dividing Out (Canceling) Common Factors When Multiplying Fractions

Before applying the Fundamental Multiplication Principle when multiplying fractions, common factors of any numerator and any denominator can be divided out. Thus,

$$\frac{1}{3} \times \frac{\overset{1}{3}}{4} = \frac{1 \times 1}{1 \times 4} = \frac{1}{4} \qquad \frac{\overset{2}{14}}{3} \times \frac{5}{\underset{1}{7}} = \frac{2 \times 5}{3 \times 1} = \frac{10}{3}$$

Expressing a Whole Number as a Fraction

A whole number may be expressed as a fraction whose numerator is the whole number and whose denominator is 1. Thus, $5 = 5/1$. Also, $1 = 1/1$ and $0 = 0/1$.

Multiplying a Whole Number by a Fraction

The product for $5 \times \frac{3}{4}$ is found by expressing 5 as 5/1.

$$\frac{5}{1} \times \frac{3}{4} = \frac{5 \times 3}{1 \times 4} = \frac{15}{4}$$

Finding the product for $5 \times \frac{3}{4}$ can be shortened as follows:

$$\frac{5 \times 3}{4} \times \frac{15}{4}$$

3.1 MULTIPLYING FRACTIONS

Multiply:

(*a*) $\frac{1}{3} \times \frac{7}{8}$ (*d*) $\frac{1}{3} \times \frac{1}{10}$ (*g*) $3 \times \frac{1}{5}$ (*j*) $\frac{4}{5} \times 14$ (*m*) $\frac{1}{2} \times \frac{1}{2} \times \frac{1}{3}$ (*p*) $8 \times \frac{1}{5} \times 4$

(*b*) $\frac{1}{4} \times \frac{5}{2}$ (*e*) $\frac{1}{7} \times \frac{1}{7}$ (*h*) $\frac{2}{7} \times 5$ (*k*) $10 \times \frac{22}{7}$ (*n*) $\frac{1}{2} \times \frac{3}{5} \times \frac{7}{4}$ (*q*) $\frac{1}{2} \times \frac{1}{5} \times 17$

(*c*) $\frac{25}{17} \times \frac{1}{2}$ (*f*) $\frac{1}{9} \times \frac{1}{9}$ (*i*) $\frac{9}{10} \times 3$ (*l*) $\frac{22}{7} \times 4$ (*o*) $4 \times 5 \times \frac{22}{7}$

Illustrative Solution

(*n*) $\frac{1}{2} \times \frac{3}{5} \times \frac{7}{4} = \frac{1 \times 3 \times 7}{2 \times 5 \times 4} = \frac{21}{40}$ *Ans.*

Ans. (a) $\dfrac{7}{24}$ (b) $\dfrac{5}{8}$ (c) $\dfrac{25}{34}$ (d) $\dfrac{1}{30}$ (e) $\dfrac{1}{49}$ (f) $\dfrac{1}{81}$ (g) $\dfrac{3}{5}$ (h) $\dfrac{10}{7}$

(i) $\dfrac{27}{10}$ (j) $\dfrac{56}{5}$ (k) $\dfrac{220}{7}$ (l) $\dfrac{88}{7}$ (m) $\dfrac{1}{12}$ (o) $\dfrac{440}{7}$ (p) $\dfrac{32}{5}$ (q) $\dfrac{17}{10}$

3.2 MULTIPLYING FRACTIONS HAVING CANCELABLE COMMON FACTORS

Multiply:

(a) $\dfrac{1}{2} \times \dfrac{2}{3}$ (b) $\dfrac{3}{5} \times \dfrac{5}{3}$ (c) $\dfrac{31}{57} \times \dfrac{57}{31}$ (d) $\dfrac{137}{250} \times \dfrac{250}{137} \times 5$ (e) $10 \times \dfrac{7}{5}$ (f) $\dfrac{8}{9} \times \dfrac{9}{16}$

(g) $\dfrac{9}{5} \times \dfrac{10}{27}$ (h) $\dfrac{3}{7} \times \dfrac{70}{9}$ (i) $\dfrac{1}{3} \times 12 \times \dfrac{5}{2}$ (j) $100 \times \dfrac{11}{20} \times \dfrac{2}{55}$ (k) $250 \times \dfrac{3}{25} \times \dfrac{9}{10} \times \dfrac{2}{27}$

Illustrative Solution   (d) Cancel common factors 137 and 250; thus,

$$\dfrac{\overset{1}{\cancel{137}}}{\underset{1}{\cancel{250}}} \times \dfrac{\overset{1}{\cancel{250}}}{\underset{1}{\cancel{137}}} \times 5 = 5 \quad Ans.$$

Ans. (a) (Cancel 2), 1/3   (f) (Cancel 8 and 9), 1/2   (j) (Cancel 11 and 20), 2
        (b) (Cancel 3 and 5), 1   (g) (Cancel 5 and 9), 2/3   (k) (Cancel 3, 9, 10, 25), 2
        (c) (Cancel 31 and 57), 1   (h) (Cancel 3 and 7), 10/3
        (e) (Cancel 5), 14   (i) (Cancel 2 and 3), 10

4. DIVIDING WITH FRACTIONS

Using a Number Line to Divide with Fractions

1. In Fig. 3-6, note that if $\dfrac{1}{4}$ is divided into 8ths, the result is 2. Hence, $\dfrac{1}{4} \div \dfrac{1}{8} = 2$.

2. Also, note that if $\dfrac{3}{4}$ is divided into 4ths, the result is 3. Hence, $\dfrac{3}{4} \div \dfrac{1}{4} = 3$.

Reciprocals or Multiplicative Inverses

Reciprocals are two numbers whose product is 1. Reciprocals are also called *multiplicative inverses*. Zero (0) has no reciprocal or multiplicative inverse.

Thus, 2/3 and 3/2 are reciprocals or multiplicative inverses of each other since

$$\dfrac{2}{3} \times \dfrac{3}{2} = 1$$

Also, 4 is the reciprocal of 1/4 since

$$\dfrac{4}{1} \times \dfrac{1}{4} = 1$$

To obtain the reciprocal of a fraction, interchange its terms. Hence, the numerator of a fraction is the denominator of its reciprocal, while the denominator of a fraction is the numerator of its reciprocal. Interchanging the terms of a fraction is called "inverting the fraction." Thus, if 2/3 is inverted, we obtain its reciprocal, 3/2.

Rule 1: To divide by a nonzero fraction, multiply by its reciprocal (invert and multiply).

Thus,
$$\frac{1}{4} \div \frac{1}{8} = \frac{1}{\underset{1}{4}} \times \frac{\overset{2}{8}}{1} = 2 \quad \text{and} \quad \frac{3}{4} \div \frac{1}{4} = \frac{3}{\underset{1}{4}} \times \frac{\overset{1}{4}}{1} = 3$$

These results agree with those found above using the number line.

Simplifying a Complex Fraction

In a *complex fraction*, one or both of the terms are themselves fractions. Thus $5/\frac{1}{2}$, $\frac{1}{4}/3$, and $\frac{1}{4}/\frac{1}{2}$ are complex fractions.

A complex fraction is *simplified* when it is changed into a common fraction of the same value. To simplify a complex fraction, apply the following Rule 2.

Rule 2: To simplify a complex fraction, multiply the numerator by the reciprocal of the denominator.

Thus,
$$\frac{10}{\frac{1}{2}} = 10 \times \frac{2}{1} = 20, \quad \frac{2/3}{4} = \frac{2}{3} \times \frac{\overset{1}{1}}{\underset{2}{4}} = \frac{1}{6}, \quad \frac{\frac{1}{4}}{\frac{1}{2}} = \frac{1}{4} \times \frac{\overset{1}{2}}{1} = \frac{1}{2}$$

Since the complex fraction $\frac{1}{4}/\frac{1}{2}$ can be taken to mean $\frac{1}{4} \div \frac{1}{2}$, Rule 2 is just a restatement of Rule 1.

4.1 RECIPROCALS OR MULTIPLICATIVE INVERSES

In (a) to (d), state the reciprocal or multiplicative inverse of each: (a) 10 (b) $\frac{1}{15}$ (c) $\frac{4}{9}$ (d) $\frac{41}{25}$. In (e) to (k), find the missing number in each:

(e) $20 \times ? = 1$ (g) $\frac{3}{8} \times ? = 1$ (i) $4 \times \frac{1}{4} \times ? = 7$ (k) $13 \times \frac{1}{9} \times ? = 13$

(f) $? \times 50 = 1$ (h) $? \times \frac{20}{21} = 1$ (j) $\frac{6}{7} \times \frac{?}{6} \times 12 = 12$

Illustrative Solutions

(a) Obtain the reciprocal of 10 or 10/1 by inverting terms. *Ans.* 1/10

(j) The product of 6/7 and ?/6 must be 1. Hence, ?/6 is the reciprocal of 6/7. *Ans.* ? = 7

Ans. (b) 15 (c) $\frac{9}{4}$ (d) $\frac{25}{41}$ (e) $\frac{1}{20}$ (f) $\frac{1}{50}$ (g) $\frac{8}{3}$ (h) $\frac{21}{20}$ (i) 7 (k) 9

4.2 DIVIDING WITH FRACTIONS

Divide and multiply as indicated:

(a) $\frac{2}{5} \div 2$ (c) $4 \div \frac{4}{9}$ (e) $\frac{7}{8} \div \frac{1}{8}$ (g) $\frac{5}{8} \div \frac{5}{2}$ (i) $\left(\frac{1}{2} \times 5\right) \div \frac{7}{2}$ (k) $\frac{22}{7} \times \left(7 \div \frac{11}{15}\right)$

(b) $\frac{5}{2} \div 10$ (d) $5 \div \frac{10}{3}$ (f) $\frac{4}{5} \div \frac{2}{5}$ (h) $\frac{1}{3} \div \frac{1}{40}$ (j) $\left(\frac{4}{9} \times \frac{9}{2}\right) \div \frac{2}{3}$

Solutions Apply Rule 1: To divide by a nonzero fraction, invert and multiply.

(a) $\frac{2}{5} \times \frac{1}{2} = \frac{1}{5}$ (d) $5 \times \frac{3}{10} = \frac{3}{2}$ (g) $\frac{5}{8} \times \frac{2}{5} = \frac{1}{4}$ (j) $\frac{4}{9} \times \frac{9}{2} \times \frac{3}{2} = 3$

(b) $\frac{5}{2} \times \frac{1}{10} = \frac{1}{4}$ (e) $\frac{7}{8} \times 8 = 7$ (h) $\frac{1}{3} \times 40 = \frac{40}{3}$ (k) $\frac{22}{7} \times 7 \times \frac{15}{11} = 30$

(c) $4 \times \frac{9}{4} = 9$ (f) $\frac{4}{5} \times \frac{5}{2} = 2$ (i) $\frac{1}{2} \times 5 \times \frac{2}{7} = \frac{5}{7}$

4.3   SIMPLIFYING COMPLEX FRACTIONS

Simplify each.

(a) $\dfrac{1/2}{2}$ (b) $\dfrac{4}{1/2}$ (c) $\dfrac{1/4}{5}$ (d) $\dfrac{10}{1/4}$ (e) $\dfrac{1/2}{1/4}$ (f) $\dfrac{2/3}{2/5}$ (g) $\dfrac{5/9}{2/9}$ (h) $\dfrac{2/5}{3/8}$ (i) $\dfrac{11/15}{22/45}$

Solutions   Apply Rule 2: Multiply the numerator by the reciprocal of the denominator.

(a) $\dfrac{1}{2} \times \dfrac{1}{2} = \dfrac{1}{4}$ (c) $\dfrac{1}{4} \times \dfrac{1}{5} = \dfrac{1}{20}$ (e) $\dfrac{1}{2} \times 4 = 2$ (g) $\dfrac{5}{9} \times \dfrac{9}{2} = \dfrac{5}{2}$ (i) $\dfrac{11}{15} \times \dfrac{45}{22} = \dfrac{3}{2}$

(b) $4 \times \dfrac{2}{1} = 8$ (d) $10 \times \dfrac{4}{1} = 40$ (f) $\dfrac{2}{3} \times \dfrac{5}{2} = \dfrac{5}{3}$ (h) $\dfrac{2}{5} \times \dfrac{8}{3} = \dfrac{16}{15}$

5.   EXPRESSING A FRACTION IN HIGHER TERMS

Expressing a fraction in higher terms means changing the fraction to an equivalent fraction, each of whose terms is a *multiple* of the corresponding term of the given fraction.

Thus, 2/3 may be expressed in higher terms as 4/6, 20/30, or 50/75. Each term of each higher fraction is a multiple of 2 or 3, the terms of the given fraction, 2/3.

Expressing a fraction in higher terms is of the greatest importance in the adding and subtracting of fractions. We will also find it useful when comparing or ordering fractions.

To Express a Fraction in Higher Terms

Express 2/3 as 30ths.

PROCEDURE	SOLUTION
1. Equate the given fraction with its required fraction:	1. $\dfrac{2}{3} = \dfrac{?}{30}$
2. Determine the needed multiplier:	2. To obtain 30, 3 is multiplied by 10.
3. Multiply each term by the multiplier found in step 2:	3. $\dfrac{2}{3} = \dfrac{2 \times 10}{3 \times 10} = \dfrac{20}{30}$ *Ans.*

To Express Fractions as Equivalent Fractions Having a Common Denominator

Express 3/4 and 7/10 as equivalent fractions having 20 as the common denominator.

PROCEDURE	SOLUTION
1. Equate each given fraction with its required fraction:	1. Since 20 is the required common denominator, $\dfrac{3}{4} = \dfrac{?}{20}$ and $\dfrac{7}{10} = \dfrac{?}{20}$
2. Determine the needed multipliers:	2. Since $4 \times 5 = 20$ and $10 \times 2 = 20$, the needed multipliers are 5 and 2.
3. Multiply the terms by the multipliers found in step 2:	3. $\dfrac{3}{4} = \dfrac{3 \times 5}{4 \times 5} = \dfrac{15}{20}$ and $\dfrac{7}{10} = \dfrac{7 \times 2}{10 \times 2} = \dfrac{14}{20}$. *Ans.* $\dfrac{15}{20}, \dfrac{14}{20}$

5.1   EXPRESSING A FRACTION IN HIGHER TERMS

Find each missing term:

(a) $\dfrac{1}{2} = \dfrac{?}{20}$ (b) $\dfrac{3}{5} = \dfrac{30}{?}$ (c) $\dfrac{7}{9} = \dfrac{?}{99}$ (d) $\dfrac{12}{5} = \dfrac{1,200}{?}$ (e) $\dfrac{75}{32} = \dfrac{?}{128}$ (f) $\dfrac{225}{490} = \dfrac{?}{9,800}$

Illustrative Solution (*d*) Since $12 \times 100 = 1,200$, multiply terms by 100:

$$\frac{12}{5} = \frac{12 \times 100}{5 \times 100} = \frac{1,200}{500} \quad Ans.$$

Ans. (*a*) 10 (*b*) 50 (*c*) 77 (*e*) 300 (*f*) 4,500

5.2 EXPRESSING A FRACTION IN HIGHER TERMS GIVEN THE NEW DENOMINATOR

(*a*) Express 3/4 as (1) 16ths, (2) 40ths, (3) 144ths, (4) 4,000ths.
(*b*) Express 21/25 as (1) 100ths, (2) 250ths, (3) 1,000ths, (4) 12,500ths.

Illustrative Solution (*a*) (3) To express 3/4 as 144ths, it may be stated as $3/4 = ?/144$. Since $4 \times 36 = 144$, multiply terms by 36. Hence,

$$\frac{3}{4} = \frac{3 \times 36}{4 \times 36} = \frac{108}{144} \quad Ans.$$

Ans. (*a*) (1) $\frac{12}{16}$ (2) $\frac{30}{40}$ (4) $\frac{3,000}{4,000}$

(*b*) (1) $\frac{84}{100}$ (2) $\frac{210}{250}$ (3) $\frac{848}{1,000}$ (4) $\frac{10,500}{12,500}$

5.3 EXPRESSING FRACTIONS AS EQUIVALENT FRACTIONS HAVING A COMMON DENOMINATOR

In each, express the given fractions as equivalent fractions having the required common denominator.

(*a*) $\frac{1}{2}$ and $\frac{3}{4}$ as 4ths   (*c*) $\frac{7}{15}$ and $\frac{12}{25}$ as 75ths   (*e*) $\frac{1}{3}$, $\frac{5}{6}$, and $\frac{8}{9}$ as 18ths

(*b*) $\frac{1}{6}$ and $\frac{1}{9}$ as 18ths   (*d*) $\frac{7}{8}$ and $\frac{5}{12}$ as 24ths   (*f*) $\frac{3}{2}$, $\frac{13}{12}$, and $\frac{11}{15}$ as 60ths

Illustrative Solution (*c*) The required common denominator is the *least common multiple* (Chapter 2, Section 14) of both denominators. Since 75 is the least common multiple of 15 and 25, multiply terms of 7/15 by 5 and terms of 12/25 by 3:

$$\frac{7}{15} = \frac{7 \times 5}{15 \times 5} = \frac{35}{75} \quad Ans. \qquad \frac{12}{25} = \frac{12 \times 3}{25 \times 3} = \frac{36}{75} \quad Ans.$$

Ans. (*a*) $\frac{2}{4}$, $\frac{3}{4}$ (*b*) $\frac{3}{18}$, $\frac{2}{18}$ (*d*) $\frac{21}{24}$, $\frac{10}{24}$ (*e*) $\frac{6}{18}$, $\frac{15}{18}$, $\frac{16}{18}$ (*f*) $\frac{90}{60}$, $\frac{65}{60}$, $\frac{44}{60}$

6. ADDING OR SUBTRACTING LIKE FRACTIONS

Like fractions are fractions having a common denominator. *Unlike fractions* are fractions having different denominators. Thus, 3/8 and 5/8 are like fractions, whereas 3/8 and 5/9 are unlike fractions. Like fractions are combined (added or subtracted) by applying the following rule:

Rule: Combine like fractions into a single fraction by keeping the common denominator and combining numerators according to the fraction sign, adding if the sign is +, and subtracting if the sign is −.

$$\text{Thus,} \quad \frac{3}{8} + \frac{4}{8} = \frac{3+4}{8} = \frac{7}{8} \quad \text{and} \quad \frac{4}{8} - \frac{3}{8} = \frac{4-3}{8} = \frac{1}{8}.$$

To Add or Subtract Like Fractions

Combine into a single fraction: $(a)\ \dfrac{9}{16}+\dfrac{3}{16}$ $(b)\ \dfrac{2}{15}+\dfrac{8}{15}-\dfrac{7}{15}$

PROCEDURE   **SOLUTIONS**

1. Add or subtract numerators according to fraction sign: $\dfrac{9+3}{16}$ $\dfrac{2+8-7}{15}$

2. Keep the denominator unchanged:

3. Reduce to lowest terms: $\dfrac{12}{16}=\dfrac{3}{4}$ *Ans.* $\dfrac{3}{15}=\dfrac{1}{5}$ *Ans.*

Understanding the Addition or Subtraction of Like Fractions

Think of 4/8 as $4 \times (1/8)$ or 4 eighths and 3/8 as $3 \times (1/8)$ or 3 eighths.

Hence, $\dfrac{4}{8}+\dfrac{3}{8}=4\text{ eighths }+3\text{ eighths }=7\text{ eighths }=\dfrac{7}{8}$

Also, $\dfrac{4}{8}-\dfrac{3}{8}=4\text{ eighths }-3\text{ eighths }=1\text{ eighth }=\dfrac{1}{8}$

Number Line Method

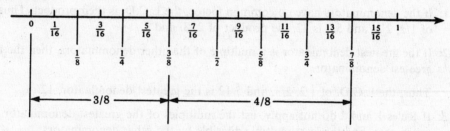

Fig. 3-7

To use a number line to add like fractions 3/8 and 4/8, combine segments having these lengths, as shown in Fig. 3-7. Hence,

$$\frac{3}{8}+\frac{4}{8}=\frac{7}{8}$$

6.1   Adding or Subtracting Two Like Fractions

In each, combine into a single fraction:

$(a)\ \dfrac{7}{8}+\dfrac{1}{8}$ $(c)\ \dfrac{15}{32}+\dfrac{9}{32}$ $(e)\ \dfrac{7}{10}+\dfrac{5}{10}$ $(g)\ \dfrac{3}{8}-\dfrac{1}{8}$ $(i)\ \dfrac{31}{32}-\dfrac{15}{32}$ $(k)\ \dfrac{13}{10}-\dfrac{7}{10}$

$(b)\ \dfrac{11}{16}+\dfrac{7}{16}$ $(d)\ \dfrac{5}{7}+\dfrac{1}{7}$ $(f)\ \dfrac{12}{25}+\dfrac{13}{25}$ $(h)\ \dfrac{11}{16}-\dfrac{5}{16}$ $(j)\ \dfrac{12}{7}-\dfrac{10}{7}$ $(l)\ \dfrac{21}{25}-\dfrac{11}{25}$

Solutions

$(a)\ \dfrac{8}{8}=1$ $(c)\ \dfrac{24}{32}=\dfrac{3}{4}$ $(e)\ \dfrac{12}{10}=\dfrac{6}{5}$ $(g)\ \dfrac{2}{8}=\dfrac{1}{4}$ $(i)\ \dfrac{16}{32}=\dfrac{1}{2}$ $(k)\ \dfrac{6}{10}=\dfrac{3}{5}$

$(b)\ \dfrac{18}{16}=\dfrac{9}{8}$ $(d)\ \dfrac{6}{7}$ $(f)\ \dfrac{25}{25}=1$ $(h)\ \dfrac{6}{16}=\dfrac{3}{8}$ $(j)\ \dfrac{2}{7}$ $(l)\ \dfrac{10}{25}=\dfrac{2}{5}$

6.2 ADDING OR SUBTRACTING MORE THAN TWO LIKE FRACTIONS

Combine into a single fraction:

(a) $\dfrac{5}{2}+\dfrac{3}{2}+\dfrac{1}{2}$ (b) $\dfrac{5}{4}+\dfrac{3}{4}-\dfrac{1}{4}$ (c) $\dfrac{5}{8}-\dfrac{3}{8}-\dfrac{1}{8}$ (d) $\dfrac{12}{11}+\dfrac{10}{11}-\dfrac{8}{11}$ (e) $\dfrac{23}{25}-\dfrac{17}{25}-\dfrac{4}{25}$

Solutions

(a) $\dfrac{5+3+1}{2}=\dfrac{9}{2}$ (b) $\dfrac{5+3-1}{4}=\dfrac{7}{4}$ (c) $\dfrac{5-3-1}{8}=\dfrac{1}{8}$ (d) $\dfrac{12+10-8}{11}=\dfrac{14}{11}$ (e) $\dfrac{23-17-4}{25}=\dfrac{2}{25}$

7. ADDING OR SUBTRACTING UNLIKE FRACTIONS

Least Common Denominator

The *least common denominator* (L.C.D.) of two or more fractions is the least common multiple (Chapter 2, Section 14) of their denominators.

Thus, 18 is the L.C.D. of 1/3, 5/6, and 7/9 since 18 is the least common multiple of the denominators 3, 6 and 9.

Finding the L.C.D.

Rule 1: If the denominators have no common factor, the L.C.D. is their product. Thus, the L.C.D. of 1/2, 2/3, and 5/7 is 42, the product of 2, 3, and 7.

Rule 2: If the greatest denominator is a multiple of the other denominators, then the L.C.D. is the greatest denominator.

Thus, the L.C.D. of 1/2, 2/3, and 5/12 is the greatest denominator, 12.

Rule 3: If Rules 1 and 2 do not apply, list the multiples of the greatest denominator in ascending order, then find the least multiple divisible by the other denominators.

Thus, to find the L.C.D. of 1/2, 2/5, 11/12, and 13/15, list the multiples of 15, which are 15, 30, 45, 60, 75, . . . , and note that 60 is the least multiple divisible by 2, 5, and 12. Hence, 60 is the L.C.D.

Method for Combining Unlike Fractions into a Single Fraction

Combine unlike fractions into a single fraction by changing the fractions into like fractions whose denominator is their L.C.D., then combining the resulting like fractions into a single fraction in lowest terms.

To Add or Subtract Unlike Fractions

Combine into a single fraction: (a) $\dfrac{1}{6}+\dfrac{3}{4}$ (b) $\dfrac{2}{3}+\dfrac{7}{12}-\dfrac{5}{9}-\dfrac{1}{36}$

PROCEDURE

1. Find the L.C.D.:

2. Change each fraction to an equivalent fraction having the L.C.D. as denominator:

3. Combine into a single fraction in lowest terms:

SOLUTIONS

L.C.D. = 12

$\dfrac{1\times2}{6\times2}+\dfrac{3\times3}{4\times3}$

$\dfrac{2}{12}+\dfrac{9}{12}$

$\dfrac{2+9}{12}$

Ans. $\dfrac{11}{12}$

L.C.D. = 36

$\dfrac{2\times12}{3\times12}+\dfrac{7\times3}{12\times3}-\dfrac{5\times4}{9\times4}-\dfrac{1}{36}$

$\dfrac{24}{36}+\dfrac{21}{36}-\dfrac{20}{36}-\dfrac{1}{36}$

$\dfrac{24+21-20-1}{36}=\dfrac{24}{36}$

Ans. $\dfrac{2}{3}$

7.1   FINDING THE LEAST COMMON DENOMINATOR (L.C.D.)

In each, find the least common denominator of the given fractions.

(a) $\dfrac{1}{2}, \dfrac{2}{3}$ (c) $\dfrac{1}{2}, \dfrac{4}{3}, \dfrac{2}{5}$ (e) $\dfrac{7}{8}, \dfrac{1}{4}$ (g) $\dfrac{5}{8}, \dfrac{7}{12}$ (i) $\dfrac{5}{6}, \dfrac{7}{9}, \dfrac{11}{15}$

(b) $\dfrac{1}{3}, \dfrac{4}{5}$ (d) $\dfrac{1}{3}, \dfrac{5}{6}$ (f) $\dfrac{3}{5}, \dfrac{1}{3}, \dfrac{8}{15}$ (h) $\dfrac{3}{4}, \dfrac{1}{6}, \dfrac{7}{9}$

Solutions

Apply Rule 1:   (a) $2 \cdot 3 = 6$ *Ans.*,   (b) $3 \cdot 5 = 15$ *Ans.*,   (c) $2 \cdot 3 \cdot 5 = 30$ *Ans.*

Apply Rule 2:   (d) 6 *Ans.*,   (e) 8 *Ans.*,   (f) 15 *Ans.*

Apply Rule 3:   (g) Multiples of 12 are 12, 24. *Ans.* 24.   (h) Multiples of 9 are 9, 18, 27, 36.
Ans. 36.   (i) Multiples of 15 are 15, 30, 45, 60, 75, 90. *Ans.* 90.

7.2   ADDING OR SUBTRACTING UNLIKE FRACTIONS

Combine into a single fraction:

(a) $\dfrac{1}{2} + \dfrac{2}{3}$ (b) $\dfrac{4}{5} - \dfrac{1}{3}$ (c) $\dfrac{1}{2} + \dfrac{4}{3} - \dfrac{2}{5}$ (d) $\dfrac{1}{3} + \dfrac{5}{6}$ (e) $\dfrac{7}{8} - \dfrac{1}{4}$

(f) $\dfrac{3}{5} + \dfrac{1}{3} - \dfrac{8}{15}$ (g) $\dfrac{5}{8} - \dfrac{7}{12}$ (h) $\dfrac{3}{4} + \dfrac{1}{6} + \dfrac{7}{9}$ (i) $\dfrac{11}{15} + \dfrac{5}{6} - \dfrac{7}{9}$

Solutions

(a) $\dfrac{3}{6} + \dfrac{4}{6} = \dfrac{7}{6}$ (d) $\dfrac{2}{6} + \dfrac{5}{6} = \dfrac{7}{6}$ (g) $\dfrac{15}{24} - \dfrac{14}{24} = \dfrac{1}{24}$

(b) $\dfrac{12}{15} - \dfrac{5}{15} = \dfrac{7}{15}$ (e) $\dfrac{7}{8} - \dfrac{2}{8} = \dfrac{5}{8}$ (h) $\dfrac{27}{36} + \dfrac{6}{36} + \dfrac{28}{36} = \dfrac{61}{36}$

(c) $\dfrac{15}{30} + \dfrac{40}{30} - \dfrac{12}{30} = \dfrac{43}{30}$ (f) $\dfrac{9}{15} + \dfrac{5}{15} - \dfrac{8}{15} = \dfrac{6}{15} = \dfrac{2}{5}$ (i) $\dfrac{66}{90} + \dfrac{75}{90} - \dfrac{70}{90} = \dfrac{71}{90}$

7.3   PROBLEM SOLVING INVOLVING ADDING FRACTIONS

The table shows the parts of three parks in a city that have been set aside for recreation, picnicking, and parking. In each, find the remaining part.

	Recreation	Picnicking	Parking	Remaining Part
(a)	3/8	1/4	3/16	?
(b)	1/3	2/9	1/18	?
(c)	2/5	1/3	1/6	?

Solutions

(a) $\dfrac{3}{8} + \dfrac{1}{4} + \dfrac{3}{16} = \dfrac{13}{16}$, Remainder $= \dfrac{3}{16}$ (b) $\dfrac{6}{18} + \dfrac{4}{18} + \dfrac{1}{18} = \dfrac{11}{18}$, Remainder $= \dfrac{7}{18}$

(c) $\dfrac{12}{30} + \dfrac{10}{30} + \dfrac{5}{30} = \dfrac{27}{30} = \dfrac{9}{10}$, Remainder $= \dfrac{1}{10}$

8.   COMPARING AND ORDERING FRACTIONS

Ordering Numbers: Ascending and Descending Order

Numbers are *ordered* when they are arranged according to size. In a *descending order*, the greatest number is listed first, while in an *ascending order*, the smallest is listed first.

Thus, 1, 2, 3 is an ascending order, and the reverse, 3, 2, 1, is a descending order.

Methods of Comparing Fractions

Three methods of comparing fractions are to be shown. Each of these methods is to be used to arrange 3/8, 5/16, and 1/4 in descending order.

Common Denominator Method

Fractions may be compared by changing them into equivalent fractions having a common denominator and then applying the following rule:

Rule: If fractions have a common denominator, the fraction with the greater numerator is the greater.

The fractions 1/4, 3/8, and 5/16 can be compared by changing them to 16ths. Thus, 1/4 = 4/16, 3/8 = 6/16. In descending order, beginning with the greatest, we have 6/16 or 3/8, then 5/16, and lastly, 4/16 or 1/4.

Cross Products Method

Fractions may be compared by using the cross products obtained by multiplying the numerator of one fraction by the denominator of the other.

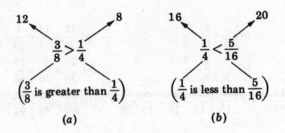

Fig. 3-8

Cross Products Rule: If cross products pointing upward are compared as shown, the greater fraction is nearer the greater cross product.

Thus, in Fig. 3-8(a), since the cross product 12 is greater than the cross product 8, the fraction 3/8 is greater than 1/4. In Fig. 3-8(b), since 20 is greater than 16, 5/16 is greater than 1/4, or 1/4 is less than 5/16.

Number Line Method

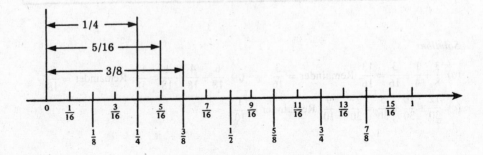

Fig. 3-9

A number line, as in Fig. 3-9 may be used to compare fractions. Comparing the lengths of segments, it can be seen that 3/8 is greater than 5/16, which in turn is greater than 1/4. The number line method is limited to those fractions that can be represented by points on the number line.

8.1   Using Common Denominators to Compare Fractions

Using the common denominator method, compare:

(a) $\dfrac{1}{3}$ and $\dfrac{4}{9}$ (b) $\dfrac{2}{3}$ and $\dfrac{7}{10}$ (c) $\dfrac{3}{2}$ and $\dfrac{17}{10}$ (d) $\dfrac{3}{4}$ and $\dfrac{11}{16}$ (e) $\dfrac{5}{4}$ and $\dfrac{27}{20}$

Solutions

(a) Since $1/3 = 3/9$, 1/3 is less than 4/9.   (b) Change to 30ths. Since $2/3 = 20/30$ and $7/10 = 21/30$, then 2/3 is less then 7/10.   (c) Since $3/2 = 15/10$, then 3/2 is less than 17/10.   (d) Since $3/4 = 12/16$, then 3/4 is greater than 11/16.   (e) Since $5/4 = 25/20$, then 5/4 is less than 27/20.

8.2   Using Cross Products to Compare Fractions

Using cross products, compare the fractions in each.

(a) $\dfrac{1}{6}$ and $\dfrac{1}{7}$ (b) $\dfrac{4}{6}$ and $\dfrac{20}{30}$ (c) $\dfrac{2}{3}$ and $\dfrac{13}{20}$ (d) $\dfrac{5}{8}$ and $\dfrac{7}{10}$

Solutions > means "is greater than," < means "is less than."

(a) $\dfrac{1}{6} ? \dfrac{1}{7}$ Since $7 > 6, \dfrac{1}{6} > \dfrac{1}{7}$.

(c) $\dfrac{2}{3} ? \dfrac{13}{20}$ Since $40 > 39, \dfrac{2}{3} > \dfrac{13}{20}$.

(b) $\dfrac{4}{6} ? \dfrac{20}{30}$ Since $120 = 120, \dfrac{4}{6} = \dfrac{20}{30}$.

(d) $\dfrac{5}{8} ? \dfrac{7}{10}$ Since $50 > 56, \dfrac{5}{8} < \dfrac{7}{10}$.

9.   EXPRESSING A WHOLE NUMBER OR A MIXED NUMBER AS AN IMPROPER FRACTION

Expressing a Whole Number as an Improper Fraction

A whole number may be expressed as an improper fraction (Section 1) having a given denominator. Thus, 3 may be expressed as 3/1, in halves as 6/2, in 3rds as 9/3, and in 4ths as 12/4.

To Express a Whole Number as an Improper Fraction

Express: (a) 3 in 4ths   (b) 5 in 7ths   (c) 12 in 10ths

PROCEDURE   SOLUTIONS

1. Obtain the product of the whole number and the given denominator: 1. $3 \cdot 4 = 12$ $5 \cdot 7 = 35$ $12 \cdot 10 = 120$

2. Place the product found in step 1 over the denominator: 2. $\dfrac{12}{4}$ *Ans.* $\dfrac{35}{7}$ *Ans.* $\dfrac{120}{10}$ *Ans.*

Expressing a Mixed Number as an Improper Fraction

A mixed number has two parts, a whole number and a fraction, and the mixed number is the sum of its two parts. Thus, $3\frac{1}{4} = 3 + \frac{1}{4}$. Hence,

$$3\frac{1}{4} = \frac{12}{4} + \frac{1}{4} = \frac{12 + 1}{4} = \frac{13}{4} \quad \text{(an improper fraction)}$$

To Express a Mixed Number as an Improper Fraction

Express as a fraction: (a) $3\frac{1}{4}$ -   (b) $5\frac{4}{7}$ (c) $12\frac{7}{10}$

PROCEDURE   SOLUTIONS

1. To the product of the whole number and   1. $3 \cdot 4 + 1 = 13$ | $5 \cdot 7 + 4 = 39$ | $12 \cdot 10 + 7 = 127$
   the denominator of the fraction, add the
   numerator:

2. Place the result found in step 1 over the   2. $\frac{13}{4}$ Ans. | $\frac{39}{7}$ Ans. | $\frac{127}{10}$ Ans.
   denominator of the fraction:

9.1   EXPRESSING A WHOLE NUMBER AS AN IMPROPER FRACTION

Express (a) 3 as 5ths,   (b) 10 as 8ths,   (c) 5 as 9ths,   (d) 12 as 6ths,   (e) 20 as 3rds.

Solutions   The numerator needed is the product of the whole number and the given denominator.

(a) $\frac{3 \cdot 5}{5} = \frac{15}{5}$ (b) $\frac{10 \cdot 8}{8} = \frac{80}{8}$ (c) $\frac{5 \cdot 9}{9} = \frac{45}{9}$ (d) $\frac{12 \cdot 6}{6} = \frac{72}{6}$ (e) $\frac{20 \cdot 3}{3} = \frac{60}{3}$

9.2   EXPRESSING A MIXED NUMBER AS AN IMPROPER FRACTION

Express as an improper fraction: (a) $3\frac{1}{2}$, (b) $5\frac{2}{5}$, (c) $8\frac{5}{7}$, (d) $10\frac{4}{9}$, (e) $100\frac{11}{20}$.

Solutions

(a) $\frac{3 \cdot 2 + 1}{2} = \frac{7}{2}$ (c) $\frac{8 \cdot 7 + 5}{7} = \frac{61}{7}$ (e) $\frac{100 \cdot 20 + 11}{20} = \frac{2,011}{20}$

(b) $\frac{5 \cdot 5 + 2}{5} = \frac{27}{5}$ (d) $\frac{10 \cdot 9 + 4}{9} = \frac{94}{9}$

9.3   MEASUREMENT PROBLEMS REQUIRING IMPROPER FRACTION EXPRESSIONS

(a) How many half-inches are there in   (1) 3 inches,   (2) $2\frac{1}{2}$ inches?
(b) How many quarter-inches are there in   (1) 5 inches,   (2) $5\frac{1}{2}$ inches?
(c) How many eighth-inches are there in   (1) 10 inches,   (2) $10\frac{5}{8}$ inches?

Solutions

(a) (1) 3 in. $= 6\left(\frac{1}{2} \text{in.}\right)$ *Ans.* 6   (2) $2\frac{1}{2} = \frac{5}{2} = 5\left(\frac{1}{2}\right)$ *Ans.* 5

(b) (1) $5 = \frac{20}{4} = 20\left(\frac{1}{4}\right)$ *Ans.* 20   (2) $5\frac{1}{2} = \frac{11}{2} = \frac{22}{4} = 22\left(\frac{1}{4}\right)$ *Ans.* 22

(c) (1) $10 = \frac{80}{8} = 80\left(\frac{1}{8}\right)$ *Ans.* 80   (2) $10\frac{5}{8} = \frac{85}{8} = 85\left(\frac{1}{8}\right)$ *Ans.* 85

10.   EXPRESSING AN IMPROPER FRACTION AS A WHOLE OR MIXED NUMBER

Forming a Fraction by Placing the Remainder over the Divisor

If 13 tennis balls are to fill cans each of which can hold 3 balls, the number of cans needed can be found by dividing 13 by 3. By dividing we find that 4 cans will be filled, with a remainder of 1 ball. If there is another can, the remaining 1 ball will occupy 1/3 of the can. Hence, we say that the number of cans needed is $4\frac{1}{3}$.

Note that the fraction 1/3 is formed by making the remainder 1, the numerator and making the divisor, 3, the denominator. We think of the fraction as "the remainder over the divisor."

Expressing an Improper Fraction as a Whole or Mixed Number

Like any fraction, an *improper fraction may mean division*. To express an improper fraction as either a whole number or a mixed number, divide the numerator by the denominator. The quotient obtained consists of a whole number and a remainder. Make "the remainder over the divisor" the fraction part of the mixed number and combine it with the whole number to form the mixed number. If the remainder is zero, then the improper fraction is equivalent to a whole number.

Thus, for the improper fraction 13/3:

$$\frac{13}{3} = 3\overline{)13}^{\,4\,R\,1} = 4\tfrac{1}{3} \text{ (a mixed number)}$$

Also, for the improper fraction 12/3:

$$\frac{12}{3} = 3\overline{)12}^{\,4\,R\,0} = 4 \text{ (a mixed number)}$$

Standard Form of a Mixed Number

To be in standard form, the fraction part of the mixed number must meet the following requirements:

1. The fraction part must be a *proper fraction*. For example, $5\tfrac{5}{9}$ is in standard form, but its equivalent $4\tfrac{14}{9}$ is not.

2. The fraction part must be *reduced to lowest terms*. Thus, $3\tfrac{1}{2}$ is in standard form, but $3\tfrac{2}{4}$ is not.

To Express an Improper Fraction as a Whole Number or a Mixed Number in Standard Form

Express as a whole or mixed number: (a) 7/4   (b) 28/7   (c) 51/15   (d) 485/25

PROCEDURE   SOLUTIONS

1. Divide numerator by denominator to obtain a whole number and a remainder:	1. $4\overline{)7}^{\,1\,R\,3}$	$7\overline{)28}^{\,4}$	$15\overline{)51}^{\,3\,R\,6}$ $\frac{45}{6}$	$25\overline{)485}^{\,19\,R\,10}$ $\frac{25}{235}$ $\frac{225}{10}$
2. Make the *remainder over the divisor* the fraction part of the mixed number:	2. $1\tfrac{3}{4}$	4	$3\tfrac{6}{15}$	$19\tfrac{10}{25}$
3. Express mixed number in standard form by reducing the fraction part to lowest terms:	3. $1\tfrac{3}{4}$ *Ans.*	4 *Ans.*	$3\tfrac{2}{5}$ *Ans.*	$19\tfrac{2}{5}$ *Ans.*

10.1 Expressing an Improper Fraction as a Whole or Mixed Number

Express each as a whole number or as a mixed number.

(a) $\dfrac{3}{2}$ (b) $\dfrac{85}{5}$ (c) $\dfrac{43}{10}$ (d) $\dfrac{355}{110}$

Solutions

(a) $2\overline{)3}^{\,1\,R\,1}$ (b) $5\overline{)85}^{\,17}$ (c) $10\overline{)43}^{\,4\,R\,3}$ (d) $110\overline{)355}^{\,3\,R\,25}$ $\frac{330}{25}$

Ans. $1\tfrac{1}{2}$ *Ans.* 17   *Ans.* $4\tfrac{3}{10}$

$$3\tfrac{25}{110} = 3\tfrac{5}{22} \text{ *Ans.*}$$

10.2 MEASUREMENT PROBLEMS REQUIRING WHOLE OR MIXED NUMBERS

How many centimeters are there in (*a*) 14 half-centimeters, (*b*) 25 quarter-centimeters, (*c*) 100 eighth-centimeters, (*d*) 175 sixteenth-centimeters, (*e*) 390 thirty-second centimeters.

Solutions

(*a*) $14\left(\dfrac{1}{2} \text{ cm.}\right) = 7 \text{ cm}$ *Ans.*   (*c*) $100\left(\dfrac{1}{8}\right) = 12\dfrac{4}{8} = 12\dfrac{1}{2}$ *Ans.*   (*e*) $390\left(\dfrac{1}{32}\right) = 12\dfrac{6}{32} = 12\dfrac{3}{16}$ *Ans.*

(*b*) $25\left(\dfrac{1}{4}\right) = 6\dfrac{1}{4}$ *Ans.*   (*d*) $175\left(\dfrac{1}{16}\right) = 10\dfrac{15}{16}$ *Ans.*

11. ADDING WITH MIXED NUMBERS

Mixed numbers are added by applying the following rule:

Rule: Add two or more mixed numbers by finding the sum of the whole number parts and the sum of the fraction parts, then adding both sums.

Horizontal Form: $1\dfrac{1}{4} + 1\dfrac{1}{2} = (1 + 1) + \left(\dfrac{1}{4} + \dfrac{1}{2}\right)$ Vertical Form: $\begin{array}{r} 1\frac{1}{4} \\ +1\frac{1}{2} \\ \hline 2\frac{3}{4} \text{ Ans.} \end{array}$

$$= \quad 2 \quad + \quad \dfrac{3}{4} = 2\dfrac{3}{4} \text{ Ans.}$$

To Add Mixed Numbers

Add: (*a*) $1\dfrac{3}{4} + 2\dfrac{5}{8}$ (*b*) $5\dfrac{2}{3} + 8\dfrac{2}{5}$ (*c*) $20\dfrac{3}{8} + 29\dfrac{5}{6}$

PROCEDURE   SOLUTIONS

1. Add the whole number parts:

$\begin{array}{r|l} 1 & \frac{3}{4} = \frac{6}{8} \\ +2 & \frac{5}{8} = \frac{5}{8} \\ \hline 3 & \frac{11}{8} = 1\frac{3}{8} \end{array}$ $\begin{array}{r|l} 5 & \frac{2}{3} = \frac{10}{15} \\ +8 & \frac{2}{5} = \frac{6}{15} \\ \hline 13 & \frac{16}{15} = 1\frac{1}{15} \end{array}$ $\begin{array}{r|l} 20 & \frac{3}{8} = \frac{9}{24} \\ +29 & \frac{5}{6} = \frac{20}{24} \\ \hline 49 & \frac{29}{24} = 1\frac{5}{24} \end{array}$

2. Add the fraction parts, expressing the sum as a mixed number:

3. Add the sums found in steps 1 and 2: $3 + 1\dfrac{3}{8}$ $13 + 1\dfrac{1}{15}$ $49 + 1\dfrac{5}{24}$

4. Express the result in standard form: *Ans.* $4\dfrac{3}{8}$ *Ans.* $14\dfrac{1}{15}$ *Ans.* $50\dfrac{5}{24}$

11.1 ADDING MIXED NUMBERS

Add:

(*a*) $4 + 3\dfrac{1}{2}$ (*b*) $5\dfrac{2}{3} + 15$ (*c*) $5\dfrac{1}{5} + \dfrac{3}{5}$ (*d*) $2\dfrac{1}{2} + 4\dfrac{3}{4} + 8\dfrac{1}{8}$ (*e*) $5\dfrac{1}{4} + 8\dfrac{2}{3} + \dfrac{7}{12}$

(*f*) $\begin{array}{r} 7\frac{5}{9} \\ +3\frac{4}{9} \end{array}$ (*g*) $\begin{array}{r} 10\frac{7}{8} \\ +5\frac{5}{8} \end{array}$ (*h*) $\begin{array}{r} 4\frac{2}{7} \\ +9\frac{6}{7} \end{array}$ (*i*) $\begin{array}{r} 3\frac{1}{5} \\ +12\frac{3}{10} \end{array}$ (*j*) $\begin{array}{r} 7\frac{1}{6} \\ +13\frac{3}{4} \end{array}$ (*k*) $\begin{array}{r} 12\frac{7}{8} \\ +25\frac{7}{12} \end{array}$

Illustrative Solution

(*i*) $\begin{array}{r|l} 3 & \frac{1}{5} = \frac{2}{10} \\ +12 & \frac{3}{10} = \frac{3}{10} \\ \hline 15 & \frac{5}{10} = \frac{1}{2} \end{array}$

Ans. $15\dfrac{1}{2}$

Ans.   (a) $7\frac{1}{2}$ (b) $20\frac{2}{3}$ (c) $5\frac{4}{5}$ (d) $15\frac{3}{8}$ (e) $14\frac{1}{2}$ (f) 11 (g) $16\frac{1}{2}$ (h) $14\frac{1}{7}$

(j) $20\frac{11}{12}$ (k) $38\frac{11}{24}$

11.2 PROBLEM SOLVING INVOLVING THE ADDITION OF MIXED NUMBERS

(a) Figure 3-10 shows the distances in miles between towns A, B, C, and D located on a straight road. What is the distance in miles (1) from A to C, (2) from B to D, and (3) from A to D?

(b) Find the total number of grams in two bags if their weights are (1) each $14\frac{1}{4}$ g, (2) $15\frac{3}{4}$ g and $20\frac{3}{8}$ g, (3) $125\frac{7}{8}$ g and $250\frac{15}{16}$ g.

Fig. 3-10

(c) Find the total number of meters in three rolls of cloth if their measurements are (1) $50\frac{1}{2}$ m, $48\frac{5}{8}$ m, and $36\frac{3}{4}$ m; (2) $156\frac{7}{8}$ m, 175 m and $182\frac{1}{4}$ m.

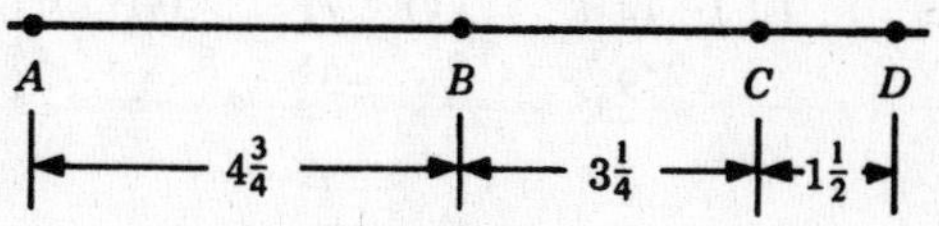

Illustrative Solution   (a) (3): $4\frac{3}{4}+3\frac{1}{4}+1\frac{1}{2}=8+1\frac{1}{2}=9\frac{1}{2}$ *Ans.*

Ans.   (a) (1) 8   (2) $4\frac{3}{4}$

(b) (1) $28\frac{1}{2}$ (2) $36\frac{1}{8}$ (3) $376\frac{13}{16}$

(c) (1) $135\frac{7}{8}$ (2) $514\frac{1}{8}$

12. SUBTRACTING WITH MIXED NUMBERS

Mixed numbers are subtracted by applying the following rule:

Rule: Subtract one mixed number from another by finding the difference between the whole number parts and the difference of the fraction parts, then adding both differences.

Horizontal Form: $2\frac{3}{4}-1\frac{1}{2}=(2-1)+(\frac{3}{4}+\frac{1}{2})$

$$= \quad 1 \quad + \quad \frac{1}{4} \quad =1\frac{1}{4}$$

Vertical Form:

$$\begin{array}{r} 2\ \ \frac{3}{4}=\frac{3}{4} \\ -1\ \ \frac{1}{2}=\frac{2}{4} \\ \hline 1\ \ \ \ \frac{1}{4} \end{array}$$

Ans. $1\frac{1}{4}$

To Subtract with Mixed Numbers

Subtract:

(a) $10\frac{7}{12}-3\frac{1}{4}$ (b) $21\frac{1}{6}-8\frac{4}{9}$

PROCEDURE

1. If the minuend's fraction part is less than the subtrahend's fraction part, transfer 1 from the minuend's whole number part to its fraction part:

2. Find the difference of the whole number parts:

3. Find the difference of the fraction parts, reducing it to lowest terms:

4. Add the differences found in steps 2 and 3:

SOLUTIONS

No transfer needed.

$$\begin{array}{r} 10\ \left|\ \frac{7}{12}=\frac{7}{12}\right. \\ -3\ \left|\ \frac{1}{4}=\frac{3}{12}\right. \\ \hline 7\ \ \ \ \left|\ \frac{4}{12}=\frac{1}{3}\right. \end{array}$$

Ans. $7\frac{1}{3}$

Transfer 1 or 6/6:

$21\frac{1}{6}=20\frac{7}{6}$

$$\begin{array}{r} 20\ \left|\ \frac{7}{6}=\frac{21}{18}\right. \\ -8\ \left|\ \frac{4}{9}=\frac{8}{18}\right. \\ \hline 12\ \ \ \left|\ \frac{13}{18}\right. \end{array}$$

Ans. $12\frac{13}{18}$

12.1 SUBTRACTION WITH MIXED NUMBERS

Subtract:

(a) $14\frac{7}{8} - 10$ (b) $25\frac{1}{6} - 17$ (c) $10 - \frac{3}{4}$ (d) $20 - 5\frac{4}{5}$ (e) $6\frac{11}{16} - 2\frac{11}{16}$

(f) $12\frac{13}{15}$ (g) $7\frac{3}{8}$ (h) $6\frac{1}{2}$ (i) $13\frac{2}{5}$ (j) $26\frac{5}{6}$ (k) $15\frac{7}{12}$ (l) $35\frac{9}{10}$
$\ -9\frac{7}{15}$ $\ -3\frac{5}{8}$ $\ -2\frac{3}{8}$ $\ -10\frac{1}{6}$ $\ -20\frac{11}{12}$ $\ -8\frac{7}{8}$ $\ -29\frac{3}{4}$

Illustrative Solution

$$(d) \qquad 20 = 19\frac{5}{5}$$
$$-5\frac{4}{5}$$
$$\overline{\ \ 14\frac{1}{5}}\ \ Ans.$$

Ans. (a) $4\frac{7}{8}$ (b) $8\frac{1}{6}$ (c) $9\frac{1}{4}$ (e) 4   (f) $3\frac{2}{5}$ (g) $3\frac{3}{4}$ (h) $4\frac{1}{8}$ (i) $3\frac{7}{30}$

(j) $5\frac{11}{12}$ (k) $6\frac{17}{24}$ (l) $6\frac{3}{20}$

12.2 PROBLEM SOLVING INVOLVING SUBTRACTION WITH MIXED NUMBERS

(a) Figure 3-11 shows the distance in kilometers between towns A, B, and C located on a straight road. What is the distance in miles from B to C if the distance d from A to C is (1) 10 km, (2) $7\frac{5}{8}$ km, (3) $7\frac{5}{16}$ km?

(b) Find the number of pounds left in a bag of 100 pounds if the number of pounds removed from it is (1) $90\frac{1}{4}$, (2) $75\frac{3}{8}$, (3) $35\frac{87}{100}$.

(c) From a roll of cloth $20\frac{1}{8}$ yards long, a length is cut. How many yards remain if the length cut off is (1) $5\frac{1}{8}$ yd, (2) $12\frac{7}{8}$ yd, (3) $18\frac{15}{16}$ yd?

(d) Normal body temperature is $98\frac{3}{5}$ degrees. How many degrees above normal is a temperature of (1) 100 degrees, (2) $101\frac{1}{2}$ degrees, (3) $103\frac{3}{10}$ degrees?

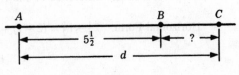

Fig. 3-11

Illustrative Solution

$$(a)\ (3):\quad 7\frac{5}{16} = 6\ \begin{array}{c|c} \frac{21}{16} = \frac{21}{16} \\ \frac{1}{2} = \frac{8}{16} \\ \hline \frac{13}{16} \end{array}$$

Ans. $1\frac{13}{16}$

Ans. (a) (1) $4\frac{1}{2}$ (2) $2\frac{1}{8}$

(b) (1) $9\frac{3}{4}$ (2) $24\frac{5}{8}$ (3) $64\frac{13}{100}$

(c) (1) 15   (2) $7\frac{1}{4}$ (3) $1\frac{3}{16}$

(d) (1) $1\frac{2}{5}$ (2) $2\frac{9}{10}$ (3) $4\frac{7}{10}$

13. MULTIPLYING OR DIVIDING WITH MIXED NUMBERS

When multiplying or dividing with mixed numbers, express each mixed number as an improper fraction, then apply the rules for multiplying fractions or for dividing fractions. Results should be expressed as mixed numbers in standard form.

To Multiply with Mixed Numbers

Multipy: (a) $\frac{5}{12} \times 3\frac{3}{5}$ (b) $5\frac{5}{6} \times 1\frac{2}{5}$

PROCEDURE   SOLUTIONS

1. Express each mixed number as an improper fraction: 1. $\frac{5}{12} \times \frac{18}{5}$ $\frac{35}{6} \times \frac{7}{5}$

2. Multiply fractions: 2. $\frac{\overset{1}{\cancel{5}}}{\underset{2}{\cancel{12}}} \times \frac{\overset{3}{\cancel{18}}}{\cancel{5}}$ $\frac{35}{6} \times \frac{7}{\cancel{5}}\overset{}{\underset{1}{}}$

3. Express the result as a mixed number in standard form: 3. $\frac{3}{2} = 1\frac{1}{2}$ *Ans.* $\frac{49}{6} = 8\frac{1}{6}$ *Ans.*

To Divide with Mixed Numbers

Divide: (a) $4\frac{1}{4} \div 4$ (b) $3\frac{2}{3} \div 2\frac{14}{15}$

PROCEDURE   SOLUTIONS

1. Express each mixed number as an improper fraction: 1. $\frac{9}{2} \div \frac{4}{1}$ $\frac{11}{3} \div \frac{44}{15}$

2. Multiply by the reciprocal of the divisor: 2. $\frac{9}{2} \times \frac{1}{4}$ $\frac{\overset{1}{\cancel{11}}}{\cancel{3}} \times \frac{\overset{5}{\cancel{15}}}{\underset{4}{\cancel{44}}}$
   (Cancel every common factor first)

3. Express the result as a mixed number in standard form: 3. $\frac{9}{8} = 1\frac{1}{8}$ *Ans.* $\frac{5}{4} = 1\frac{1}{4}$ *Ans.*

13.1 MULTIPLYING WITH MIXED NUMBERS

Multipy:

(a) $3\frac{1}{2} \times 2$ (b) $3\frac{1}{4} \times 8$ (c) $40 \times 5\frac{3}{4}$ (d) $10 \times 4\frac{2}{5}$ (e) $50 \times 3\frac{7}{10}$

(f) $5\frac{1}{16} \times 8$ (g) $4\frac{3}{8} \times 7$ (h) $2\frac{2}{5} \times \frac{1}{6}$ (i) $3\frac{3}{4} \times \frac{2}{5}$ (j) $\frac{9}{25} \times 4\frac{1}{6}$

(k) $4\frac{1}{2} \times 3\frac{1}{3}$ (l) $3\frac{1}{7} \times \frac{3}{4} \times 14$ (m) $10 \times 2\frac{1}{4} \times \frac{4}{5}$ (n) $2\frac{5}{6} \times 1\frac{1}{11} \times 3\frac{2}{3}$ (o) $1\frac{3}{4} \times 2\frac{1}{2} \times 3\frac{1}{7}$

Illustrative Solutions

(k) $4\frac{1}{2} \times 3\frac{1}{3} = \frac{9}{2} \times \overset{5}{\underset{1}{\cancel{\frac{10}{3}}}} = \frac{15}{1} = 15$ *Ans.*   (l) $3\frac{1}{7} \times \frac{3}{4} \times 14 = \overset{11}{\underset{1}{\cancel{\frac{22}{7}}}} \times \frac{3}{4} \times \overset{\overset{1}{\cancel{2}}}{\underset{1}{\cancel{\frac{14}{1}}}} = 33$ *Ans.*

Ans.   (a) 7   (c) 230   (e) 185   (g) $30\frac{5}{8}$ (i) $1\frac{1}{2}$ (m) 18   (o) $13\frac{3}{4}$

(b) 26   (d) 44   (f) $40\frac{1}{2}$ (h) 2/5   (j) $1\frac{1}{2}$ (n) $11\frac{1}{3}$

13.2 PROBLEM SOLVING INVOLVING MULTIPLICATION OF MIXED NUMBERS

(a) A piecrust requires $3\frac{3}{8}$ cups of flour, $1\frac{1}{4}$ teaspoons of salt, and 5/6 cup of water. How much of each ingredient is needed for 8 piecrusts?

(b) A metal rod weighs $2\frac{1}{3}$ pounds per foot. How many pounds does a rod weigh if its length is (1) 9 ft, (2) 12 ft, (3) $5\frac{1}{2}$ ft?

(c) The cost of a meter of cloth is $\$2\frac{1}{2}$. Find the cost of (1) 10 m, (2) $8\frac{1}{2}$ m, (3) $10\frac{2}{5}$ m.

(d) At a rate of $14\frac{2}{3}$ miles per gallon of gas, how many miles can a car go if the number of gallons in the tank is (1) 3, (2) $5\frac{1}{4}$, (3) $10\frac{1}{2}$?

(e) If a worker gets time and a half for overtime work, his hourly rate of pay is multiplied by $1\frac{1}{2}$ or 3/2. What is the hourly rate for overtime work if the worker earns (1) \$12 per hour, (2) $\$19\frac{1}{2}$ per hour, (3) $\$12\frac{4}{5}$ per hour?

Illustrative Solution

(c) (3): $2\frac{1}{2} \times 10\frac{2}{5} = \overset{1}{\underset{1}{\cancel{\frac{5}{2}}}} \times \overset{26}{\underset{1}{\cancel{\frac{52}{5}}}} = \frac{26}{1} = 26$ *Ans.* \$26

Ans.   (a) (1) 27 cups of flour, 10 teaspoons of salt, $6\frac{2}{3}$ cups of water

(b) (1) 21   (10 28   (3) $12\frac{5}{6}$

(c) (1) \$25   (2) $\$21\frac{1}{4}$

(d) (1) 44   (2) 77   (3) 154

(e) (1) \$18 per hour   (2) $\$28\frac{1}{4}$ per hour   (3) $\$19\frac{1}{5}$ per hour

13.3 DIVISION WITH MIXED NUMBERS

Divide:

(a) $3\frac{1}{2} \div 7$ (d) $8 \div 1\frac{1}{2}$ (g) $\frac{3}{4} \div 2\frac{2}{5}$ (j) $2\frac{1}{4} \div 3\frac{1}{5}$ (m) $12\frac{1}{4} \div 2\frac{3}{16}$

(b) $6\frac{2}{3} \div 5$ (e) $16 \div 1\frac{3}{4}$ (h) $\frac{7}{12} \div 1\frac{3}{4}$ (k) $5\frac{1}{2} \div 3\frac{2}{3}$

(c) $13\frac{1}{3} \div 10$ (f) $6 \div 3\frac{3}{5}$ (i) $\frac{15}{16} \div 2\frac{1}{2}$ (l) $1\frac{5}{6} \div 3\frac{1}{7}$

Illustrative Solution

$$(k)\ 5\tfrac{1}{2} \div 3\tfrac{2}{3} = \frac{\overset{1}{\cancel{11}}}{2} \times \frac{3}{\underset{1}{\cancel{11}}} = \frac{3}{2} = 1\tfrac{1}{2}\ \ Ans.$$

Ans. (a) $\dfrac{1}{2}$ (b) $1\tfrac{1}{3}$ (c) $1\tfrac{1}{3}$ (d) $5\tfrac{1}{3}$ (e) $9\tfrac{1}{7}$ (f) $1\tfrac{2}{3}$ (g) $\dfrac{5}{16}$ (h) $\dfrac{1}{3}$

(i) $\dfrac{3}{8}$ (j) $\dfrac{45}{64}$ (l) $\dfrac{7}{12}$ (m) $5\tfrac{3}{5}$

13.4   PROBLEM SOLVING INVOLVING DIVISION WITH MIXED NUMBERS

(a) How many half-inches are in a line whose length in inches is (1) 30, (2) $5\tfrac{1}{2}$, (3) $34\tfrac{1}{2}$?

(b) How many brooms, each $1\tfrac{1}{4}$ feet long, can be cut from a pole whose length in feet is (1) 10, (2) 15, (3) 25?

(c) A trip is to be completed in $2\tfrac{1}{2}$ days. What average rate in miles per day must be driven to travel a distance of (1) 400 mi, (2) 550 mi, (3) 725 mi?

(d) The diameter of a circle is found approximately by dividing the circumference, which is the distance around the circle, by $3\tfrac{1}{7}$. What is the diameter of a circle whose circumference is (1) 33 feet, (2) 11/14 yards, (3) $9\tfrac{5}{8}$ inches?

Illustrative Solution

$$(a)\ (2):\ 5\tfrac{1}{2} \div \tfrac{1}{2} = \frac{11}{2} \times \frac{\overset{2}{\cancel{2}}}{\underset{1}{\cancel{1}}} = 11\ \ Ans.$$

Ans. (a) (1) 60   (3) 69
      (b) (1) 8   (2) 12   (3) 20
      (c) (1) 160   (2) 220   (3) 290
      (d) (1) $10\tfrac{1}{2}$ feet   (2) 1/4 yard   (3) $3\tfrac{1}{16}$ inches

Supplementary Problems

3.1.   Express each of the following as a fraction or a mixed number: (a) one-seventh,   (1.1)

(b) seven-ninths,   (c) three-twentieths,   (d) three and one-fifth,   (e) 3 divided by 4,

(f) $5 \div 7$,   (g) $7 \div 5$,   (h) 6 out of 13,   (i) $9:11$ (j) $11:9$,   (k) $5 \times \tfrac{1}{8}$,   (l) $7 + \tfrac{2}{9}$,

(m) $3 + \tfrac{4}{5}$,   (n) $15 + 9 \times \tfrac{1}{17}$.

Ans. (a) $\dfrac{1}{7}$ (c) $\dfrac{3}{20}$ (e) $\dfrac{3}{4}$ (g) $\dfrac{7}{5}$ (i) $\dfrac{9}{11}$ (k) $\dfrac{5}{8}$ (m) $3\tfrac{4}{5}$

(b) $\dfrac{7}{9}$ (d) $3\tfrac{1}{5}$ (f) $\dfrac{5}{7}$ (h) $\dfrac{6}{13}$ (j) $\dfrac{11}{9}$ (l) $7\tfrac{2}{9}$ (n) $15\tfrac{9}{17}$

3.2. State the fractions that indicate the shaded and unshaded sections in each part of Fig.   **(1.2)**
3-12.

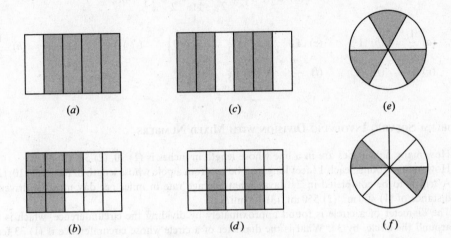

Fig. 3-12

Ans.

	Shaded	Unshaded
(a)	4/5	1/5
(b)	7/15	8/15
(c)	4/6 or 2/3	2/6 or 1/3

	Shaded	Unshaded
(d)	4/15	11/15
(e)	3/6 or 1/2	3/6 or 1/2
(f)	2/8 or 1/4	6/8 or 3/4

3.3. (a) If 5 of 12 pencils are used, what part of the entire number is used? unused?   **(1.3)**
  (b) If out of 15 games played, a team wins 8, what part of the games played are won? lost?
  (c) In a club of 18 members, 11 are girls. What part of the membership is girls? boys?
  (d) If 25¢ of a dollar is spent, what part of the dollar is spent? unspent?
  (e) If 2 of a dozen rolls are eaten, what part of the dozen is eaten? uneaten?
  (f) At 6 A.M., what part of a day has elapsed? remains?
  (g) In a set of 10 crayons, 4 are red. What part of the set are red? not red?

Ans. (a) 5/12, 7/12   (d) 25/100 or 1/4, 75/100 or 3/4   (g) 4/10 or 2/5, 6/10 or 3/5
   (b) 8/15, 7/15   (e) 2/12 or 1/6, 10/12 or 5/6
   (c) 11/18, 7/18   (f) 6/24 or 1/4, 18/24 or 3/4

3.4. (a) Express each of the following as a fraction: **(1.4)**

   (1) 13 divided by 17   (2) 14 ÷ 9   (3) 4)‾5   (4) 5)‾4   (5) 25)‾1

  (b) Express each fraction as a division using the division symbol)‾

   (1) $\frac{7}{10}$ (2) $\frac{10}{7}$ (3) $\frac{21}{100}$ (4) $\frac{100}{21}$

(a) (1) $\dfrac{13}{17}$ (2) $\dfrac{14}{9}$ (3) $\dfrac{5}{4}$ (4) $\dfrac{4}{5}$ (5) $\dfrac{1}{25}$

(b) (1) $10\overline{)7}$ (2) $7\overline{)10}$ (3) $100\overline{)21}$ (4) $21\overline{)100}$

3.5. (a) The value of a nickel is what part of the value of a dime? a quarter?   **(1.5)**
  (b) The value of a quarter is what part of the value of a half-dollar? a dollar?
  (c) A half-year is what part of a decade? a century?
  (d) A foot is what part of a yard? a mile?
  (e) A second is what part of a minute? an hour?
  (f) A day is what part of a week? three weeks?
  (g) A gram is what part of a kilogram? a hectogram?

  Ans. (a) 1/2, 1/5   (c) 1/20, 1/200   (e) 1/60, 1/3,600
    (b) 1/2, 1/4   (d) 1/3, 1/5,280 (f) 1/7, 1/21
    (g) 1/1,000, 1/100

3.6. Given $f = 1\tfrac{4}{7}$ $g = \dfrac{7}{11}$ $h = \dfrac{11}{7}$ $j = 7\tfrac{4}{11}$ $k = \dfrac{11}{81}$ $m = \dfrac{81}{11}$ **(1.6)**
  Which of the given numbers are (a) proper fractions, (b) improper fractions, (c) mixed numbers?

  Ans. (a) g, k (b) h, m (c) f, j

3.7. (a) Evaluate: (1) 0/3, (2) 3/0, (3) 0/1,000, (4) 1,000/0.   **(1.7)**
  (b) For what value of n is the fraction equal to 0?

$$(1)\ \dfrac{n}{4}\quad (2)\ \dfrac{n-2}{5}\quad (3)\ \dfrac{5-n}{2}\quad (4)\ \dfrac{2n}{5}$$

  Ans. (a) (1) 0 (2) no value (3) 0 (4) no value
    (b) (1) 0 (2) 2 (3) 5 (4) 0

3.8. Change each of the following to equivalent fractions by multiplying the numerator   **(2.1)**
  and denominator by 2, 5, 10, and 25: (a) 1/7, (b) 3/4, (c) 9/10, (d) 11/15.

(a) $\dfrac{1}{7} = \dfrac{2}{14} = \dfrac{5}{35} = \dfrac{10}{70} = \dfrac{25}{175}$ (c) $\dfrac{9}{10} = \dfrac{18}{20} = \dfrac{45}{50} = \dfrac{90}{100} = \dfrac{225}{250}$

(b) $\dfrac{3}{4} = \dfrac{6}{8} = \dfrac{15}{20} = \dfrac{30}{40} = \dfrac{75}{100}$ (d) $\dfrac{11}{15} = \dfrac{22}{30} = \dfrac{55}{75} = \dfrac{110}{150} = \dfrac{275}{375}$

3.9. Change each to equivalent fractions by dividing its numerator and denominator by 2, 3, **(2.2)**
  4, 6, and 12: (a) 12/24, (b) 36/48, (c) 48/120.

  Ans. (a) $\dfrac{12}{24} = \dfrac{6}{12} = \dfrac{4}{8} = \dfrac{3}{6} = \dfrac{2}{4} = \dfrac{1}{2}$ (b) $\dfrac{36}{48} = \dfrac{18}{24} = \dfrac{12}{16} = \dfrac{9}{12} = \dfrac{6}{8} = \dfrac{3}{4}$

    (c) $\dfrac{48}{120} = \dfrac{24}{60} = \dfrac{16}{40} = \dfrac{12}{30} = \dfrac{8}{20} = \dfrac{4}{10}$

3.10. In each, show how the first fraction is changed into the equivalent second fraction: **(2.3)**

(a) $\dfrac{9}{12}=\dfrac{3}{4}$ (b) $\dfrac{9}{12}=\dfrac{180}{240}$ (c) $\dfrac{75}{125}=\dfrac{3}{5}$ (d) $\dfrac{3}{5}=\dfrac{75}{125}$ (e) $\dfrac{4}{5}=\dfrac{800}{1,000}$ (f) $\dfrac{63}{105}=\dfrac{3}{5}$

Ans. (a) $\dfrac{9\div 3}{12\div 3}$ (b) $\dfrac{9\times 20}{12\times 10}$ (c) $\dfrac{75\div 25}{125\div 25}$ (d) $\dfrac{3\times 25}{5\times 25}$ (e) $\dfrac{4\times 200}{5\times 200}$ (f) $\dfrac{63\div 21}{105\div 21}$

3.11. Reduce each fraction to lowest terms: **(2.4)**

(a) $\dfrac{7}{28}$ (c) $\dfrac{15}{225}$ (e) $\dfrac{350}{525}$ (g) $\dfrac{3,003}{7,007}$ (i) $\dfrac{45}{75}$ (k) $\dfrac{36}{90}$ (m) $\dfrac{280}{770}$ (o) $\dfrac{143}{2,002}$

(b) $\dfrac{9}{81}$ (d) $\dfrac{25}{750}$ (f) $\dfrac{1,234}{4,936}$ (h) $\dfrac{15}{55}$ (j) $\dfrac{750}{450}$ (l) $\dfrac{90}{54}$ (n) $\dfrac{91}{1,001}$ (p) $\dfrac{495}{770}$

Ans. (a) $\dfrac{1}{4}$ (c) $\dfrac{1}{15}$ (e) $\dfrac{2}{3}$ (g) $\dfrac{3}{7}$ (i) $\dfrac{3}{5}$ (k) $\dfrac{2}{5}$ (m) $\dfrac{4}{11}$ (o) $\dfrac{1}{14}$

(b) $\dfrac{1}{9}$ (d) $\dfrac{1}{30}$ (f) $\dfrac{1}{4}$ (h) $\dfrac{3}{11}$ (j) $\dfrac{5}{3}$ (l) $\dfrac{5}{3}$ (n) $\dfrac{1}{11}$ (p) $\dfrac{9}{14}$

3.12. Multiply: **(3.1)**

(a) $\dfrac{1}{2}\times\dfrac{5}{7}$ (c) $\dfrac{1}{2}\times\dfrac{1}{11}$ (e) $\dfrac{1}{20}\times\dfrac{11}{13}$ (g) $\dfrac{5}{9}\times 17$ (i) $\dfrac{10}{17}\times 15$ (k) $\dfrac{1}{3}\times\dfrac{2}{3}\times 10$

(b) $\dfrac{1}{3}\times\dfrac{8}{5}$ (d) $\dfrac{1}{9}\times\dfrac{1}{9}$ (f) $4\times\dfrac{2}{3}$ (h) $\dfrac{12}{5}\times 6$ (j) $\dfrac{22}{7}\times 30$ (l) $\dfrac{1}{4}\times\dfrac{7}{5}\times\dfrac{11}{10}$

Ans. (a) $\dfrac{5}{14}$ (c) $\dfrac{1}{22}$ (e) $\dfrac{11}{260}$ (g) $\dfrac{85}{9}$ (i) $\dfrac{150}{17}$ (k) $\dfrac{20}{9}$

(b) $\dfrac{8}{15}$ (d) $\dfrac{1}{81}$ (f) $\dfrac{8}{3}$ (h) $\dfrac{72}{5}$ (j) $\dfrac{660}{7}$ (l) $\dfrac{77}{200}$

3.13. Multiply: **(3.2)**

(a) $\dfrac{1}{3}\times\dfrac{3}{5}$ (c) $\dfrac{11}{17}\times\dfrac{17}{11}$ (e) $\dfrac{50}{9}\times\dfrac{9}{5}$ (g) $\dfrac{50}{7}\times\dfrac{70}{5}$ (i) $\dfrac{3}{7}\times 21$ (k) $\dfrac{5}{21}\times\dfrac{41}{43}\times\dfrac{21}{5}$

(b) $\dfrac{4}{5}\times\dfrac{5}{12}$ (d) $\dfrac{159}{997}\times\dfrac{997}{159}$ (f) $\dfrac{3}{130}\times\dfrac{13}{3}$ (h) $15\times\dfrac{2}{5}$ (j) $\dfrac{3}{7}\times\dfrac{7}{3}\times 148$ (l) $\dfrac{9}{8}\times\dfrac{4}{45}$

Ans. In each, the cancelable factors are shown in parentheses after the answer.

(a) $\dfrac{1}{5}$ (3) (c) 1 (11, 17) (e) 10 (5, 9) (g) 100 (5, 7) (i) 9 (7) (k) $\dfrac{41}{43}$ (5, 21)

(b) $\dfrac{1}{3}$ (4, 5) (d) 1 (159, 997) (f) $\dfrac{1}{10}$ (3, 13) (h) 6 (5) (j) 148 (3, 7) (l) $\dfrac{1}{10}$ (4, 9)

3.14. In (a) to (d), state the reciprocal or multiplicative inverse of each: **(4.1)**

(a) 15 (b) $\dfrac{1}{20}$ (c) $\dfrac{3}{14}$ (d) $\dfrac{15}{11}$

In (e) to (l), find the missing number in each:

(e) $10\times ?=1$ (g) $\dfrac{11}{15}\times ?=1$ (i) $7\times\dfrac{1}{7}\times ?=5$ (k) $\dfrac{2}{3}\times\dfrac{5}{7}\times ?=\dfrac{5}{7}$

(f) $?\times 25=1$ (h) $?\times\dfrac{28}{27}=1$ (j) $\dfrac{2}{3}\times\dfrac{5}{7}\times ?=\dfrac{2}{3}$ (l) $\dfrac{1}{2}\times\dfrac{1}{4}\times 2\times 4=?$

Ans. (a) $\dfrac{1}{15}$ (b) 20 (c) $\dfrac{14}{3}$ (d) $\dfrac{11}{15}$ (e) $\dfrac{1}{10}$ (f) $\dfrac{1}{25}$ (g) $\dfrac{15}{11}$ (h) $\dfrac{27}{28}$ (i) 5 (j) $\dfrac{7}{5}$

(k) $\dfrac{3}{2}$ (l) 1

3.15. Divide and multiply as indicated: **(4.2)**

(a) $\dfrac{3}{5} \div 3$ (c) $12 \div \dfrac{3}{4}$ (e) $\dfrac{3}{4} \div 12$ (g) $\dfrac{7}{3} \div \dfrac{28}{15}$ (i) $\dfrac{1}{12} \div \dfrac{2}{5}$ (k) $49 \times \dfrac{27}{20} \div \dfrac{27}{10}$

(b) $\dfrac{3}{5} \div 5$ (d) $12 \div \dfrac{4}{3}$ (f) $\dfrac{4}{3} \div 12$ (h) $\dfrac{1}{5} \div \dfrac{1}{12}$ (j) $\dfrac{4}{9} \times \dfrac{27}{8} \div \dfrac{27}{10}$ (l) $49 \times \dfrac{10}{7} \div \dfrac{7}{10}$

Ans. (a) $\dfrac{3}{5} \times \dfrac{1}{3} = \dfrac{1}{5}$ (d) $12 \times \dfrac{3}{4} = 9$ (g) $\dfrac{7}{3} \times \dfrac{15}{28} = \dfrac{5}{4}$ (j) $\dfrac{4}{9} \times \dfrac{27}{8} \times \dfrac{10}{27} = \dfrac{5}{9}$

(b) $\dfrac{3}{5} \times \dfrac{1}{5} = \dfrac{3}{25}$ (e) $\dfrac{3}{4} \times \dfrac{1}{12} = \dfrac{1}{16}$ (h) $\dfrac{1}{5} \times 12 = \dfrac{12}{5}$ (k) $49 \times \dfrac{27}{20} \times \dfrac{10}{27} = \dfrac{49}{2}$

(c) $12 \times \dfrac{4}{3} = 16$ (f) $\dfrac{4}{3} \times \dfrac{1}{12} = \dfrac{1}{9}$ (i) $\dfrac{1}{12} \times \dfrac{5}{2} = \dfrac{5}{24}$ (l) $49 \times \dfrac{10}{7} \times \dfrac{10}{7} = 100$

3.16. Simplify each: **(4.3)**

(a) $\dfrac{1/3}{2}$ (c) $\dfrac{1/5}{10}$ (e) $\dfrac{1/5}{1/10}$ (g) $\dfrac{3/8}{3/8}$ (i) $\dfrac{4/7}{8/7}$ (k) $\dfrac{10/11}{50/77}$

(b) $\dfrac{4}{1/3}$ (d) $\dfrac{10}{1/5}$ (f) $\dfrac{3/4}{3/8}$ (h) $\dfrac{3/8}{8/3}$ (j) $\dfrac{7/9}{2/3}$ (l) $\dfrac{33/50}{22/75}$

Ans. (a) $\dfrac{1}{3} \times \dfrac{1}{2} = \dfrac{1}{6}$ (d) $10 \times 5 = 50$ (g) $\dfrac{3}{8} \times \dfrac{8}{3} = 1$ (j) $\dfrac{7}{9} \times \dfrac{3}{2} = \dfrac{7}{6}$

(b) $4 \times 3 = 12$ (e) $\dfrac{1}{5} \times 10 = 2$ (h) $\dfrac{3}{8} \times \dfrac{3}{8} = \dfrac{9}{64}$ (k) $\dfrac{10}{11} \times \dfrac{77}{50} = \dfrac{7}{5}$

(c) $\dfrac{1}{5} \times \dfrac{1}{10} = \dfrac{1}{50}$ (f) $\dfrac{3}{4} \times \dfrac{8}{3} = 2$ (i) $\dfrac{4}{7} \times \dfrac{7}{8} = \dfrac{1}{2}$ (l) $\dfrac{33}{50} \times \dfrac{75}{22} = \dfrac{9}{4}$

3.17. Find each missing term: **(5.1)**

(a) $\dfrac{1}{3} = \dfrac{?}{33}$ (b) $\dfrac{2}{5} = \dfrac{24}{?}$ (c) $\dfrac{8}{9} = \dfrac{?}{72}$ (d) $\dfrac{12}{7} = \dfrac{360}{?}$ (e) $\dfrac{25}{36} = \dfrac{?}{108}$ (f) $\dfrac{107}{71} = \dfrac{5,350}{?}$

Ans. (a) 11   (b) 60   (c) 64   (d) 210   (e) 75   (f) 3,550

3.18. (a) Express 2/5 as (1) 10ths, (2) 30ths, (3) 150ths, (4) 1,000ths.   **(5.2)**
(b) Express 7/20 as (1) 100ths, (2) 180ths, (3) 1,000ths, (4) 4,440ths.

Ans. (a) (1) $\dfrac{4}{10}$ (2) $\dfrac{12}{30}$ (3) $\dfrac{60}{150}$ (4) $\dfrac{400}{1,000}$

(b) (1) $\dfrac{35}{100}$ (2) $\dfrac{63}{180}$ (3) $\dfrac{350}{1,000}$ (4) $\dfrac{1,554}{4,440}$

3.19. In each, express the given fractions as equivalent fractions having a required common   **(5.3)**
denominator.

(a) $\dfrac{2}{5}$ and $\dfrac{5}{9}$ as 9ths   (d) $\dfrac{7}{12}$ and $\dfrac{9}{16}$ as 48ths   (g) $\dfrac{5}{6}, \dfrac{11}{15}, \dfrac{17}{30},$ and $\dfrac{31}{45}$ as 90ths

(b) $\dfrac{3}{4}$ and $\dfrac{5}{6}$ as 12ths   (e) $\dfrac{1}{2}, \dfrac{1}{4},$ and $\dfrac{3}{8}$ as 8ths

(c) $\dfrac{11}{50}$ and $\dfrac{12}{75}$ as 150ths   (f) $\dfrac{2}{5}, \dfrac{7}{10},$ and $\dfrac{14}{25}$ as 50ths

Ans. (a) $\dfrac{6}{9}, \dfrac{5}{9}$ (c) $\dfrac{33}{150}, \dfrac{24}{150}$ (e) $\dfrac{4}{8}, \dfrac{2}{8}, \dfrac{3}{8}$ (g) $\dfrac{75}{90}, \dfrac{66}{90}, \dfrac{51}{90}, \dfrac{62}{90}$

(b) $\dfrac{9}{12}, \dfrac{10}{12}$ (d) $\dfrac{28}{48}, \dfrac{27}{48}$ (f) $\dfrac{20}{50}, \dfrac{35}{50}, \dfrac{28}{50}$

3.20. In each, combine into a single fraction: **(6.1)**

(a) $\dfrac{5}{8}+\dfrac{3}{8}$ (c) $\dfrac{15}{32}+\dfrac{1}{32}$ (e) $\dfrac{3}{20}+\dfrac{7}{20}$ (g) $\dfrac{37}{100}-\dfrac{17}{100}$ (i) $\dfrac{17}{18}-\dfrac{5}{18}$ (k) $\dfrac{33}{25}-\dfrac{13}{25}$

(b) $\dfrac{11}{16}+\dfrac{9}{16}$ (d) $\dfrac{5}{7}+\dfrac{4}{7}$ (f) $\dfrac{29}{50}+\dfrac{11}{50}$ (h) $\dfrac{7}{10}-\dfrac{3}{10}$ (j) $\dfrac{15}{7}-\dfrac{8}{7}$

Ans. (a) 1   (c) $\dfrac{1}{2}$ (e) $\dfrac{1}{2}$ (g) $\dfrac{1}{5}$ (i) $\dfrac{2}{3}$ (k) $\dfrac{4}{5}$

(b) $\dfrac{5}{4}$ (d) $\dfrac{9}{7}$ (f) $\dfrac{4}{5}$ (h) $\dfrac{2}{5}$ (j) 1

3.21. In (a) to (e), combine into a single fraction: **(6.2)**

(a) $\dfrac{7}{2}+\dfrac{3}{2}+\dfrac{1}{2}$ (b) $\dfrac{7}{4}+\dfrac{3}{4}-\dfrac{5}{4}$ (c) $\dfrac{13}{8}+\dfrac{5}{8}-\dfrac{3}{8}$ (d) $\dfrac{15}{13}+\dfrac{2}{13}-\dfrac{4}{13}$ (e) $\dfrac{17}{25}+\dfrac{7}{25}-\dfrac{6}{25}-\dfrac{3}{25}$

In (f) to (i), if $n = 20/21$, $p = 11/21$, $q = 8/21$, and $r = 1/21$, find

(f) $n+p+q$ (g) $p+q-r$ (h) $n-p-q$ (i) $n+p-r$

Ans. (a) $\dfrac{11}{2}$ (c) $\dfrac{15}{8}$ (e) $\dfrac{3}{5}$ (g) $\dfrac{6}{7}$ (i) $\dfrac{10}{7}$

(b) $\dfrac{5}{4}$ (d) 1   (f) $\dfrac{13}{7}$ (h) $\dfrac{1}{21}$

3.22. In each, find the least common denominator of the given fractions: **(7.1)**

(a) $\dfrac{2}{3}, \dfrac{1}{4}$ (c) $\dfrac{1}{3}, \dfrac{3}{5}, \dfrac{4}{7}$ (e) $\dfrac{1}{5}, \dfrac{7}{10}$ (g) $\dfrac{1}{4}, \dfrac{5}{6}$ (i) $\dfrac{2}{3}, \dfrac{1}{9}, \dfrac{7}{12}$

(b) $\dfrac{2}{5}, \dfrac{3}{7}$ (d) $\dfrac{3}{4}, \dfrac{7}{8}$ (f) $\dfrac{1}{4}, \dfrac{2}{3}, \dfrac{5}{12}$ (h) $\dfrac{1}{2}, \dfrac{3}{4}, \dfrac{4}{5}$

Ans. (a) 12   (b) 35   (c) 105   (d) 8   (e) 10   (f) 12   (g) 12   (h) 20   (i) 36

3.23. In (a) through (i), combine into a single fraction: **(7.2)**

(a) $\dfrac{2}{3}+\dfrac{1}{4}$ (c) $\dfrac{3}{4}+\dfrac{7}{8}$ (e) $\dfrac{3}{4}-\dfrac{1}{6}$ (g) $\dfrac{3}{4}-\dfrac{2}{5}-\dfrac{1}{10}$ (i) $\dfrac{7}{12}+\dfrac{1}{9}-\dfrac{2}{3}$

(b) $\dfrac{3}{7}-\dfrac{2}{5}$ (d) $\dfrac{7}{10}-\dfrac{1}{5}$ (f) $\dfrac{1}{2}+\dfrac{3}{4}-\dfrac{3}{5}$ (h) $\dfrac{2}{3}+\dfrac{2}{9}-\dfrac{7}{12}$

In (j) through (m), if $p = 11/12$, $q = 7/8$, and $r = 1/4$, find

(j) $p+q+r$ (k) $p+q-r$ (l) $p+r-q$ (m) $r+q-p$

Ans. (a) $\dfrac{11}{12}$ (c) $\dfrac{13}{8}$ (e) $\dfrac{7}{12}$ (g) $\dfrac{1}{4}$ (i) $\dfrac{1}{36}$ (k) $\dfrac{37}{24}$ (m) $\dfrac{5}{24}$

(b) $\dfrac{1}{35}$ (d) $\dfrac{1}{2}$ (f) $\dfrac{13}{20}$ (h) $\dfrac{11}{36}$ (j) $\dfrac{49}{24}$ (l) $\dfrac{7}{24}$

3.24. Table 3-1 shows the parts of five family budgets that have been set aside for three kinds **(7.3)** of food. In each, find the remaining part for baked products.

Table 3-1

Family	Groceries	Fruit and Produce	Meats, Fish, and Dairy	Baked Products
(a)	3/10	1/5	2/5	?
(b)	9/20	1/10	3/10	?
(c)	7/20	1/10	7/20	?
(d)	2/5	1/4	3/10	?
(e)	9/20	6/25	1/4	?

Ans. (a) 1/10   (b) 3/20   (c) 1/5   (d) 1/20   (e) 3/50

3.25. Using the common denominator method, compare the fractions in each: **(8.1)**

(a) $\frac{3}{4}$ and $\frac{12}{16}$ (b) $\frac{2}{3}$ and $\frac{3}{5}$ (c) $\frac{6}{5}$ and $\frac{16}{15}$ (d) $\frac{5}{9}$ and $\frac{16}{27}$ (e) $\frac{7}{10}$ and $\frac{37}{50}$

Ans. (a) $\frac{3}{4} = \frac{12}{16}$ (b) $\frac{2}{3} > \frac{3}{5}$ (c) $\frac{6}{5} > \frac{16}{15}$ (d) $\frac{5}{9} < \frac{16}{27}$ (e) $\frac{7}{10} < \frac{37}{50}$

3.26. Using cross products, compare the fractions in each. **(8.2)**

(a) $\frac{1}{8}$ and $\frac{2}{15}$ (b) $\frac{5}{8}$ and $\frac{7}{11}$ (c) $\frac{9}{20}$ and $\frac{21}{50}$ (d) $\frac{6}{11}$ and $\frac{6}{13}$ (e) $\frac{11}{12}$ and $\frac{23}{25}$

Ans. (a) $15 < 16$; hence, $\frac{1}{8} < \frac{2}{15}$ (c) $450 > 420$; hence, $\frac{9}{20} > \frac{21}{50}$ (e) $275 < 276$; hence, $\frac{11}{12} < \frac{23}{25}$

(b) $55 < 56$; hence, $\frac{5}{8} < \frac{7}{11}$ (d) $78 > 66$; hence, $\frac{6}{11} > \frac{6}{13}$

3.27. Express (a) 5 as 4ths, (b) 3 as 8ths, (c) 4 as 16ths, (d) 10 as 32nds, (e) 7 as 9ths **(9.1)**
Ans. (a) $\frac{20}{4}$ (b) $\frac{24}{8}$ (c) $\frac{64}{16}$ (d) $\frac{320}{32}$ (e) $\frac{63}{9}$

3.28. Express as an improper fraction: (a) $5\frac{1}{2}$, (b) $3\frac{2}{3}$, (c) $4\frac{2}{7}$, (d) $8\frac{5}{9}$, (e) $20\frac{7}{10}$. **(9.2)**
Ans. (a) $\frac{11}{2}$ (b) $\frac{11}{3}$ (c) $\frac{30}{7}$ (d) $\frac{77}{9}$ (e) $\frac{207}{10}$

3.29. (a) How many half-inches are there in (1) 5 inches, (2) $4\frac{1}{2}$ inches? **(9.3)**
(b) How many quarter-inches are there in (1) 7 inches, (2) $9\frac{1}{4}$ inches?
(c) How many eighth-inches are there in (1) 10 inches, (2) $8\frac{7}{8}$ inches?
(d) How many sixteenth-inches are there in (1) 3 inches, (2) $4\frac{1}{4}$ inches?

Ans. (a) (1) 10   (2) 9   (c) (1) 80   (2) 71
(b) (1) 28   (2) 37   (d) (1) 48   (2) 68

3.30. Express each as a whole number or as a mixed number.   **(10.1)**

(a) $\dfrac{7}{2}$ (b) $\dfrac{39}{3}$ (c) $\dfrac{52}{5}$ (d) $\dfrac{75}{10}$ (e) $\dfrac{150}{36}$

Ans.   (a) $3\frac{1}{2}$ (b) 13   (c) $10\frac{2}{5}$ (d) $7\frac{1}{2}$ (e) $4\frac{1}{6}$

3.31. How many meters are in (a) 20 half-meters,   (b) 42 quarter-meters,   (c) 144   **(10.2)**
eighth-meters,   (d) 84 sixteenth-meters,   (e) 327 thirty-second-meters

Ans.   (a) 10   (b) $10\frac{1}{2}$ (c) 18   (d) $5\frac{1}{4}$ (e) $10\frac{7}{32}$

3.32. Add: **(11.1)**

(a) $5 + 7\frac{1}{2}$ (b) $8\frac{3}{4} + 26$ (c) $12\frac{3}{5} + \dfrac{2}{5}$ (d) $\dfrac{7}{8} + 19\frac{1}{8}$ (e) $3\frac{3}{5} + 5\frac{4}{5}$

(f) $\begin{array}{r} 10\frac{3}{7} \\ +7\frac{4}{7} \end{array}$ (g) $\begin{array}{r} 12\frac{5}{8} \\ +16\frac{7}{8} \end{array}$ (h) $\begin{array}{r} 21\frac{5}{9} \\ +15\frac{8}{9} \end{array}$ (i) $\begin{array}{r} 31\frac{11}{16} \\ +14\frac{13}{16} \end{array}$ (j) $\begin{array}{r} 25\frac{2}{5} \\ +13\frac{1}{2} \end{array}$ (k) $\begin{array}{r} 43\frac{3}{4} \\ +25\frac{1}{8} \end{array}$ (l) $\begin{array}{r} 56\frac{5}{6} \\ +27\frac{2}{9} \end{array}$

Ans.   (a) $12\frac{1}{2}$ (c) 13   (e) $9\frac{2}{5}$ (g) $29\frac{1}{2}$ (i) $79\frac{1}{2}$ (k) $68\frac{7}{8}$

       (b) $34\frac{3}{4}$ (d) 20   (f) 18   (h) $37\frac{4}{9}$ (j) $38\frac{9}{10}$ (l) $84\frac{1}{18}$

3.33. Add: **(11.1)**

(a) $\dfrac{1}{4} + 4\frac{3}{4} + 5\frac{1}{4}$ (c) $10\frac{7}{8} + 20\frac{5}{8} + 25\frac{1}{8}$ (e) $21\frac{1}{2} + 17\frac{1}{3} + 38\frac{3}{5}$

(b) $7\frac{2}{5} + 3\frac{4}{5} + 8\frac{1}{5}$ (d) $12\frac{2}{3} + 15\frac{1}{6} + 30\frac{5}{12}$ (f) $51\frac{1}{4} + 46\frac{3}{8} + 29\frac{5}{12}$

Ans.   (a) $10\frac{1}{4}$ (b) $19\frac{2}{5}$ (c) $56\frac{5}{8}$ (d) $58\frac{1}{4}$ (e) $77\frac{13}{30}$ (f) $127\frac{1}{24}$

3.34.   (a) Figure 3-13 shows the distances in kilometers between towns *A, B, C, D* and *E*   **(12.2)**
located on a straight road. Find the distance in kilometers (1) from *A* to *C*, (2) from *B* to *D*, (3) from *B* to *E*, (4) from *A* to *E*.

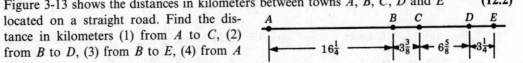

Fig. 3-13

    (b) Find the total number of pounds in two boxes if their weights in pounds are: (1) each $35\frac{1}{4}$, (2) $35\frac{1}{6}$ and $40\frac{5}{6}$, (3) $18\frac{2}{3}$ and $25\frac{5}{9}$.

    (c) Find the total number of yards in three rolls of cloth if their measurements in yards are (1) each $24\frac{1}{2}$; (2) $24\frac{3}{8}$, $24\frac{5}{8}$, and $24\frac{7}{16}$; (3) $75\frac{1}{8}$, $49\frac{15}{16}$, and $103\frac{1}{4}$.

Ans.   (a) (1) $19\frac{5}{8}$ (2) 10   (3) $13\frac{1}{4}$ (4) $29\frac{1}{2}$

       (b) (1) $70\frac{1}{2}$ (2) 76   (3) $44\frac{2}{9}$

       (c) (1) $73\frac{1}{2}$ (2) $73\frac{7}{16}$ (3) $228\frac{5}{16}$

3.35. Subtract: **(12.1)**

(a) $17\frac{11}{15} - 12$ (b) $24\frac{5}{13} - 15$ (c) $12 - \dfrac{5}{8}$ (d) $25 - 5\frac{3}{7}$ (e) $18\frac{17}{20} - 9\frac{17}{20}$

(f) $\begin{array}{r} 8\frac{7}{9} \\ -4\frac{5}{9} \end{array}$ (g) $\begin{array}{r} 9\frac{5}{9} \\ -6\frac{7}{9} \end{array}$ (h) $\begin{array}{r} 10\frac{1}{4} \\ -4\frac{5}{8} \end{array}$ (i) $\begin{array}{r} 24\frac{3}{5} \\ -17\frac{1}{6} \end{array}$ (j) $\begin{array}{r} 30\frac{3}{4} \\ -15\frac{5}{6} \end{array}$ (k) $\begin{array}{r} 85\frac{7}{12} \\ -71\frac{7}{8} \end{array}$ (l) $\begin{array}{r} 111\frac{1}{6} \\ -85\frac{3}{8} \end{array}$

Ans.   (a) $5\frac{11}{15}$ (c) $11\frac{3}{8}$ (e) 9   (g) $2\frac{7}{9}$ (i) $7\frac{13}{30}$ (k) $13\frac{17}{24}$

       (b) $9\frac{5}{13}$ (d) $19\frac{4}{7}$ (f) $4\frac{2}{9}$ (h) $5\frac{5}{8}$ (j) $14\frac{11}{12}$ (l) $25\frac{19}{24}$

3.36. (a) Figure 4-14 shows the distances in miles between towns A, B, and C located on a   **(12.2)**
straight road. Find the distance in
miles from B to C if the distance d
from A to B is (1) $5\frac{1}{4}$, (2) $12\frac{5}{11}$,
(3) $16\frac{14}{17}$.

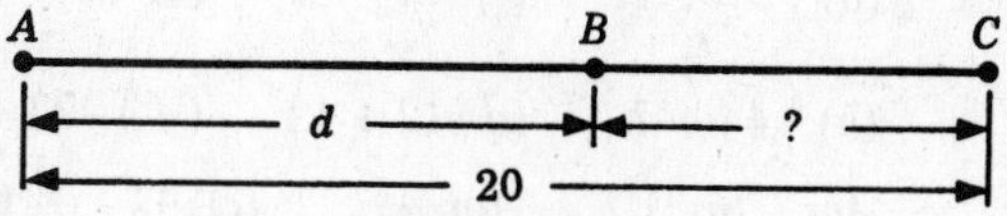

(b) Find the number of grams removed
from a box weighing 150 grams if the
remaining weight in grams is (1) $85\frac{1}{4}$, (2) $63\frac{5}{8}$, (3) $28\frac{17}{50}$.

Fig. 3-14

(c) From a roll of cloth $30\frac{1}{4}$ yards long, a length is cut. How many yards are cut if the number of
yards remaining is (1) $5\frac{1}{8}$, (2) $14\frac{5}{16}$, (3) $26\frac{13}{32}$?

(d) What is the rise in temperature when the temperature changes from an initial reading of $65\frac{1}{2}$
degrees to a new reading in degrees of (1) 74, (2) $83\frac{1}{4}$, (3) $102\frac{3}{5}$?

Ans. (a) (1) $14\frac{3}{4}$ (2) $7\frac{6}{11}$ (3) $3\frac{3}{17}$

(b) (1) $64\frac{3}{4}$ (2) $86\frac{3}{8}$ (3) $121\frac{33}{50}$

(c) (1) $25\frac{1}{8}$ (2) $15\frac{15}{16}$ (3) $3\frac{27}{32}$

(d) (1) $8\frac{1}{2}$ (2) $17\frac{3}{4}$ (3) $37\frac{1}{10}$

3.37. Multiply: **(13.1)**

(a) $12\frac{1}{2} \times 2$ (d) $40 \times 2\frac{3}{8}$ (g) $9\frac{1}{8} \times 5$ (j) $\frac{7}{15} \times 2\frac{1}{7}$ (m) $3\frac{3}{10} \times 1\frac{9}{11}$

(b) $8\frac{1}{4} \times 12$ (e) $48 \times 5\frac{3}{8}$ (h) $3\frac{1}{5} \times \frac{1}{4}$ (k) $5\frac{1}{3} \times 1\frac{1}{2}$ (n) $10\frac{2}{3} \times 3\frac{3}{4}$

(c) $15 \times 7\frac{2}{5}$ (f) $8\frac{3}{16} \times 4$ (i) $\frac{2}{7} \times 4\frac{2}{3}$ (l) $7\frac{1}{2} \times 5\frac{1}{5}$ (o) $8\frac{7}{11} \times 6\frac{3}{5}$

Ans. (a) 25   (c) 111   (e) 258   (g) $45\frac{5}{8}$ (i) $1\frac{1}{3}$ (k) 8   (m) 6   (o) 57

(b) 99   (d) 95   (f) $32\frac{3}{4}$ (h) 4/5   (j) 1   (l) 39   (n) 40

3.38. Multiply: **(13.1)**

(a) $3\frac{1}{7} \times \frac{4}{5} \times 35$ (c) $3\frac{3}{4} \times 2\frac{1}{7} \times \frac{14}{25}$ (e) $5\frac{5}{7} \times 2\frac{1}{3} \times 4\frac{1}{2}$

(b) $20 \times 3\frac{1}{3} \times \frac{3}{5}$ (d) $4\frac{1}{5} \times 3\frac{1}{7} \times 1\frac{9}{11}$ (f) $1\frac{1}{4} \times 1\frac{4}{7} \times 2\frac{4}{5}$

Ans. (a) 88   (b) 40   (c) $4\frac{1}{2}$ (d) 24   (e) 60   (f) $5\frac{1}{2}$

3.39. (a) A rod is $7\frac{1}{2}$ meters long. How much does it weigh if each meter weighs (1) 4 kg,   **(13.2)**
(2) $3\frac{1}{3}$ kg, (3) $2\frac{4}{5}$ kg?

(b) Find the cost of $4\frac{3}{8}$ yards of cloth if each yard costs (1) $8, (2) $7\frac{1}{5}$, (3) $3\frac{9}{25}$.

(c) At a rate of 45 miles per hour, how many miles can a car go if the time taken to travel is (1)
$3\frac{1}{5}$ hours, (2) $2\frac{2}{3}$ hours, (3) $1\frac{1}{15}$ hours?

(d) If a man is paid at the rate of $12\frac{4}{5}$ per hour, how much does he earn in (1) 5 hours, (2) 10
hours, (3) $3\frac{3}{4}$ hours?

Ans. (a) (1) 30 kg   (2) 25 kg   (3) 21 kg

(b) (1) $35   (2) $31\frac{1}{2}$ (3) $14\frac{7}{10}$

(c) (1) 144   (2) 120   (3) 48

(d) (1) $64   (2) $128   (3) $48

3.40. Divide: **(13.3)**

(a) $5\frac{1}{2} \div 11$ (c) $14\frac{1}{4} \div 3$ (e) $15 \div 3\frac{1}{3}$ (g) $1\frac{2}{3} \div 2\frac{7}{9}$ (i) $12\frac{1}{4} \div 2\frac{11}{12}$ (k) $3\frac{1}{7} \div 2\frac{5}{14}$

(b) $4\frac{2}{3} \div 7$ (d) $12 \div 1\frac{1}{5}$ (f) $22 \div 1\frac{5}{6}$ (h) $\dfrac{11}{12} \div 7\frac{1}{3}$ (j) $2\frac{3}{16} \div 2\frac{1}{10}$ (l) $5\frac{1}{2} \div 3\frac{1}{7}$

Ans. (a) 1/2 (c) $4\frac{3}{4}$ (e) $4\frac{1}{2}$ (g) 3/5 (i) $4\frac{1}{5}$ (k) $1\frac{1}{3}$

(b) 2/3 (d) 10 (f) 12 (h) 1/8 (j) $1\frac{1}{24}$ (l) $1\frac{3}{4}$

3.41. (a) How many quarter-inch pieces can be cut from a board whose length in inches is **(13.4)**
(1) 15, (2) $18\frac{1}{2}$, (3) $25\frac{3}{4}$?

(b) At a rate of $2\frac{2}{3}$ miles per minute, how many minutes does it take a racing car to travel (1) $3\frac{5}{9}$ miles, (2) $35\frac{5}{9}$ miles, (3) 144 miles?

(c) How many poles, each $1\frac{1}{2}$ meters long, can be cut from a longer pole whose length in meters is (1) $22\frac{1}{2}$, (2) 30, (3) $37\frac{1}{2}$?

(d) At what hourly rate is a worker paid if he earns \$36 in (1) $4\frac{1}{2}$ hr, (2) $4\frac{4}{5}$ hr, (3) $6\frac{2}{3}$ hr?

Ans. (a) (1) 60 (2) 74 (3) 103
(b) (1) $1\frac{1}{3}$ min (2) $13\frac{1}{3}$ min (3) 54 min
(c) (1) 15 (2) 20 (3) 25
(d) (1) \$8 per hr (2) $7\frac{1}{2}$ per hr (3) $5\frac{2}{5}$ per hr

Chapter 4

Decimals

1. UNDERSTANDING DECIMALS AND DECIMAL FRACTIONS

The use of decimals is fundamental in our monetary system. To understand decimals, you must know the difference between decimals, common fractions, and decimal fractions, as shown in Table 4-1.

Table 4-1   United States Coinage Table

Coin	Number of Cents	Fractional Part of a Dollar		Decimals
		Common Fraction	Decimal Fraction	
Half-dollar	50	1/2	5/10 or 50/100	.5 or .50
Quarter	25	1/4	25/100	.25
Dime	10	1/10	1/10 or 10/100	.1 or .10
Nickel	5	1/20	5/100	.05
Penny	1	1/100	1/100	.01

Decimal Fraction and Decimal

A *decimal fraction* is a common fraction whose denominator is 10, 100, 1,000, or any other place value. Note the decimal fractions in Table 4-1.

A *decimal* is a symbol for an equivalent decimal fraction. Note the equivalency in Table 4-1. The use of decimals greatly simplifies writing and computation. The metric system is based on the use of the decimal.

Thus, .5 is the symbol for the decimal fraction 5/10, and .05 is the symbol for the decimal fraction 5/100.

Expressing a Decimal Fraction as a Decimal, and Vice Versa

Rule: The number of decimal places of the decimal equals the number of 0s in the denominator of the decimal fraction.

Thus, 3/10 = .3 shows that a decimal fraction having 1 zero in the denominator is expressed as a decimal of 1 decimal place. Also, 705/100 = 7.05 shows that a decimal fraction having 2 zeros in the denominator is expressed as a decimal having 2 decimal places.

To express a decimal fraction as a decimal, or a decimal as a decimal fraction, match the number of 0s in the denominator of the decimal fraction with the number of decimal places of the decimal.

Thus, express 25/1,000 as .025 by matching the 3 zeros in the denominator with 3 decimal places. Also, express .00475 as 475/100,000 by matching the 5 decimal places with 5 zeros in the denominator.

Expressing a Common Fraction, a Whole Number, or a Mixed Number in Decimal Form

To express a common fraction in decimal form, first express the common fraction as a decimal fraction; then express the decimal fraction as a decimal. Thus, to express 1/2 in decimal form, first express 1/2 as 5/10, then express 5/10 as .5.

To express a whole number in decimal form, place a decimal point after the whole number. Thus, 32. is the decimal form of 32.

To express a whole number in decimal form, first express the fraction part as a decimal, then combine the whole number part and the decimal. Thus, to express $3\frac{1}{4}$ in decimal form, first express $\frac{1}{4}$ as .25, then combine with 3 to obtain the decimal equivalent 3.25.

Reading a Decimal

The two methods used to read decimals are the following:

Method 1: Read the decimal in the same way as the equivalent decimal fraction.

Thus, read .35 as "thirty-five hundredths," exactly as you would read 35/100.

Important Notes:

(1) When reading a decimal, end with the place value of the rightmost digit.

Thus, .00025 is read "twenty-five hundred-thousandths," since the rightmost digit 5 is in hundred-thousandths place.

(2) Use "and" only when naming the decimal point.

Thus, .702 is read "seven hundred two thousandths." The reading "seven hundred and two thousandths" may be mistaken for 700.002.

Method 2: Read the decimal by saying the digits in order from left to right, as one reads telephone numbers. Use "point" to name the decimal point.

Thus, .00025 is read "point zero zero zero two five."

Caution: Learn to use Method 1 before using Method 2. An understanding of Method 1 is valuable and essential to an understanding of decimals.

Place Value Relationships among Whole Numbers and Decimals

In Table 4-2, note the following:

(1) From left to right, each place value, whether whole number or decimal, is ten times the next place value. Thus, $1,000 = 10 \times 100$ and $.1 = 10 \times .01$.

(2) From right to left, each place value is one-tenth the next place value. Thus, $10 = \frac{1}{10} \times 100$ and $.001 = \frac{1}{10} \times .01$.

Table 4-2

	Whole Number Place Values			Decimal Point	Decimal Place Values		
	Thousands	Hundreds	Tens	Units	Tenths	Hundredths	Thousandths
	1,000	100	10	1	.1 or $\frac{1}{10}$	.01 or $\frac{1}{100}$	.01 or $\frac{1}{1,000}$
(a) 99.9			9	9 .	0		
(b) 707.07		7	0	7 .	0	7	
(c) 3,003.003	3	0	0	3 .	0	0	3

(a) 99.9 is read "ninety nine *and* nine tenths." The expanded form of 99.9 is $99.9 = 90 + 9 + .9$. Note that the decimal point is read "and."

(b) 707.07 is read "seven hundred seven *and* seven hundredths." The expanded form of 707.07 is $700 + 7 + .07$. Include an addend for each nonzero digit.

(c) 3,003.003 is read "three thousand three *and* three thousandths." The expanded form of 3,003.003 is $3,000 + 3 + .003$. Use "and" only for the decimal point.

Reading Decimals on a Number Line in Tenths

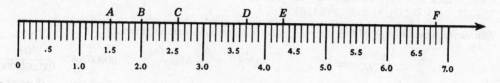

Fig. 4-1

The number line, Fig. 4-1, is divided into tenths. Hence the coordinates of the lettered points shown are 1.5 for *A*, 2.0 or 2 for *B*, 2.6 for *C*, 3.7 for *D*, 4.3 for *E*, and 6.8 for *F*.

Reading Decimals on a Number Line in Hundredths

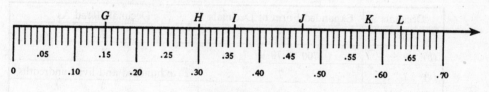

Fig. 4-2

The number line, Fig. 4-2, is divided into hundredths. Hence, the coordinates of the lettered points shown are .15 for *G*, .30 for *H*, .36 for *I*, .47 for *J*, .58 for *K*, and .63 for *L*.

1.1 EXPRESSING UNITED STATES MONEY IN DECIMAL FORM

Express each as a decimal in dollars: (a) 12 pennies; (b) 7 nickels; (c) 5 dimes; (d) 3 quarters; (e) 2 dollars and 4 pennies; (f) 1 quarter, 2 dimes, and 3 pennies; (g) 3 dollars, 7 dimes, and 3 nickels.

Illustrative Solution (f) Since 1 quarter = \$.25, 2 dimes = \$.20, and 3 pennies = \$.03, then

$$1 \text{ quarter} + 2 \text{ dimes} + 3 \text{ pennies} = \$.25 + \$.20 + \$.03 = \$.48 \quad Ans.$$

Ans. (a) \$.12 (b) \$.35 (c) \$.50 (d) \$.75 (e) \$2.04 (g) \$3.85

1.2 EXPRESSING DECIMAL FRACTIONS AS DECIMALS

Express each as a decimal:

(a) $\dfrac{9}{10}, \dfrac{77}{10}, \dfrac{185}{10}$ (b) $\dfrac{5}{100}, \dfrac{56}{100}, \dfrac{567}{100}$ (c) $\dfrac{3}{1,000}, \dfrac{33}{1,000}, \dfrac{3,333}{1,000}$ (d) $\dfrac{666}{1,000,000}$

Illustrative Solution (*d*) Match zeros in the denominator and decimal places. In 1,000,000, there are 6 zeros. Hence, the decimal has 6 places:

$$\frac{666}{1,000,000} = .000666 \quad Ans.$$

$$\text{(6 zeros)} \qquad \text{(6 places)}$$

Ans. (*a*) .9, 7.7, 18.5 (*b*) .05, .56, 5.67 (*c*) .003, .033, 3.333

1.3 EXPRESSING DECIMALS AS DECIMAL FRACTIONS

Express each as a decimal fraction: (*a*) .06, .006, .00006; (*b*) 3.09, 30.9, .00000309; (*c*) .003, 3.0003, 33.0333; (*d*) 812.03, 81.203, .00081203.

Illustrative Solution (*b*) Match decimal places with zeros in the denominator.

$$3.09 = \frac{309}{100} \; Ans. \qquad 30.9 = \frac{309}{10} \; Ans. \qquad .00000309 = \frac{309}{100,000,000} \; Ans.$$

$$\text{(2 places)} \quad \text{(2 zeros)} \qquad \text{(1 place)} \quad \text{(1 zero)} \qquad \text{(8 places)} \quad \text{(8 zeros)}$$

Ans. (*a*) $\frac{6}{100}, \frac{6}{1,000}, \frac{6}{100,000}$ (*c*) $\frac{3}{1,000}, \frac{30,003}{10,000}, \frac{330,333}{10,000}$ (*d*) $\frac{81,203}{100}, \frac{81,203}{1,000}, \frac{81,203}{100,000,000}$

1.4 RELATING DECIMALS, EXPANDED FORM OF DECIMALS, AND NAMES

Complete Table 4-3.

Table 4-3

Decimals	Expanded Form of Decimals	Decimals Read As
(*a*) 88.08	?	?
(*b*) ?	700 + 70 + .7	?
(*c*) ?	?	Five hundred and five hundredths
(*d*) 40.004	?	?
(*e*) ?	6,000 + 60 + .00006	?
(*f*) ?	?	One million one and one tenth

Illustrative Solution (*a*) In expanded form, use an addend for each nonzero digit:

$$88.08 = 80 + 8 + .08$$

When reading a decimal, use "and" only for the decimal point: eighty eight *and* eight hundredths.
Ans. See Table 4-4.

Table 4-4

Decimals	Expanded Form of Decimals	Decimals Read As
(*b*) 770.7	700 + 70 + .7	Seven hundred seventy and seven tenths
(*c*) 500.05	500 + .05	Five hundred and five hundredths
(*d*) 40.004	40 + .004	Forty and four thousandths
(*e*) 6,060.00006	6,000 + 60 + .00006	Six thousand sixty and six hundred-thousandths
(*f*) 1,000,001.1	1,000,000 + 1 + .1	One million one and one tenth

1.5 **EXPRESSING PARTS OF A FIGURE AS A DECIMAL**

In Fig. 4-3, state the decimals that indicate the shaded and unshaded parts.

Illustrative Solution   (c) Out of 100 squares, 44 are shaded and the remaining 56 are unshaded. Hence, the shaded part occupies 44/100 or .44 and the unshaded part 56/100 or .56 of the figure.

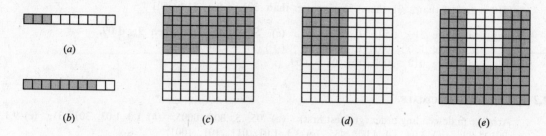

(a)   (b)   (c)   (d)   (e)

Fig. 4-3

Ans.

	Shaded Part	Unshaded Part
(a)	.3	.7
(b)	.8	.2
(c)	.44	.56

	Shaded Part	Unshaded Part
(d)	.25	.75
(e)	.76	.24

2. ORDERING AND COMPARING DECIMALS

Believe it or not, .7000 and .7 are equivalent decimals. The reason is that

$$\frac{7,000}{10,000} = \frac{7}{10}$$

Rule: The value of a decimal does not change when zeros are annexed to the right.

Ordering Decimals

Observe in the following example how decimals are ordered by changing them to equivalent decimals having the same number of decimal places.

To Order Decimals

Arrange in descending order (greatest first):

(a) .05, .5, .4899, .501   (b) .15, .0051, .105, .051

PROCEDURE   SOLUTIONS

1. Express the given decimals as equivalent decimals having the same number of decimal places by annexing zeros to the right:

2. Ignoring the decimal point, the greater the number, the greater the decimal:

1. .0500   .1500
    .5000   .0051
    .4899   .1050
    .5010   .0510

2. In descending order: In descending order:
    .5010 = .501   .1500 = .15
    .5000 = .5   .1050 = .105
    .4899 = .4899   .0510 = .051
    .0500 = .05   .0051 = .0051

2.1 COMPARING DECIMALS

Compare the decimals in each pair, using the symbol =, >, or <:

(a) .51, 1.5   (c) .15, .051   (e) .9, 1.09
(b) .51, .501   (d) .01, .009   (f) .51, .51000

Illustrative Solution (b) To compare .51 and .501, express .51 as .510 by adding a zero to the right. Since .510 is greater than .501, then, .51 is greater than .501.   Ans. .51 > .501

Ans. (a) Since .51 < 1.50, .51 < 1.5.   (e) Since .90 < 1.09, then .9 < 1.09.
      (c) Since .150 > .051, .15 > .051.   (f) .51 = .51000
      (d) Since .010 > .009, then .01 > .009.

2.2 ORDERING DECIMALS

Arrange in descending order (greatest first): (a) .05, .5, .005, .0005; (b) 1.3, 1.03, .301, .31; (c) 9.1, .19, .091, 1.09; (d) 3.04, 3.4, 4.003 .43; (e) 1.1, 1.01, .011, .101, .1001.

Illustrative Solution (b) Express each decimal in 3 decimal places for comparison purposes. Hence, add 2 zeros to 1.3, 1 zero to 1.03 and .31. Comparing 1.300, 1.030, .301, and .310, the arrangement in descending order is 1.300, 1.030, .310, 301.   Ans. 1.3, 1.03, .31, .301

Ans. (a) .5, .05, .005, .0005   (d) 4.003, 3.4, 3.04, .43
      (c) 9.1, 1.09, .19, .091   (e) 1.1, 1.01, .101, .1001, .011

3. ROUNDING DECIMALS

Rounding a Decimal to the Nearest Unit

To learn how to round a decimal to the nearest unit, note how an amount of money is rounded to the nearest dollar. Consider amounts of money between $61 and $62.
(1) Amounts greater than $61.00 and less than $61.50 are rounded to $61.
(2) Amounts equal to or greater than $61.50 are rounded to $62.
Thus, $61.49 is rounded to $61. Also, $61.50 and $61.75 are rounded to $62.
Besides the units place, any other decimal place may be chosen for rounding, as described below.

General Method for Rounding a Decimal to a Specific Place

(1) Note the digit to the right of the digit in the specific place.
(2) Add 1 to the digit in the specific place if the digit to its right is 5 or more; keep the digit in the specific place if the digit to its right is less than 5.
(3) Eliminate all digits to the right of the digit in the specific place.

To Round a Decimal to Hundredths

Round to hundredths: (a) 8.523   (b) 8.5279

PROCEDURE	SOLUTIONS	
1. Note the digit to the right of the *hundredths* digit:	1. Note 3 to the right of 2, the hundredths digit.	Note 7 to the right of 2, the hundredths digit.
2. Add 1 to the *hundredths* digit if the digit to its right is 5 or more; keep the *hundredths* digit if the noted digit is less than 5:	2. Since 3 is less than 5, keep 2.	Since 7 is more than 5, add 1 to 2, making it 3.
3. Eliminate all digits to the right of the *hundredths* digit:	3. Eliminate 3. Ans. 8.52	Eliminate 79. Ans. 8.53

3.1   USING A NUMBER LINE TO ROUND DECIMALS

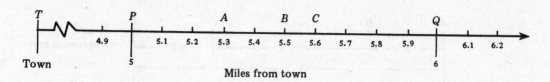

Fig. 4-4

In Fig. 4-4, each point on segment $\overline{PQ}$ indicates a distance between 5 and 6 miles from town T. To the nearest mile, find the distance from town T of   (a) a home at A,   (b) a road junction at B,   (c) a factory at C.

Illustrative Solution   (b) The halfway distance of 5.5 miles is considered to be nearer to 6 miles.   *Ans.* 6 miles

Ans.   (a) 5 miles   (c) 6 miles

3.2   ROUNDING AMOUNTS OF MONEY TO THE NEAREST DOLLAR OR NEAREST CENT

Round to the nearest dollar: (a) $5.60,   (b) $25.40,   (c) $15.50,   (d) $54.49,   (e) $45.51.
Round to the nearest cent: (f) $8.547,   (g) $5.624,   (h) $1.765.

Illustrative Solutions
   (d)   Since $54.49 is less than $54.50, round to $54.
   (f)   Since $8.547 is more than $8.545, round to $8.55.

Ans.   (a) $6   (b) $25   (c) $16   (e) $46   (g) $5.62   (h) $1.77

3.3   ROUNDING DECIMALS TO A GIVEN NUMBER OF DECIMAL PLACES

Round
to tenths: (a) .72,   (b) .75,   (c) 2.76,   (d) 2.7489.
to hundredths: (e) 0.234,   (f) 0.239,   (g) 0.235,   (h) 0.25333.
to thousandths: (i) 4.5005,   (j) 4.5001,   (k) 4.50048,   (l) 4.499499.
to ten-thousandths: (m) 1.23456,   (n) 1.23465,   (o) 1.234506.

Illustrative Solution   (d) In 2.7489, note 4 to the right of 7 in *tenths* place. Since 4 is less than 5, keep 7. Eliminate 489 to the right of 7.   *Ans.* 2.7

Ans.   (a) .7   (c) 2.8   (f) 0.24   (h) 0.25   (j) 4.500   (l) 4.499   (n) 1.2347
      (b) .8   (e) 0.23   (g) 0.24   (i) 4.501   (k) 4.500   (m) 1.2346   (o) 1.2345

3.4   FINDING SETS OF DECIMALS WHICH ROUND TO A GIVEN NUMBER

Find the set of decimals which round   (a) to 9 and are expressed in tenths,   (b) to .4 and are expressed in hundredths,   (c) to .57 and are expressed in thousandths,   (d) to 4.037 and are expressed in ten-thousandths.

Illustrative Solution   (b) Since the decimals round to .4, then 10ths place must be either .3 or .4. If .3, then 100ths place is 5 or more. If .4, then 100ths place is less than 5. The desired set is thus:

$$\{.35, .36, .37,\ldots, .43, .44\}; \text{ that is, } \{.35, .36, .37, .38, .39, .40, .41, .42, .43, .44\}$$

Ans.   (a) $\{8.5, 8.6, 8.7,\ldots, 9.3, 9.4\}$ (d) $\{4.0365, 4.0366, 4.0367,\ldots 4.0373, 4.0374\}$
      (c) $\{.565, .566, .567,\ldots, .573, .574\}$

4. ADDING DECIMALS

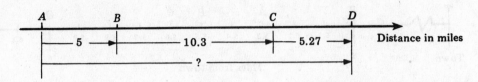

Fig. 4-5

Homes A, B, C, and D, Fig. 4-5, lie along a straight road. The number of miles in the distance from A to D is the sum of 5, 10.3, and 5.27. That sum, 20.57, may be found using the following procedure:

To Add Decimals

	(a) $5 + 10.3 + 5.27$	(b) $3 + 5.61 + 0.904$

Add:

PROCEDURE   SOLUTIONS

1. Express any whole number as a decimal: 1. Express 5 as 5..   Express 3 as 3..

2. Arrange the decimals vertically with deci-   2. 5.   3.
   mal points aligned: 10.3   5.61
                                                     5.27   0.904

3. Add the decimals exactly as whole num-   3. 5.00*   3.000*
   bers are added, aligning the decimal point   10.30   5.610
   of the sum under the other decimal   5.27   0.904
   points: 20.57 *Ans.*   9.514 *Ans.*

*Zeros may be added to equalize the number of decimal places of each addend.

Understanding the Addition of Decimals

To understand the addition of decimals, express 5.00, 10.30, and 5.27 as the like decimal fractions 500/100, 1030/100, and 527/100. Hence,

$$5.00 + 10.30 + 5.27 = \frac{500}{100} + \frac{1030}{100} + \frac{527}{100} = \frac{500 + 1030 + 527}{100}$$

$$= \frac{2057}{100} \quad \text{Express as a decimal, } 20.57 \quad Ans.$$

4.1 ADDING AMOUNTS OF MONEY

Add:

(a) $\$.25 + \$.50$ (d) $\$455.67 + \$3 + \$19.05 + \$93.10$

(b) $\$1.10 + \$2.01$ (e) $\$635 + \$79.24 + \$17 + \$200 + \$1,483.56$

(c) $\$30.40 + \$50 + \$125.90$

> ***Illustrative Solution*** (c) $ \quad 30.40
> \qquad\qquad\qquad\qquad\qquad 50.00
> \qquad\qquad\qquad\qquad\quad\; \underline{125.90}
> \qquad\qquad\qquad\qquad \$\,\overline{206.30} \quad Ans.$

Ans. (a) $\$.75$ (b) $\$3.11$ (d) $\$570.82$ (e) $\$2,414.80$

4.2 ADDING DECIMALS

Add: (a) $.5 + .7 + .9$,   (b) $.52 + .82 + .35$,   (c) $3.07 + 4.90 + 11.60$,   (d) $1.2 + .5 + .63$,
(e) $.84 + 17 + .4$,   (f) $.84 + 7.3 + .2 + 9 + 8.49$,

(g)	4.74	(h)	7.276	(i)	0.5684	(j)	812.79	(k)	0.84	(l)	129.75
	9.60		1.751		1.9478		932.67		32.7		14.
	8.51		8.003		3.7253		108.95		0.5		652.14
	7.19		5.930		0.6407		633.02		2.143		9.

Illustrative Solution (e) .84
                              17.00
                               .40
                             ─────
                             18.24   *Ans.*

Ans. (a) 2.1   (c) 19.57   (f) 25.83   (h) 22.960   (j) 2,487.43   (l) 804.89
        (b) 1.69   (d) 2.33   (g) 30.04   (i) 6.8822   (k) 36.183

4.3 PROBLEM SOLVING INVOLVING ADDING DECIMALS

(a) Find the total number of kilometers flown for successive flights of (1) 150, 344.7, and 290.3 km; (2) 1,230, 340, 789.2, and 465.9 km.

(b) Find the total number of dollars spent for purchases amounting to (1) $12.63, $18, $40.65, and $8.50; (2) $155, $250.95, $318.40, and $67.36.

(c) Find the total number of liters consumed for separate consumptions of (1) 14.2, 17.3, and 9.5 liters; (2) 20.5, 22.8, 19.3, and 14 liters.

(d) Find the length and width of the object shown in Fig. 4-6 if dimensions are in inches.

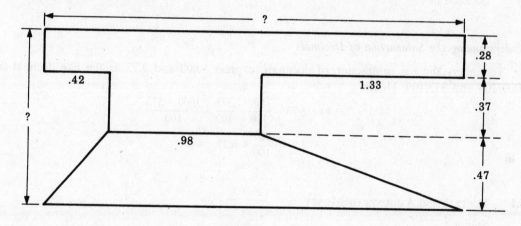

Fig. 4-6

Ans. (a) (1) 785   (2) 2,825.1   (c) (1) 41   (2) 76.6
        (b) (1) $79.78   (2) $791.71   (d) length = 2.73 in.   width = 1.12 in.

5. SUBTRACTING DECIMALS

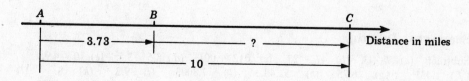

Fig. 4-7

Towns A, B, and C, Fig. 4-7, lie along a straight road. The number of miles in the distance from B to C is found by subtracting 3.73 from 10. The difference, 6.27, may be found using the following procedure:

To Subtract Decimals

Subtract: $10 - 3.73$.

PROCEDURE

1. Express any whole number as a decimal:
2. Arrange the decimals vertically with decimal points aligned:
3. Subtract the decimals exactly as whole numbers are subtracted, aligning the decimal point of the difference under the other decimal points:

SOLUTION

Express 10 as 10..

$$10. \\ - \ 3.73$$

Method 1

$$
\begin{array}{r}
9\ \ 9 \\
1^{1}0.\ {}^{1}0\ {}^{1}0* \\
-\ \ \ \ 3.\ 7\ 3 \\
\hline
6.\ 2\ 7 \quad Ans.
\end{array}
$$

Method 2

$$
\begin{array}{r}
1\ {}^{1}0.\ {}^{1}0\ {}^{1}0* \\
-\ \ \ 3.\ 7\ 3 \\
{}_{1}\ \ {}_{1}\ \ {}_{1} \\
\hline
6.\ 2\ 7 \quad Ans.
\end{array}
$$

*Zeros should be added to equalize the number of decimal places.

Important Note: In Method 1, minuend digits are decreased by 1, while in Method 2, subtrahend digits are increased by 1.

Understanding the Subtraction of Decimals

To understand the subtraction of decimals, express 10.00 and 3.73 as the like decimal fractions 1000/100 and 373/100. Hence,

$$10.00 - 3.73 = \frac{1000}{100} - \frac{373}{100} = \frac{1000 - 373}{100}$$

$$= \frac{627}{100} = 6.27 \quad Ans.$$

5.1 SUBTRACTING AMOUNTS OF MONEY

Subtract:
(a) $\$.42 - \$.27$ (d) $\$125.55 - \$90.75$ (g) $\$45.15 - \$27.80 - \$5.67 - \$4.89$
(b) $\$1.75 - \$.99$ (e) $\$50 - \$30 - \$15.25$
(c) $\$5.50 - \$2.80$ (f) $\$100 - \$5.55 - \$90.75$

Illustrative Solution

$$
(d) \quad
\begin{array}{r}
\overset{4}{\$1^{1}2\ \overset{1}{5}.\ {}^{1}5\ 5} \\
-\quad\ \ 9\ 0.\ 7\ 5 \\
\hline
\$\ \ 3\ 4.\ 8\ 0 \quad Ans.
\end{array}
\quad or \quad
\begin{array}{r}
\$1^{1}2\ 5.\ {}^{1}5\ 5 \\
-\quad\ \ 9\ 0.\ 7\ 5 \\
{}_{1}\ \ {}_{1} \\
\hline
\$\ \ 3\ 4.\ 8\ 0 \quad Ans.
\end{array}
$$

Ans. (a) $.15   (b) $.76   (c) $2.70   (e) $4.75   (f) $3.70   (g) $6.79

5.2 SUBTRACTING DECIMALS

Subtract: (a) $.7 - .3$,   (b) $.47 - .23$,   (c) $.012 - .001$,   (d) $8.5 - 3.5$,   (e) $10 - 4.9$,

$$
(f) \quad
\begin{array}{r}
.431 \\
-.320 \\
\hline
\end{array}
\qquad
(g) \quad
\begin{array}{r}
6.75 \\
-3.8 \\
\hline
\end{array}
\qquad
(h) \quad
\begin{array}{r}
52.48 \\
-48.68 \\
\hline
\end{array}
\qquad
(i) \quad
\begin{array}{r}
7.8003 \\
-6.0061 \\
\hline
\end{array}
\qquad
(j) \quad
\begin{array}{r}
9.3 \\
-\ .593 \\
\hline
\end{array}
\qquad
(k) \quad
\begin{array}{r}
8. \\
-1.702 \\
\hline
\end{array}
\qquad
(l) \quad
\begin{array}{r}
4.6708 \\
-\ \ .98 \\
\hline
\end{array}
$$

(m) Subtract .65 from 1.1.   (n) From 46.9 take 45.1.   (o) Decrease 25 by 19.79.

Illustrative Solution

$$
\begin{array}{r}
7\ 9 \\
(i)\quad 7.8\,^1 0\,^1 0\ 3 \\
6.0\ 0\ 6\ 1 \\
\hline
1.7\ 9\ 4\ 2 \quad Ans.
\end{array}
\qquad
\begin{array}{r}
7.8\,^1 0\,^1 0\ 3 \\
or \quad 6.0\ 0\ 6\ 1 \\
^1\ ^1 \\
\hline
1.7\ 9\ 4\ 2 \quad Ans.
\end{array}
$$

Ans. (a) .4   (c) 0.011   (e) 5.1   (g) 2.95   (j) 8.707   (l) 3.6908   (n) 1.8
     (b) .24   (d) 5   (f) .111   (h) 3.80   (k) 6.298   (m) .45   (o) 5.21

5.3   PROBLEM SOLVING INVOLVING SUBTRACTION OF DECIMALS

(a) How much change from $20 should be given if the purchase is (1) $2.50, (2) $9.75, (3) $15.48?

(b) How much money is left in a bank account if from a balance of $295 there is a withdrawal of (1) $205.30, (2) $198.55, (3) $46.74?

(c) How many kilometers were covered on a trip if at the beginning of the trip the odometer read 4,761.3 and at the end the reading was (1) 5,000.0, (2) 6,500.5, (3) 10,890.6?

(d) How many miles per hour faster is a plane traveling at 545 miles per hour than a plane traveling at (1) 400.7 miles per hour, (2) 528.9 miles per hour, (3) 389.4 miles per hour?

Illustrative Solution

$$
\begin{array}{r}
1\,^1 8\,^1 4\ 9 \\
(b)\ (2)\quad \$2\ 9\ 5.\,^1 0\,^1 0 \\
-\quad 1\ 9\ 8.\ 5\ 5 \\
\hline
\$\ \ 9\ 6.\ 4\ 5 \quad Ans.
\end{array}
\qquad
\begin{array}{r}
\$2\,^1 9\,^1 5.\,^1 0\,^1 0 \\
or \quad -\quad 1\ 9\ 8.\ 5\ 5 \\
^1\ ^1\ ^1\ ^1 \\
\hline
\$\ \ 9\ 6.\ 4\ 5 \quad Ans.
\end{array}
$$

Ans. (a) (1) $17.50   (2) $10.25   (3) $4.52
     (b) (1) $89.70   (3) $248.26
     (c) (1) 238.7   (2) 1,739.2   (3) 6,129.3
     (d) (1) 144.3   (2) 16.1   (3) 155.6

6.   MULTIPLYING DECIMALS

Since one-tenth of a dime is a penny, it follows that 1/10 of .1 equals .01, or .1 × .1 = .01. Observe that the number of decimal places of the product, .01, equals the sum of the number of decimal places of the factors, .1 and .1.

Rule 1: The number of decimal places in the product of two or more factors is the sum of the numbers of decimal places of the factors.

Thus: .3 × .05 = .015   2.4 × .001 = .0024   .2 × .04 × .0001 = .0000008

$$
\begin{array}{r}
.3 \text{ has 1 place} \\
\times\ .05 \text{ has \underline{2 places}} \\
\hline
.015 \text{ has 3 places}
\end{array}
\qquad
\begin{array}{r}
2.4 \text{ has 1 place} \\
\times\ .001 \text{ has \underline{3 places}} \\
\hline
.0024 \text{ has 4 places}
\end{array}
\qquad
\begin{array}{r}
.2 \text{ has 1 place} \\
\times\ .04 \text{ has 2 places} \\
\times\ .0001 \text{ has \underline{4 places}} \\
\hline
.0000008 \text{ has 7 places}
\end{array}
$$

To Multiply Decimals

Multiply: (a) 5.6×0.2 (b) $1.24 \times .003$

PROCEDURE   SOLUTIONS

1. Multiply as whole numbers, ignoring   1.   56   124
   the decimal points: $\times$ 2 $\times$ 3
                                                   ───   ───
                                                   112   372

2. Place the decimal point of the product   2.   5.6 has 1 place   1.24 has 2 places
   so that the number of decimal places $\times$ .02 has 2 places $\times$ .003 has 3 places
   of the product is the sum of the   ─────────────────   ──────────────────
   number of decimal places of the   .112 has 3 places   .00372 has 5 places
   factors:
                                                   Ans. .112   *Ans.* .00372

Understanding Multiplying Decimals

To understand the procedure for multiplying decimals such as 5.6 and .02, in (a) above, multiply their equivalents:

$$\frac{56}{10} \times \frac{2}{100} = \frac{56 \times 2}{10 \times 100}$$

Multiplying 56×2 is done in step 1 of the above procedure. Multiplying 10×100, the resulting product of 1,000 becomes the denominator of a decimal fraction which is equivalent to a 3-place decimal.

Multiplying a Whole Number by a Decimal

If the cost of 4 bottles of soda at \$.80 a bottle is \$3.20, it follows that $4 \times .80 = 3.20$. In this case, the number of decimal places of the product, 3.20, equals the number of decimal places of the decimal factor, .80.

Rule 2: If a whole number is multiplied by a decimal, the number of places of the product is the same as the number of decimal places of the decimal.

Thus: $4 \times .22 = .88$ $3.2 \times 4 = 12.8$ $100 \times 2.1374 = 213.7400$

         (Both .22 and .88   (Both 3.2 and 12.8   (Both 2.1374 and 213.7400
         have 2 places.)   have 1 place.)   have 4 places.)

Understanding Multiplying a Whole Number and a Decimal

Express the whole number as a 0-place decimal by placing a decimal point after the last digit; then apply Rule 1, as follows:

         $4 \times .22 = .88$ $100 \times 2.1374 = 213.7400$

              4. has 0 places   100. has 0 places
         $\times$ .22 has 2 places $\times$ 2.1374 has 4 places
         ──────────────   ──────────────────
              .88 has 2 places   213.7400 has 4 places

Multiplying a Decimal by 10, 100, 1,000, or Any Place Value

Rule 3: To multiply a decimal by 10, 100, 1,000, or any place value, move the decimal point as many places to the *right* as there are zeros in the place value.

| To multiply 2.1374 by 10, move decimal point 1 place to the right, as follows: 2.1.374 *Ans.* 21.374 | To multiply 2.1374 by 100, move decimal point 2 places to the right, as follows: 2.13.74 *Ans.* 213.74 | To multiply 2.1374 by 1000, move decimal point 3 places to the right, as follows: 2.137.4 *Ans.* 2137.4 |

6.1   MULTIPLYING A DECIMAL BY A DECIMAL

Multiply: (*a*) .2 × .5, (*b*) .02 × .005, (*c*) .5 × .05, (*d*) .05 × .0005, (*e*) .875 × .8, (*f*) 9.5 × 3.4, (*g*) .95 × .0034, (*h*) 15.6 × 2.75, (*i*) 4.056 × .06, (*j*) .116 × .145, (*k*) 7.004 × 1.0055.

Illustrative Solutions   Apply Rule 1.

(*d*) Multiply whole numbers: 5 × 5 = 25. Mark off 2 + 4 or 6 places. *Ans.* .000025
(*i*) Multiply whole numbers: 4,056 × 6 = 24,336. Mark off 3 + 2 or 5 places. *Ans.* .24336

Ans.   (*a*) .10   (*b*) .00010   (*c*) .025   (*e*) .7000   (*f*) 32.30   (*g*) .003230   (*h*) 42.900
         (*j*) .016820   (*k*) 7.0425220

6.2   MULTIPLYING A WHOLE NUMBER BY A DECIMAL

Multiply: (*a*) 20 × .7, (*b*) 150 × .03, (*c*) 2,000 × .005, (*d*) 12 × .08, (*e*) .04 × 240, (*f*) .003 × 45,000, (*g*) 6.075 × 3,000, (*h*) 6,075 × .0003, (*i*) 60.75 × 30.

Illustrative Solution   Apply Rule 2.

(*b*) Multiply whole numbers: 150 × 3 = 450. Mark off 2 places. *Ans.* 4.50

Ans.   (*a*) 14.0   (*c*) 10.000   (*d*) .96   (*e*) 9.60   (*f*) 135.000   (*g*) 18,225.000
         (*h*) 1.8225   (*i*) 1,822.50

6.3   MULTIPLYING BY 10, 100, 1,000, OR ANY PLACE VALUE

(*a*) Multiply by 10: (1) 3.4, (2) 4.36, (3) 5.6709.
(*b*) Multiply by 100: (1) 4.5, (2) 5.86, (3) 45.0783.
(*c*) Multiply by 1,000: (1) 5.8, (2) 70.46, (3) 568.00034.
(*d*) Multiply (1) 5.9 × 100,000, (2) 1,000,000 × 45.68, (3) 100,000,000 × .40036.

Solutions   Apply Rule 3.

(*a*) To multiply by 10, move the decimal point 1 place to the right:

(1) 34.   (2) 43.6   (3) 56.709

(*b*) To multiply by 100, move the decimal point 2 places to the right:

(1) 450.   (2) 586.   (3) 4,507.83

(*c*) To multiply by 1,000, move the decimal point 3 places to the right:

(1) 5,800.   (2) 70,460.   (3) 568,000.34

(*d*) (1) 590,000.   (2) 45,680,000.   (3) 40,036,000.

6.4   PROBLEM SOLVING INVOLVING MULTIPLICATION OF DECIMALS

(*a*) How much is earned for 40 hours of work at the rate of (1) $7.00 an hour, (2) $8.50 an hour, (3) $30.72 an hour?

(*b*) To the nearest cent, find the cost of 15.5 gallons of gas at (1) 42¢ a gallon, (2) 44.9¢ a gallon, (3) 61.5¢ a gallon.

(c) How many miles can a car travel on 20.5 gallons of gas if for each gallon, it travels (1) 15 miles, (2) 16.4 miles, (3) 20.45 miles?

(d) Find the value of 4.5 ounces of gold if each ounce is worth (1) $125, (2) $126.50, (3) $158.40.

(e) Find the yearly cost of each service, given each of the following average monthly costs: (1) telephone, $25.40 per month; (2) gas, $12.51 per month; (3) electricity, $34.17 per month.

(f) Find the total earnings, counting a rate of pay of time and a half for hours over 40 hours:

	(1)	(2)	(3)	(4)	(5)
Number of hours	52	45	50	54	60
Hourly rate	$7.00	$8.40	$11.70	$12.24	$8.88

Illustrative Solutions

(b) (2) Multiply: $15.5 \times 44.9¢ = 695.95¢$. To the nearest cent, $695.95¢ = 696¢$ or $6.96 *Ans.*

(f) (1) The number of hours overtime is 12, since $52 - 40 = 12$. At time and a half, the worker is paid for an extra $\frac{1}{2}(12)$ or 6 hours, thus making a total of 58 hours. Multiplying $58 \times 7.00, earnings equal $406 *Ans.*

Ans.   (a) (1) $140   (2) $170   (3) $614.40

        (b) (1) $6.51   (3) $9.53

        (c) (1) 307.5   (2) 336.20   (3) 419.225

        (d) (1) $562.50   (2) $569.25   (3) $712.80

        (e) (1) $304.80   (2) $150.12   (3) $410.04

        (f) (2) $399   (3) $643.50   (4) $746.64

            (5) $621.60

7. DIVIDING DECIMALS

The division fact $.72 \div 6 = .12$ can be used to show that if 6 bottles of soda cost $.72, then each bottle costs $.12. The method for dividing a decimal by a whole number is shown in the following Procedure A:

To Divide a Decimal by a Whole Number

Divide:

PROCEDURE A

	(a) $.72 \div 6$	(b) $4,006 \div 4$	(c) $3.14 \div 25$
		SOLUTIONS	

1. Align the decimal point in the quotient with the decimal point in the dividend:

   1. $6\overline{).72}$ $4\overline{)4006.}$ $25\overline{)3.14}$

2. Divide as whole numbers, ignoring the decimal point. Annex zeros to the dividend as needed:

   2. $\dfrac{.1\ 2}{6)7^12}$ *Ans.* $\dfrac{1001.\ 5}{4)4006.^20}$ *Ans.* $\dfrac{.1256}{25)3.1400}$ *Ans.*

                          (Annex 1 zero to dividend.)   (Annex 2 zeros to dividend.)

Note: In (c), the complete division is left to be done by the student.

Checking a Division Problem Involving Decimals

To check a division problem, use the rule DIVISOR × QUOTIENT = DIVIDEND. Thus, to check $.72 \div 6 = .12$, multiply 6 by .12 to obtain .72, the dividend.

Dividing a Decimal by 10, 100, 1,000, or Any Place Value

Rule: To divide a decimal by 10, 100, 1,000, or any place value, move the decimal point as many places to the *left* as there are zeros in the place value.

To divide 21.374 by 10, move decimal point 1 place to the left, as follows:
  2.1̸374   *Ans.* 2.1374

To divide 21.374 by 100, move decimal point 2 places to the left, as follows:
  .21̸374   *Ans.* .21374

To divide 21.374 by 1,000, move decimal point 3 places to the left, as follows:
  .021̸374   *Ans.* .021374

(Annex zeros as needed to provide places to the left.)

Pricing Goods by 100s or by 1,000s

If a quantity of goods is priced in terms of 100s, 1,000s, or other counting unit of measure, the number of such units is needed to find the cost of the quantity. The number of 100s is found by dividing the number of items by 100. Similarly, the number of 1,000s is found by dividing the number of items by 1,000. In either case, move the decimal point to the left; 2 places if dividing by 100, and 3 places if dividing by 1,000.

To Find the Cost of Goods Priced by 100s or by 1,000s

Find each cost:

(*a*) 1,350 items at $4 per 100

(*b*) 1,350 items at $4 per 1000

PROCEDURE   SOLUTIONS

1. Find the number of 100s or 1000s by moving the decimal point to the left 2 or 3 places, respectively:

   1. 1350. = 13.5̸0̸100

   1350. = 1.3̸5̸0̸1000

2. Find the cost by multiplying the result in step 1 by the price of the counting unit:

   2. Cost = 13.5 × $4
           = $54   *Ans.*

   Cost = 1.35 × $4
        = $5.40   *Ans.*

Dividing a Decimal by a Decimal

The division fact $.10 \div .05 = 2$ can be used to show that the value of a dime is twice the value of a nickel. To divide a decimal by a decimal, the first step, as shown below in Procedure B, is to move the decimal point of both terms to the right as many places as is needed to make the divisor a whole number. Once this is done, Procedure A, in which a decimal is divided by a whole number, can be followed:

To Divide a Decimal by a Decimal

Divide:

(*a*) $.072 \div .6$

(*b*) $40.06 \div .04$

(*c*) $.314 \div 2.5$

PROCEDURE B   SOLUTIONS

1. Move the decimal point of the divisor to the right as many places as is needed to make the divisor a whole number and, at the same time, move the decimal point of the dividend to the right the *same number of places*:
   (Use a caret, ∧, to indicate the new position of each decimal point.)

   .6)̅.0̲ 72
      ∧   ∧

   (Move decimal points 1 place to the right.)

   .04)̅40.06
       ∧   ∧

   (Move decimal points 2 places to the right.)

   2.5)̅.3̲ 14
       ∧   ∧

   (Move decimal points 1 place to the right.)

2. Having made the divisor a whole number, follow the two steps of Procedure A:

   .12
   6.)̅0.72
      ∧ ∧

   Ans. .12

   1001.5
   4.)̅4006.0
      ∧   ∧

   Ans. 1001.5

   .1256
   25.)̅3.1400
       ∧   ∧

   Ans. .1256

Understanding the Division of a Decimal by a Decimal

To understand what happens when the decimal points of the divisor and the dividend are moved the same number of places, first express the division as a fraction, then multiply both terms by the place value needed. Thus,

$$.072 \div .6 = \frac{.072}{.6} = \frac{.072 \times 10}{.6 \times 10} = \frac{0.72}{6.} \quad Ans. \ .12$$

(multiply
both terms
by 10)

Dividing a Decimal by a Whole Number to Obtain a Rounded Quotient

In each of the following division problems involving decimals, the instruction indicates the number of places to which the result is to be rounded, such as, "Express answer to nearest tenth, to nearest hundredth, etc."

To express a quotient to the nearest tenth, which means a 1-place answer, carry out the quotient to 2 places. Also, to express a quotient to the nearest hundredth, which means a 2-place answer, carry out the quotient to 3 places. In general, carry out the division to *one decimal place more* than the number of places needed for the answer.

To Divide a Decimal by a Whole Number to Obtain a Rounded Quotient

Express each answer to the nearest tenth: (a) $.8 \div 3$ (b) $25 \div 9$

Express answer to nearest hundredth: (c) $125.05 \div 15$

PROCEDURE C		SOLUTIONS	
1. By annexing zeros, obtain a dividend having one decimal place more than the number of places needed for the answer	(a) $3\overline{).80}$ (Annex 1 zero to .8)	(b) $9\overline{)25.00}$ (Annex 2 zeros to 25.)	(c) $15\overline{)125.050}$ (Annex 1 zero to 125.05.)
2. Align the decimal point of the quotient and the decimal point of the dividend:	$\overset{.}{3\overline{).80}}$	$\overset{.}{9\overline{)25.00}}$	$\overset{.}{15\overline{)125.050}}$
3. Divide as whole numbers, ignoring the decimal point:	$\overset{.26*}{3\overline{).80}}$	$\overset{2.77*}{9\overline{)25.00}}$	$\overset{8.336*}{15\overline{)125.050}}$
4. Round the quotient to the needed number of decimal places:	Round .26 to .3 *Ans.*	Round 2.77 to 2.8 *Ans.*	Round 8.336 to 8.34 *Ans.*

*Disregard the final remainder, since it does not affect the answer.

Dividing a Decimal by a Decimal to Obtain a Rounded Quotient

To divide a decimal by a decimal, the first step, as shown below in Procedure D, is to move the decimal point of both terms to the right as many places as is needed to make the divisor a whole number. Once this is done, Procedure C, in which a decimal is divided by a whole number, can be followed:

To Divide a Decimal by a Decimal to Obtain a Rounded Quotient

Express each answer to nearest tenth: (a) $.08 \div .3$ (b) $.25 \div .09$

Express answer to nearest hundredth: (c) $.12505 \div .015$

PROCEDURE D

1. Move the decimal point of the divisor to the right as many places as is needed to make the divisor a whole number and, at the same time, move the decimal point to the dividend to the right the same number of places:

   (Use of a caret, ∧, to indicate the new position of each decimal point.)

2. Having made the divisor a whole number, follow the four steps of Procedure C:

SOLUTIONS

(a) $.3\overline{)\,.0\,8}$ (b) $.09\overline{)\,.25}$ (c) $.015\overline{)\,.125\,05}$

(Making the divisor a whole number transforms these three divisions into the divisions used to illustrate Procedure C. Refer to Procedure C for solutions.)

$$
\begin{array}{r}
.26 \\
3.\overline{)0.80} \\
\end{array}
\qquad
\begin{array}{r}
2.77 \\
9.\overline{)25.00} \\
\end{array}
\qquad
\begin{array}{r}
8.336 \\
15.\overline{)125.050} \\
\end{array}
$$

Round .26 to .3 *Ans.*   Round 2.77 to 2.8 *Ans.*   Round 8.336 to 8.34 *Ans.*

7.1 **DIVIDING A DECIMAL BY A WHOLE NUMBER**

Divide:

(a) $12\overline{)9.3}$ (b) $15\overline{)\,.375}$ (c) $150.\overline{)375}$ (d) $125\overline{)35}$ (e) $32\overline{)296}$

Solutions Apply Procedure A.

(a)
$$
\begin{array}{r}
.775 \\
12\overline{)9.300} \\
\underline{8\,4} \\
90 \\
\underline{84} \\
60 \\
\underline{60} \\
\end{array}
$$

(b)
$$
\begin{array}{r}
.025 \\
15\overline{)\,.375} \\
\underline{30} \\
75 \\
\underline{75} \\
\end{array}
$$

(c)
$$
\begin{array}{r}
.0025 \\
150\overline{)\,.3750} \\
\underline{300} \\
750 \\
\underline{750} \\
\end{array}
$$

(d)
$$
\begin{array}{r}
.28 \\
125\overline{)35.00} \\
\underline{25\,0} \\
10\,00 \\
\underline{10\,00} \\
\end{array}
$$

(e)
$$
\begin{array}{r}
9.25 \\
32\overline{)296.00} \\
\underline{288} \\
8\,0 \\
\underline{6\,4} \\
1\,60 \\
\underline{1\,60} \\
\end{array}
$$

Ans. .775   *Ans.* .025   *Ans.* .0025   *Ans.* .28   *Ans.* 9.25

7.2 **DIVIDING BY A PLACE VALUE: 10, 100, 1,000, ETC.**

(a) Divide by 10: (1) 4.5, (2) 25.67, (3) .897.

(b) Divide by 100: (1) 4.5, (2) 0.59, (3) .0072.

(c) Divide by 1,000: (1) 315., (2) 5.23, (3) 2,570.6.

(d) Divide: (1) 64.2 by 100,000, (2) 165.3 by 1,000,000, (3) 56.187 by 100,000,000.

Solutions Apply the rule for dividing by a place value: Move the decimal point to the left as many places as there are 0s in the divisor.

(a) To divide by 10, move the decimal point 1 place to the left:

(1) .45 (2) 2.567 (3) .0897

(b) To divide by 100, move the decimal point 2 places to the left:

(1) .045 (2) .0059 (3) .000072

(c) To divide by 1,000, move the decimal point 3 places to the left:

(1) .315 (2) .00523 (3) 2.5706

(d) (1) .000642 (2) .0001653 (3) .00000056187

7.3   DIVIDING A DECIMAL BY A DECIMAL

Divide:

(a) $.3\overline{)6.}$ (e) $.8\overline{).056}$ (i) $.025\overline{)8.5}$ (m) $.48\overline{)5.4}$ (q) $6.04\overline{)19.5998}$

(b) $.03\overline{)90.}$ (f) $.08\overline{)560.}$ (j) $.16\overline{)88.}$ (n) $7.5\overline{).654}$ (r) $.3128\overline{)1.59528}$

(c) $.5\overline{).0055}$ (g) $2.5\overline{)75.}$ (k) $1.8\overline{).45}$ (o) $3.14\overline{)23.55}$

(d) $.005\overline{).0055}$ (h) $2.5\overline{)8.5}$ (l) $.64\overline{).0576}$ (p) $2.75\overline{)220.}$

Illustrative Solution   Apply Procedure B.

$$
\begin{array}{r}
70\ 00 \\
.08\ \overline{)560.00} \\
\text{\textasciicircum} \qquad \text{\textasciicircum}
\end{array}
$$

(f)   *Ans.* 7,000

Ans.   (a) 20   (c) .011   (e) .07   (h) 3.4   (j) 550   (l) .09   (n) .0872   (p) 80

       (b) 3,000   (d) 1.1   (g) 30   (i) 340   (k) .25   (m) 11.25   (o) 7.5   (q) 3.245

                                                                            (r) 5.1

7.4   DIVIDING A DECIMAL BY A WHOLE NUMBER TO OBTAIN A ROUNDED QUOTIENT

Express each answer to the nearest tenth: (a) $1.3 \div 3$,   (b) $23.5 \div 7$.

Express each answer to the nearest hundredth: (c) $31.3 \div 12$,   (d) $14.58 \div 22$.

Solutions   Apply Procedure C.

(a)
$$
\begin{array}{r}
.4\ 3 \\
3\overline{)1.3^10}
\end{array}
$$
Ans. .4

(b)
$$
\begin{array}{r}
3.\ 3\ 5 \\
7\overline{)23.^25^40}
\end{array}
$$
Ans. 3.4

(c)
$$
\begin{array}{r}
2.608 \\
12\overline{)31.300} \\
\underline{24} \\
7\ 3 \\
\underline{7\ 2} \\
100 \\
\underline{96} \\
4
\end{array}
$$
Ans. 2.61

(d)
$$
\begin{array}{r}
.662 \\
22\overline{)14.580} \\
\underline{13\ 2} \\
1\ 38 \\
\underline{1\ 32} \\
60 \\
\underline{44} \\
16
\end{array}
$$
Ans. .66

7.5   DIVIDING A DECIMAL BY A DECIMAL TO OBTAIN A ROUNDED QUOTIENT

Express each answer to the nearest tenth and also to the nearest hundredth:

(a) $.3\overline{).1}$ (c) $.03\overline{)11.}$ (e) $1.5\overline{).61}$ (g) $4.05\overline{)185.}$ (i) $1.472\overline{)355.44}$

(b) $.6\overline{).5}$ (d) $.7\overline{)13.}$ (f) $.15\overline{)61.6}$ (h) $.039\overline{)4.755}$ (j) $24.8104\overline{)34.69103}$

Illustrative Solution   Apply Procedure D.

(b)
$$
\begin{array}{r}
.8\ 3 \\
6.\overline{)5.0^20} \\
\text{\textasciicircum}\ \text{\textasciicircum}
\end{array}
\qquad
\begin{array}{r}
.8\ 3\ 3 \\
6.\overline{)5.0^20^20} \\
\text{\textasciicircum}\ \ \text{\textasciicircum}
\end{array}
$$

Ans.   .8 to nearest tenth, .83 to nearest hundredth

Ans.   (a) .3, .33   (e) .4, .41   (h) 121.9, 121.92

        (c) 366.7, 366.67   (f) 410.7, 410.67   (i) 241.5, 241.47

        (d) 18.6, 18.57   (g) 45.7, 45.68   (j) 1.4, 1.40

7.6   PROBLEM SOLVING INVOLVING DIVISION OF DECIMALS

(a) At $5 per 100, find the cost of (1) 200 items, (2) 460 items, (3) 2,350 items.

(b) At $6.50 per 1,000, find the cost of (1) 3,000 items, (2) 4,200 items, (3) 20,000 items.

(c) Find the cost of 2,400 items at (1) $8 per 100, (2) $12 per 1,000, (3) $4.65 per 100.

(d) Complete the following purchase list by finding the cost of each group of articles purchased and the total cost:

No.	Kind	Price	Cost
550	#3 black pencils	$ 7 per 100	
480	ink erasers	$32 per 100	
2500	small pads	$55 per 1,000	

Total:

(e) At $3.75 per hour, how many hours will it take to earn (1) $22.50, (2) $150, (3) $180?

(f) A plane burns up 46.2 gallons of gas for each hour of flight. In how many hours will it burn up (1) 693 gallons, (2) 591.36 gallons, (3) 519.75 gallons?

(g) The telephone rate between two zones is $.45 for the first three minutes and $.12 for each additional minute. How many minutes was a call which cost (1) $1.29, (2) $2.85, (3) $3.69?

Illustrative Solutions

(a) (3) To divide by 100, move the decimal point 2 places to the left. Hence, 2,350 items = 23.50/100. At $5 per 100, the cost is 23.50($5) = $117.50. *Ans.*

(g) (2) Find the number of additional minutes by subtracting $.45 from $2.85 and then dividing the difference by $.12. Hence,

$$\text{Number of additional minutes} = \frac{\$2.85 - \$.45}{\$.12} = 20$$

Therefore, the number of minutes in all is 20 + 3 = 23.   *Ans.*

Ans. (a) (1) $10   (2) $23
       (b) (1) $19.50 (2) $27.30 (3) $130
       (c) (1) $192   (2) $28.80 (3) $111.60
       (d) cost of pencils = $38.50, cost of erasers = $153.60, cost of pads = $137.50, total cost = $329.60
       (e) (1) 6   (2) 40   (3) 48
       (f) (1) 15   (2) 12.8   (3) 11.25
       (g) (1) 10   (3) 30

8. EXPRESSING COMMON FRACTIONS AS DECIMALS

Any common fraction can be converted into a decimal by dividing the numerator by the denominator. Before dividing, place a decimal point after the numerator, annexing zeros as needed, as in the following:

Convert to decimal form: (a) 1/4,   (b) 5/8,   (c) 13/20.

(a) $\dfrac{.25}{4)1.0^20}$ *Ans.*   (b) $\dfrac{.605}{8)5.0^20^40}$ *Ans.*   (c) $\dfrac{.65}{20)13.00}$ *Ans.*
                                                                                              $\underline{12\ 0}$
                                                                                              $1\ 00$
                                                                                              $\underline{1\ 00}$

Expressing a Common Fraction as a Rounded Decimal

Using division, a common fraction can be expressed as a decimal, rounded to the nearest tenth, the nearest hundredth, etc. Keep in mind that the division is carried out to one decimal place more than the number of places needed for rounding.

Thus, to round to one decimal place, carry out the division to two decimal places.

To Express a Common Fraction as a Rounded Decimal

Express as a decimal to the nearest tenth:　(a) 7/11　(b) 25/7　(c) 51/16

PROCEDURE　　　　　　　　　　　　　　　　　SOLUTIONS

1. Divide the numerator by the denominator, carrying the division to one decimal place more than the needed number of decimal places:

$$\begin{array}{r} .63 \\ 11\overline{)7.00} \\ \underline{6\,6} \\ 40 \\ \underline{33} \\ 7 \end{array} \qquad \begin{array}{r} 3.57 \\ 7\overline{)25.00} \\ \underline{21} \\ 4\,0 \\ \underline{3\,5} \\ 50 \\ \underline{49} \\ 1 \end{array} \qquad \begin{array}{r} 3.18 \\ 16\overline{)51.00} \\ \underline{48} \\ 3\,0 \\ \underline{1\,6} \\ 1\,40 \\ \underline{1\,28} \\ 12 \end{array}$$

2. Round the quotient obtained to the required number of decimal places:

Round .63 to .6　*Ans.*　　Round 3.57 to 3.6　*Ans.*　　Round 3.18 to 3.2　*Ans.*

Expressing a Common Fraction as a Mixed Decimal

Each fraction, previously expressed as a rounded decimal, can also be expressed as a *mixed decimal*, a decimal with a fraction at the end of it. Form the end fraction by placing a remainder over the original denominator. If this is done, then

$$(1)\ \frac{7}{11} = .6\frac{4}{11} \qquad (2)\ \frac{25}{7} = 3.5\frac{4}{7} \qquad (3)\ \frac{51}{16} = 3.1\frac{14}{16}\ \text{or}\ 3.1\frac{7}{8}$$
$$= .63\frac{7}{11} \qquad\qquad = 3.57\frac{1}{7} \qquad\qquad = 3.18\frac{12}{16}\ \text{or}\ 3.18\frac{3}{4}$$

In rounding, if the end fraction is less than one-half, keep the last decimal place. However, if the end fraction is one-half or more, increase the last decimal place by one. For example, since $7/11 = .6\frac{4}{11}$, then to the nearest tenth, $7/11 = .6$. Also, since $7/11 = .63\frac{7}{11}$, then to the nearest hundredth, $7/11 = .64$.

Expressing a Common Fraction as a Repeating Decimal

Note that dividing 1 by 3 leads to a remainder of 1 at each step:

$$\begin{array}{r} .3\,3\,3\,3\,3\,3\ldots \\ 3\overline{)1.0^10^10^10^10^10^1} \end{array}$$

The repetition of 1 continues endlessly, no matter how many zeros are annexed. The three dots indicate that the division may be extended endlessly. Hence, 1/3 is equal to the repeating decimal .333333 . . . .

Similarly, to express 5/11 as a decimal, we divide 5 by 11 and find that the pair of digits, 45, repeats endlessly:

$$\begin{array}{r} .4\,5\,4\,5\,4\,5\ldots \\ 11\overline{)5.0^60^50^60^50^60^5} \end{array}$$

Hence, 5/11 = .454545 . . . .

A *repeating decimal* is a decimal having a digit or a group of digits that repeats endlessly. The writing of a repeating decimal may be simplified by placing a bar over the digit or group of digits that repeats. Thus,

$$\frac{1}{3} = .\overline{3} = .333\ldots \qquad\qquad \frac{5}{11} = .\overline{45} = .454545\ldots$$

A *terminating decimal* does not repeat endlessly but terminates or ends because of a remainder of 0. Thus,

$$\frac{1}{4} = .25 \qquad\qquad \frac{3}{8} = .375 \qquad\qquad \frac{3}{16} = .1875$$

A common fraction must be expressible as either a repeating decimal or a terminating decimal.

8.1   EXPRESSING COMMON FRACTIONS AS TERMINATING DECIMALS

Express each as a decimal:

(a) $\dfrac{7}{10}$ (d) $\dfrac{13}{1,000}$ (g) $\dfrac{4,509}{1,000,000}$ (j) $\dfrac{3}{8}$ (m) $\dfrac{9}{2,500}$ (p) $\dfrac{11}{200}$

(b) $\dfrac{7}{100}$ (e) $\dfrac{157}{10,000}$ (h) $\dfrac{3}{5}$ (k) $\dfrac{3}{80}$ (n) $\dfrac{613}{500}$ (q) $\dfrac{11}{20,000}$

(c) $\dfrac{13}{100}$ (f) $\dfrac{457}{100,000}$ (i) $\dfrac{3}{50}$ (l) $\dfrac{9}{25}$ (o) $\dfrac{613}{5,000}$

Illustrative Solutions

(e) Since there are 4 zeros in the denominator, 10,000, 4 decimal places are needed.   *Ans.*   .0157

(k) Divide the numerator, 3, by the denominator, 80:

$$
\begin{array}{r}
.0375 \quad Ans. \\
80\overline{)3.0000} \\
2\ 40 \\
\hline
600 \\
560 \\
\hline
400 \\
400 \\
\hline
\end{array}
$$

Ans.   (a) .7   (d) .013   (h) .6   (l) .36   (o) .1226

(b) .07   (f) .00457   (i) .06   (m) .0036   (p) .055

(c) .13   (g) .004509   (j) .375   (n) 1.226   (q) .00055

**8.2   EXPRESSING COMMON FRACTIONS AS ROUNDED DECIMALS
AND MIXED DECIMALS**

(a) Express to the nearest tenth and also as a mixed decimal having one decimal place: (1) 2/3, (2) 3/7, (3) 10/11, (4) 13/16.

(b) Express to the nearest hundredth and also as a mixed decimal having two decimal places: (1) 7/18, (2) 16/21, (3) 29/24, (4) 50/27.

Illustrative Solutions

(a) (4)
$$
\begin{array}{r}
.81 \\
16\overline{)13.00} \\
12\ 8 \\
\hline
20 \\
16 \\
\hline
4 \\
4 \\
\end{array}
$$

(b) (2)
$$
\begin{array}{r}
.761 \\
21\overline{)16.000} \\
14\ 7 \\
\hline
1\ 30 \\
1\ 26 \\
\hline
40 \\
21 \\
\hline
\end{array}
$$

Ans.   .8, .8$\frac{2}{16}$ or .8$\frac{1}{8}$

Ans.   .76, .76$\frac{4}{21}$

Ans.   (a) (1) .7, .6$\frac{2}{3}$ (2) .4, .4$\frac{2}{7}$ (3) .9, .9$\frac{1}{11}$

(b) (1) .39, .38$\frac{8}{9}$ (3) 1.21, 1.20$\frac{20}{24}$ or 1.20$\frac{5}{6}$ (4) 1.85, 1.85$\frac{5}{27}$

8.3   EXPRESSING A COMMON FRACTION AS A REPEATING DECIMAL

Express each as a repeating decimal. Also, show the repeating decimal in simplified form.

(a) $\dfrac{7}{9}$ (b) $\dfrac{1}{18}$ (c) $\dfrac{4}{33}$ (d) $\dfrac{91}{111}$

Solutions   To simplify, place a bar over digits that repeat.

(a)
$$
\begin{array}{r}
.7\ 7\ 7\ldots \\
9\overline{)7.0\,{}^{7}0\,{}^{7}0\ldots}
\end{array}
$$

(b)
$$
\begin{array}{r}
.0\ 5\ 5\ldots \\
18\overline{)1.0\,{}^{10}0\,{}^{10}0\ldots}
\end{array}
$$

Ans.   .777... = .$\overline{7}$

Ans.   .055... = .0$\overline{5}$

(c) $\quad$ $\overset{.1\ 2\ 1\ 2\ \ldots}{33\overline{)4.0^7 0^4 0^7 0 \ldots}}$ $\qquad$ (d) $\quad$ $\overset{.8\ \ 1\ \ 9\ \ 8\ \ 1\ \ 9 \ldots}{111\overline{)91.0^{22}0^{109}0^9 1^{22}0^{109}0 \ldots}}$

Ans. $\quad$ .1212... = $.\overline{12}$ $\qquad\qquad$ *Ans.* $\quad$ .819819... = $.\overline{819}$

9. EXPRESSING DECIMALS AS COMMON FRACTIONS

The value of a nickel, $.05, is 1/20 of the value of a dollar. To express the decimal .05 as a common fraction, first express .05 as the decimal fraction 5/100, then reduce to lowest terms. Hence,

$$.05 = \frac{5}{100} = \frac{1}{20}$$

To Express a Decimal as a Common Fraction

Express as a common fraction: $\qquad$ (a) .85, $\qquad$ (b) .125

PROCEDURE $\qquad\qquad\qquad\qquad\qquad\qquad\qquad$ SOLUTIONS

1. Express the decimal as a decimal fraction : $\quad$ 1. $\dfrac{85}{100}$ $\qquad\qquad$ $\dfrac{125}{1000}$

2. Reduce the decimal fraction to lowest terms: $\qquad$ 2. $\dfrac{17 \times \cancel{5}}{20 \times \cancel{5}}$ $\qquad\qquad$ $\dfrac{\overset{1}{\cancel{125}}}{\underset{1}{\cancel{125}} \times 8}$

$\qquad\qquad\qquad\qquad\qquad\qquad\qquad\qquad$ $\dfrac{17}{20}$ *Ans.* $\qquad\qquad$ $\dfrac{1}{8}$ *Ans.*

Expressing a Mixed Decimal as a Common Fraction

A mixed decimal, such as $.06\frac{1}{4}$, may be expressed as a common fraction by using division, as follows:

$$.06\tfrac{1}{4} = 6\tfrac{1}{4} \div 100 = \frac{25}{4} \div \frac{100}{1} = \frac{\overset{1}{\cancel{25}}}{4} \times \frac{1}{\underset{4}{\cancel{100}}} = \tfrac{1}{16} \quad Ans.$$

9.1 EXPRESSING A DECIMAL AS A COMMON FRACTION

Express as a common fraction:

(a) .8 $\quad$ (c) .04 $\quad$ (e) .71 $\quad$ (g) .68 $\quad$ (i) .143 $\quad$ (k) 2.125 $\quad$ (m) .0065
(b) .60 $\quad$ (d) 1.5 $\quad$ (f) .75 $\quad$ (h) 3.45 $\quad$ (j) .375 $\quad$ (l) .0625

Solutions $\quad$ First express as a decimal fraction, then reduce to lowest terms.

(a) $\dfrac{8}{10} = \dfrac{4}{5}$ $\quad$ (d) $\dfrac{15}{10} = \dfrac{3}{2}$ $\quad$ (g) $\dfrac{68}{100} = \dfrac{17}{25}$ $\quad$ (j) $\dfrac{375}{1,000} = \dfrac{3}{8}$ $\quad$ (m) $\dfrac{65}{10,000} = \dfrac{13}{2,000}$

(b) $\dfrac{60}{100} = \dfrac{3}{5}$ $\quad$ (e) $\dfrac{71}{100}$ $\quad$ (h) $\dfrac{345}{100} = \dfrac{69}{20}$ $\quad$ (k) $\dfrac{2,125}{1,000} = \dfrac{17}{8}$

(c) $\dfrac{4}{100} = \dfrac{1}{25}$ $\quad$ (f) $\dfrac{75}{100} = \dfrac{3}{4}$ $\quad$ (i) $\dfrac{143}{1,000}$ $\quad$ (l) $\dfrac{625}{10,000} = \dfrac{1}{16}$

9.2 EXPRESSING A MIXED DECIMAL AS A COMMON FRACTION

Express as a common fraction: $\quad$ (a) $.6\frac{2}{3}$, $\quad$ (b) $.5\frac{5}{6}$, $\quad$ (c) $.12\frac{1}{2}$, $\quad$ (d) $.062\frac{1}{2}$, $\quad$ (e) $.13\frac{1}{3}$.

Solutions   Divide, as shown, by 10, 100, 1,000, etc.

(a) $\dfrac{20}{3} \div 10$ (b) $\dfrac{35}{6} \div 10$ (c) $\dfrac{25}{2} \div 100$ (d) $\dfrac{125}{2} \div 1,000$ (e) $\dfrac{40}{3} \div 100$

$\dfrac{20}{3} \times \dfrac{1}{10}$ $\dfrac{35}{6} \times \dfrac{1}{10}$ $\dfrac{25}{2} \times \dfrac{1}{100}$ $\dfrac{125}{2} \times \dfrac{1}{1,000}$ $\dfrac{40}{3} \times \dfrac{1}{100}$

$\dfrac{2}{3}$ *Ans.* $\dfrac{7}{12}$ *Ans.* $\dfrac{1}{8}$ *Ans.* $\dfrac{1}{16}$ *Ans.* $\dfrac{2}{15}$ *Ans.*

10. DECIMALS AND CALCULATORS

As in Chapter 2, the calculator can be used to perform the computations of this chapter.

10.1 CALCULATORS AND SIMPLE DECIMAL COMPUTATIONS

Perform the indicated operations:

(a) Add $21.6 + 431.2 + 201.067$ (b) Multiply 21.46×361.73 (c) Divide $.4\overline{)\,.103}$

Solutions

(a) 653.867   (b) 7762.7258   (c) .2575

10.2 CALCULATORS AND REPEATING DECIMALS

Express as a repeating decimal:

(a) $\dfrac{5}{9}$ (b) $\dfrac{5}{11}$ (c) $\dfrac{56}{99}$

Solutions

(a) $.\overline{5}$ (b) $.\overline{45}$ (c) $.\overline{56}$

Supplementary Problems

4.1. Express each as a decimal in dollars: (a) 3 pennies;   (b) 3 dimes;   (c) 3 half-dollars;   **(1.1)**
(d) 3 dimes and 2 nickels;   (e) 3 nickels and 2 dimes;   (f) 2 dollars and 3 quarters;   (g) 3
dollars and 2 half-dollars;   (h) 4 dollars, 2 quarters, and 3 dimes;   (i) 4 half-dollars, 3 quarters,
and 2 nickels.

Ans.   (a) $.03   (b) $.30   (c) $1.50   (d) $.40   (e) $.35   (f) $2.75   (g) $4.00   (h) $4.80   (i) $2.85

4.2. Express each as a decimal: **(1.2)**

(a) $\dfrac{3}{10}, \dfrac{34}{10}, \dfrac{345}{10}$ (c) $\dfrac{7}{1,000}, \dfrac{76}{1,000}, \dfrac{765}{1,000}$ (e) $\dfrac{555}{1,000,000}, \dfrac{66,666}{1,000,000,000}$

(b) $\dfrac{13}{100}, \dfrac{135}{100}, \dfrac{1,357}{100}$ (d) $\dfrac{975}{10,000}, \dfrac{9,753}{100,000}$

Ans.   (a) .3, 3.4, 34.5   (c) .007, .076, .765   (e) .000555, .000066666
         (b) .13, 1.35, 13.57   (d) .0975, .09753

4.3. Express each as a decimal fraction: **(1.3)**

(*a*) .07, .0007, .00007   (*c*) 3.3, 30.3, 300.03   (*e*) .0432, 43.02, 4.0302

(*b*) .123, 1.23, .00123   (*d*) .0812, 8.012, 81.002   (*f*) .000005, 5.00005

Ans. (*a*) $\dfrac{7}{100}, \dfrac{7}{10,000}, \dfrac{7}{100,000}$ (*c*) $\dfrac{33}{10}, \dfrac{303}{10}, \dfrac{30,003}{100}$ (*e*) $\dfrac{432}{10,000}, \dfrac{4,302}{100}, \dfrac{40,302}{10,000}$

       (*b*) $\dfrac{123}{1,000}, \dfrac{123}{100}, \dfrac{123}{100,000}$ (*d*) $\dfrac{812}{10,000}, \dfrac{8,012}{1,000}, \dfrac{81,002}{1,000},$ (*f*) $\dfrac{5}{1,000,000}, \dfrac{500,005}{100,000}$

4.4. Complete Table 4-5.   **(1.4)**

Table 4-5

	Decimals	Expanded Form of Decimals	Decimals Read As
(*a*)	6.6	?	?
(*b*)	?	$30 + 3 + .3$	?
(*c*)	?	?	Twenty two and two hundredths
(*d*)	20.202	?	?
(*e*)	?	$500 + 5 + .0005$	?
(*f*)	?	?	One and one ten-thousandth
(*g*)	100.001	?	?
(*h*)	?	$1,000 + 1 + .1 + .01$	?
(*i*)	?	?	One million one hundred and one tenth

Ans. See Table 4-6.

Table 4-6

	Decimals	Expanded Form of Decimals	Decimals Read As
(*a*)	6.6	$6 + .6$	Six and six tenths
(*b*)	33.3	$30 + 3 + .3$	Thirty three and three tenths
(*c*)	22.02	$20 + 2 + .02$	Twenty two and two hundredths
(*d*)	20.202	$20 + .2 + .002$	Twenty and two hundred two thousandths
(*e*)	505.0005	$500 + 5 + .0005$	Five hundred five and five ten-thousandths
(*f*)	1.0001	$1 + .0001$	One and one ten-thousandth
(*g*)	100.001	$100 + .001$	One hundred and one thousandth
(*h*)	1,001.11	$1,000 + 1 + .1 + .01$	One thousand one and eleven hundredths
(*i*)	1,000,100.1	$1,000,000 + 100 + .1$	One million one hundred and one tenth

4.5.   State the decimals that indicate the shaded and unshaded parts in Fig. 4-8.   **(1.5)**

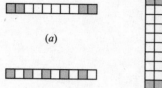

(a)

(c)

(d)

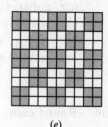

(e)

(b)

Fig. 4-8

Ans.

	Shaded	Unshaded
(a)	.4	.6
(b)	.5	.5

	Shaded	Unshaded
(c)	.36	.64
(d)	.74	.26
(e)	.52	.48

4.6.   Relate the given decimals in each using the symbol =, >, or <:   **(2.1)**

(a) .14, .041   (c) .40301, .4031   (e) 1.001, .9
(b) .401, .41   (d) .1, .0999   (f) .1203, .120300

Ans.   (a) .14 > .041   (c) .40301 < .4031   (e) 1.001 > .9
         (b) .401 < .41   (d) .1 > .0999   (f) .1203 = .120300

4.7.   Arrange in descending order (greatest first):   (a) .7, .007, .07, .0007;   (b) .502, .52,   **(2.2)**
.205, .25;   (c) 6.1, 10.06, 1.6, 1.0601;   (d) 4.4, .44, 4.04, .4004;   (e) .857, .0875, .08075, .78005,
.8705.

Ans.   (a) .7, .07, .007, .0007   (d) 4.4, 4.04, .44, .4004
         (b) .52, .502, .25, .205   (e) .8705, .857, .78005, .0875, .08075
         (c) 10.06, 6.1, 1.6, 1.0601

4.8.   To the nearest kilometer, find the distance from town T, Fig. 4-9, to each of the   **(3.1)**
following: (a) a factory at A,   (b) a church at B,   (c) a school at C,   (d) a brook at
D,   (e) a home at E.

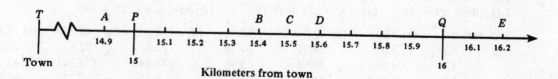

Fig. 4-9

Ans.   (a) 15 km   (b) 15 km   (c) 16 km   (d) 16 km   (e) 16 km

4.9.   Round each to the nearest   **(3.2)**

(*a*)   dollar: (1) $6.54, (2) $6.45, (3) $6.50, (4) $65.49, (5) $69.50

(*b*)   cent: (1) $.495, (2) $.594, (3) $4.995

Ans.   (*a*) (1) $7   (2) $6   (3) $7   (4) $65   (5) $70
         (*b*) (1) $.50   (2) $.59   (3) $5.00

4.10.   Round each   **(3.3)**

(*a*)   to tenths: (1) .64, (2) .65, (3) .6449, (4) .645

(*b*)   to hundredths: (1) 0.257, (2) 0.275, (3) 0.2705, (4) 0.2547

(*c*)   to thousandths: (1) 8.3529, (2) 8.3259, (3) 8.2395, (4) 8.2995

(*d*)   to ten-thousandths: (1) 3.45678, (2) 3.46785, (3) 3.67854, (4) 3.67995

Ans.   (*a*) (1) .6   (2) .7   (3) .6   (4) .6
         (*b*) (1) 0.26   (2) 0.28   (3) 0.27   (4) 0.25
         (*c*) (1) 8.353   (2) 8.326   (3) 8.240   (4) 8.300
         (*d*) (1) 3.4568   (2) 3.4679   (3) 3.6785   (4) 3.6800

4.11.   Find the set of decimals which round   (*a*) to 7 and are expressed in tenths,   (*b*) to .3   **(3.4)**
and are expressed in hundredths,   (*c*) to .30 and are expressed in thousandths,   (*d*) to 5.002
and are expressed in ten-thousandths.

Ans.   (*a*) {6.5, 6.6, 6.7, . . . , 7.4}   (*c*) {.295, .296, 297, . . . , .304}
         (*b*) {.25, .26, .27, . . . , .34}   (*d*) {5.0015, 5.0016, 5.0017, . . . , 5.0024}

4.12.   Add: **(4.1)**

(*a*) $.10 + $25   (*e*) $20.60 + $85.10   (*i*) $80 + $1.05 + $3.04

(*b*) $.25 + $.30   (*f*) $2 + $3 + $5.15   (*j*) $1.25 + $34.82 + $79.70 + $100

(*c*) $.85 + $1.15   (*g*) $20 + $3 + $51.15   (*k*) $12.50 + $348.20 + $7.97 + $1,000

(*d*) $8 + $10.30   (*h*) $8 + $10.50 + $30.40

Ans.   (*a*) $25.10   (*c*) $2   (*e*) $105.70   (*g*) $74.15   (*i*) $4.89   (*k*) $1,368.67
         (*b*) $.55   (*d*) $18.30   (*f*) $10.15   (*h*) $48.90   (*j*) $215.77

4.13.   Add: **(4.2)**

(*a*)   .3 + .6 + .8   (*e*)   3.25 + 4.02 + 6.38   (*i*) 25 + 2.5 + .25

(*b*)   1.3 + 2.7 + 3.6   (*f*) 12 + 45.03 + 60.71   (*j*) 1.23 + 12.3 + .123

(*c*)   5 + 4.8 + 3.7   (*g*)   9 + .9 + .09   (*k*) 8 + .8 + .08 + .008

(*d*)   .08 + .29 + .32   (*h*)   35 + .35 + .0035   (*l*) 85 + 8.5 + .85 + .085

(*m*) 5.4	(*n*) 5.04	(*o*) 12.04	(*p*) 12.4	(*q*) 452.03	(*r*) 45.203	(*s*) 7.8
3.7	3.07	24.31	24.31	72.8	72.8	25.06
4.9	4.09	30.96	3.96	3074.145	374.145	342.719
2.6	2.06	51.02	51.2	6.32	6.32	60.09

Ans.   (*a*) 1.7   (*e*) 13.65   (*i*) 27.75   (*m*) 16.6   (*q*) 3,605.295
         (*b*) 7.6   (*f*) 117.74   (*j*) 13.653   (*n*) 14.26   (*r*) 498.468
         (*c*) 13.5   (*g*) 9.99   (*k*) 8.8888   (*o*) 118.33   (*s*) 435.669
         (*d*) .69   (*h*) 35.3535   (*l*) 94.435   (*p*) 91.87

4.14. (*a*) Find the total number of miles in successive trips of (1) 37.4, 154, and   **(4.3)**
103.2 miles; (2) 50.6, 273.4, and 100.7 miles.

(*b*) Find the total sales in dollars for separate sales amounting to (1) $10.45, $27, $148.16, and
$113.50; (2) $472, $560.95, $483.17, and $940.

(*c*) Find the number of tons of coal carried by a group of trucks if the separate weights carried
are (1) 5.72, 6.08, and 7.37 tons; (2) 1.5, 6.93, 10.42, and 3.08 tons.

Ans. (*a*) (1) 294.6   (2) 424.7   (*c*) (1) 19.17   (2) 21.93
(*b*) (1) $299.11   (2) $2.456.12

4.15. Subtract: **(5.1)**

(*a*) $.55 − $.32   (*f*) $10 − $4.63   (*k*) $150 − $78.25
(*b*) $.52 − $.35   (*g*) $20 − $12.07   (*l*) $254.15 − $159.05
(*c*) $.93 − $.45   (*h*) $2.91 − $1.37   (*m*) $1,000 − $258.74
(*d*) $3.65 − $1   (*i*) $15.75 − $8.58   (*n*) $30 − $2.50 − $10.40
(*e*) $3 − $1.65   (*j*) $175.25 − $120   (*o*) $6,250 − $4,690 − $305.95 − $67.85

Ans. (*a*) $.23   (*d*) $2.65   (*g*) $7.93   (*j*) $55.25   (*m*) $741.26
(*b*) $.17   (*e*) $1.35   (*h*) $1.54   (*k*) $71.75   (*n*) $17.10
(*c*) $.48   (*f*) $5.37   (*i*) $7.17   (*l*) $95.10   (*o*) $1,186.20

4.16. Subtract: **(5.2)**

(*a*) .9 − .2   (*b*) 1.8 − .5   (*c*) 1.5 − .6   (*d*) 12.2 − 10   (*e*) 12 − 10.2
(*f*) 20.4 − 13   (*g*) 20 − 13.4   (*h*) 30.12 − 10.05   (*i*) 30.15 − 10.5   (*j*) 30.5 − 10.15

$$(k) \quad \begin{array}{r} 13.7 \\ -\ 8.3 \end{array} \quad (l) \quad \begin{array}{r} 1.37 \\ -\ .83 \end{array} \quad (m) \quad \begin{array}{r} 13.3 \\ -\ 8.7 \end{array} \quad (n) \quad \begin{array}{r} 13.03 \\ -\ 8.07 \end{array} \quad (o) \quad \begin{array}{r} 12.58 \\ -\ 6.73 \end{array} \quad (p) \quad \begin{array}{r} 42.7 \\ -16.58 \end{array} \quad (q) \quad \begin{array}{r} 105.32 \\ -\ 75.405 \end{array}$$

(*r*) Subtract 1.25 from 50.   (*s*) From 83.6 take 29.   (*t*) Decrease 1,000 by 57.471.

Ans. (*a*) .7   (*d*) 2.2   (*g*) 6.6   (*j*) 20.35   (*m*) 4.6   (*p*) 26.12   (*s*) 54.6
(*b*) 1.3   (*e*) 1.8   (*h*) 20.07   (*k*) 5.4   (*n*) 4.96   (*q*) 29.915   (*t*) 942.529
(*c*) .9   (*f*) 7.4   (*i*) 19.65   (*l*) .54   (*o*) 5.85   (*r*) 48.75

4.17. (*a*) How much change from $10 should be given if the purchase is (1) $3.75, (2) $3.57,   **(5.3)**
(3) $7.43?

(*b*) How much money is left in a bank account if there is a withdrawal of $43.65 from a balance
of (1) $100, (2) $120.27, (3) $140.72?

(*c*) How many miles were covered on a trip if, at the start of the trip, the mileage meter read
2,315.7 and at the end of the trip the reading was (1) 3,000.0, (2) 2,800.9, (3) 2.602.5?

(*d*) How many kilometers per hour slower is a plane traveling at 345.1 kmph than a plane
traveling at (1) 600.5 kmph (2) 525.7 kmph (3) 446.8 kmph?

Ans. (*a*) (1) $6.25   (2) $6.43   (3) $2.47
(*b*) (1) $56.35   (2) $76.62   (3) $97.07
(*c*) (1) 684.3   (2) 485.2   (3) 286.8
(*d*) (1) 255.4   (2) 181.6   (3) 101.7

4.18. Multiply: **(6.1)**

(*a*) .3 × .2   (*d*) 1.1 × .7   (*g*) .011 × 1.2   (*j*) 13.57 × .24   (*m*) .425 × 16.4
(*b*) .3 × .04   (*e*) .11 × .07   (*h*) 8.5 × 2.5   (*k*) 1.357 × .024   (*n*) 156.2 × .0000345
(*c*) .003 × .0006   (*f*) 1.1 × .12   (*i*) .85 × .25   (*l*) 42.5 × .164

Ans. (a) .06   (d) .77   (g) .0132   (j) 3.2568   (m) 6.9700
      (b) .012   (e) .0077   (h) 21.25   (k) .032568   (n) .00538890
      (c) .0000018   (f) .132   (i) .2125   (l) 6.9700

4.19. Multiply: **(6.2)**

(a) $30 \times .2$ (c) $300 \times .002$ (e) $25 \times .0006$ (g) $1,020 \times .031$ (i) $.564 \times 25,800$
(b) $3,000 \times 2$ (d) $250 \times .06$ (f) 102×3.1 (h) 5.64×258 (j) $23,456 \times 45.3$

Ans. (a) 6.0   (c) .600   (e) .0150   (g) 31.620   (i) 14,551.200
       (b) 600.0   (d) 15.00   (f) 316.2   (h) 1,455.12   (j) 1,062,556.8

4.20. (a) Multiply by 10: (1) .27, (2) 1.384, (3) 93.71.   **(6.3)**
          (b) Multiply by 100: (1) 5.605, (2) .3984, (3) 74.2815.
          (c) Multiply by 1,000: (1) 5.1, (2) 60.723, (3) 8.7564.
          (d) Multiply 6.73 by (1) 10,000, (2) 1,000,000, (3) 1,000,000,000.

Ans. (a) (1) 2.7   (2) 13.84   (3) 937.1
       (b) (1) 560.5   (2) 39.84   (3) 7,428.15
       (c) (1) 5,100   (2) 60,723   (3) 8,756.4
       (d) (1) 67,300   (2) 6,730,000 (3) 6,730,000,000

4.21. (a) At the rate of $4.50 an hour, how much is earned if the number of hours of work is **(6.4)**
          (1) 10, (2) 16, (3) 36.5?
          (b) At 50.5¢ a liter, what is the cost to the nearest cent of (1) 10 liters, (2) 12.4 liters, (3) 23.5 liters?
          (c) At the rate of 15.2 miles per gallon, how many miles can a car travel on (1) 10 gal, (2) 100 gal, (3) 35.4 gal?
          (d) At $48.56 per ounce of gold, what is the value in dollars if the number of ounces of gold is (1) 11, (2) 16.8, (3) 185.9?
          (e) Find the semiannual cost of each service, given each of the following average costs: (1) gas, $17.65 per month; (2) electricity, $26.48 per month; (3) telephone, $43.85.
          (f) Find the weekly earnings, counting a rate of pay of time and a half for hours over 40 hours:

	(1)	(2)	(3)	(4)	(5)	(6)
Number of hours	50	56	42	48	55	64
Hourly rate	$4	$3.50	$4.40	$4.75	$4.82	$5.85

Ans. (a) (1) $45.00   (2) $72.00   (3) $164.25
       (b) (1) 505¢ or $5.05   (2) 626¢ or $6.26   (3) 1,187¢ or $11.87
       (c) (1) 152   (2) 1,520   (3) 538.08
       (d) (1) $534.16   (2) $815.808   (3) $9,027.304
       (e) (1) $105.90   (2) $158.88   (3)$263.10
       (f) (1) $220   (2) $224   (3) $189.20   (4) $247   (5) $301.25   (6) $444.60

4.22. Divide: **(7.1)**

(a) $6\overline{)14.4}$ (c) $4\overline{)3.4}$ (e) $8\overline{).52}$ (g) $12\overline{)28.2}$ (i) $120\overline{).0282}$ (k) $250\overline{)2.4}$
(b) $60\overline{)14.4}$ (d) $40\overline{)3.4}$ (f) $800\overline{)5.2}$ (h) $120\overline{)2.82}$ (j) $25\overline{).02}$ (l) $2500\overline{).0046}$

Ans. (a) 2.4   (c) .85   (e) .065   (g) 2.35   (i) .000235   (k) .0096
       (b) .24   (d) .085   (f) .0065   (h) .0235   (j) .0008   (l) .00000184

4.23.   (*a*)   Divide by 10: (1) 3.7, (2) 38.42, (3) .968, (4) 4,573.6.   **(7.2)**
       (*b*)   Divide by 100: (1) .37, (2) 384.2, (3) 96.8, (4) 4.5736.
       (*c*)   Divide by 1,000: (1) 370, (2) 38.42, (3) 9,680, (4) 457.36.
       (*d*)   Divide 45,384 by (1) 10,000, (2) 100,000, (3) 10,000,000.

      Ans.   (*a*) (1) .37   (2) 3.842   (3) .0968   (4) 457.36
             (*b*) (1) .0037   (2) 3.842   (3) .968   (4) .045736
             (*c*) (1) .370   (2) .03842   (3) 9.680   (4) .45736
             (*d*) (1) 4.5384   (2) .45384   (3) .0045384

4.24.   Divide: **(7.3)**
      (*a*) $.5\overline{)1.5}$ (*b*) $.05\overline{)1.5}$ (*c*) $.005\overline{).0015}$ (*d*) $.8\overline{)1.28}$ (*e*) $.08\overline{).128}$ (*f*) $.08\overline{)12.8}$
      (*g*) $1.5\overline{)2.55}$ (*h*) $.15\overline{).0255}$ (*i*) $.015\overline{)25.5}$ (*j*) $.32\overline{)7.68}$ (*k*) $3.2\overline{).0768}$ (*l*) $.032\overline{)7,680}$

      (*m*) $2.88 \div .24$ (*n*) $.288 \div 2.4$ (*o*) $.576 \div .32$ (*p*) $.0576 \div 3.2$ (*q*) $\dfrac{9.45}{2.1}$ (*r*) $\dfrac{945}{.021}$

                    (*s*) $\dfrac{20.655}{1.35}$ (*t*) $\dfrac{2.0655}{13.5}$ (*u*) $\dfrac{163.834}{270.8}$ (*v*) $\dfrac{1.63834}{27.08}$

      Ans.   (*a*) 3   (*b*) 30   (*c*) .3   (*d*) 1.6   (*e*) 1.6   (*f*) 160   (*g*) 1.7   (*h*) .17   (*i*) 1,700   (*j*) 24
            (*k*) .024   (*l*) 240,000   (*m*) 12   (*n*) .12   (*o*) 1.8   (*p*) .018   (*q*) 4.5   (*r*) 45,000
            (*s*) 15.3   (*t*) .153   (*u*) .605   (*v*) .0605

4.25.   Express each answer to the nearest tenth: (*a*) $2.4 \div 7$,   (*b*) $24 \div 7$,   (*c*) $4.06 \div 11$,   **(7.4)**
      (*d*) $40.6 \div 11$. Express each answer to the nearest hundredth: (*e*) $.572 \div 12$,   (*f*) $52.7 \div 12$,
      (*g*) $2.574 \div 13$ (*h*) $257.4 \div 31$.

      Ans.   (*a*) .3   (*b*) 3.4   (*c*) .4   (*d*) 3.7   (*e*) .05   (*f*) 4.39   (*g*) .20   (*h*) 8.30

4.26.   Express each answer to the nearest tenth and also to the nearest hundredth: **(7.5)**
      (*a*) $.3\overline{).2}$ (*c*) $.06\overline{).5}$ (*e*) $.7\overline{)2.5}$ (*g*) $1.1\overline{)24}$ (*i*) $.75 \div .033$ (*k*) $52.1 \div 1.23$
      (*b*) $.3\overline{)2}$ (*d*) $.06\overline{)5}$ (*f*) $.7\overline{).25}$ (*h*) $.11\overline{).24}$ (*j*) $750 \div 3.3$ (*l*) $5.21 \div 12.3$

      Ans.   (*a*) .7, .67   (*d*) 83.3, 83.33   (*g*) 21.8, 21.82   (*j*) 227.3, 227.27
            (*b*) 6.7, 6.67   (*e*) 3.6, 3.57   (*h*) 2.2, 2.18   (*k*) 42.4, 42.36
            (*c*) 8.3, 8.33   (*f*) .4, .36   (*i*) 22.7, 22.73   (*l*) .4, .42

4.27.   (*a*)   At \$6 per 100, find the cost of (1) 400 items, (2) 250 items, (3) 575 items.   **(7.6)**
       (*b*)   At \$10.25 per 1,000, find the cost of (1) 2,000 items, (2) 3,400 items, (3) 14,200 items.
       (*c*)   Find the cost of 3,500 items at (1) \$5 per 100, (2) \$3 per 1,000, (3) \$3.70 per 1,000.
       (*d*)   Complete the following purchase list by finding the cost of each group of articles purchased
           and the total cost:

No.	Kind	Price	Cost
400	#2 red pencil	$ 7.50 per 100	
360	Scotch tape	$35 per 100	
5,500	Ball-point pens	$45 per 1,000	

Total:

(e) A metal rod, 36 m in length, is cut into rods of equal length. How many rods were cut if the length of each rod is (1) 3.6 m, (2) .36 m, (3) 2.4 m?

(f) If the flight time of a plane is 14.8 hours, how many gallons of gas are consumed per hour on the average if total gas consumption is (1) 614.2 gal, (2) 550.56 gal, (3) 485.44 gal?

Ans. (a) (1) \$24   (2) \$15   (3) \$34.50
     (b) (1) \$20.50   (2) \$34.85   (3) \$145.55
     (c) (1) \$175   (2) \$10.50   (3) \$12.95
     (d) pencils, \$30; tape, \$126; pens, \$247.50; total, \$403.50
     (e) (1) 10   (2) 100   (3) 15
     (f) (1) 41.5   (2) 37.2   (3) 32.8

4.28. Express each as a decimal: **(8.1)**

(a) $\dfrac{9}{10}$ (d) $\dfrac{29}{1,000}$ (g) $\dfrac{9}{5}$ (j) $\dfrac{7}{800}$ (m) $\dfrac{21}{25}$ (p) $\dfrac{631}{500}$

(b) $\dfrac{9}{100}$ (e) $\dfrac{413}{10,000}$ (h) $\dfrac{9}{50}$ (k) $\dfrac{17}{16}$ (n) $\dfrac{21}{250}$ (q) $\dfrac{309}{200}$

(c) $\dfrac{29}{100}$ (f) $\dfrac{413}{100,000}$ (i) $\dfrac{7}{8}$ (l) $\dfrac{17}{1,600}$ (o) $\dfrac{631}{5}$ (r) $\dfrac{309}{2,000}$

Ans. (a) .9   (d) .029   (g) 1.8   (j) .00875   (m) .84   (p) 1.262
     (b) .09   (e) .0413   (h) .18   (k) 1.0625   (n) .084   (q) 1.545
     (c) .29   (f) .00413   (i) .875   (l) .010625   (o) 126.2   (r) .1545

4.29. (a) Express to the nearest tenth and also as a mixed decimal having one decimal place: **(8.2)**
     (1) 2/7,   (2) 5/9,   (3) 7/11,   (4) 51/16.

     (b) Express to the nearest hundredth and also as a mixed decimal having two decimal places:
     (1) 8/11, (2) 11/16, (3) 23/18, (4) 50/19.

Ans. (a) (1) .3, $.2\frac{6}{7}$ (2) .6, $.5\frac{5}{9}$ (3) .6, $.6\frac{4}{11}$ (4) 3.2, $3.1\frac{7}{8}$
     (b) (1) .73, $.72\frac{8}{11}$ (2) .69, $.68\frac{3}{4}$ (3) 1.28, $1.27\frac{7}{9}$ (4) 2.63, $2.63\frac{3}{19}$

4.30. Express each as a repeating decimal. Also, show the repeating decimal in simplified form. **(8.3)**

(a) $\dfrac{5}{9}$ (b) $\dfrac{5}{12}$ (c) $\dfrac{7}{11}$ (d) $\dfrac{5}{33}$ (e) $\dfrac{71}{99}$ (f) $\dfrac{5}{111}$ (g) $\dfrac{29}{333}$ (h) $\dfrac{11}{37}$ (i) $\dfrac{19}{300}$

Ans. (a) $.555\ldots = .\overline{5}$ (d) $.151515\ldots = .\overline{15}$ (g) $.087087087\ldots = .\overline{087}$
     (b) $.416666\ldots = .41\overline{6}$ (e) $.717171\ldots = .\overline{71}$ (h) $.297297297\ldots = .\overline{297}$
     (c) $.636363\ldots = .\overline{63}$ (f) $.045045045\ldots = .\overline{045}$ (i) $.06333 = .06\overline{3}$

4.31. Express as a common fraction: (a) .5, (b) .05, (c) .005, (d) 2.2, (e) .22, **(9.1)**
(f) .022,   (g) 1.25,   (h) .125,   (i) 11.1,   (j) 1.11,   (k) .111,   (l) 2.73,   (m) .0273,
(n) .12,   (o) .123,   (p) .12345.

Ans. (a) $\dfrac{1}{2}$ (c) $\dfrac{1}{200}$ (e) $\dfrac{11}{50}$ (g) $\dfrac{5}{4}$ (i) $\dfrac{111}{10}$ (k) $\dfrac{111}{1,000}$ (m) $\dfrac{273}{10,000}$ (o) $\dfrac{123}{1,000}$

     (b) $\dfrac{1}{20}$ (d) $\dfrac{11}{5}$ (f) $\dfrac{11}{500}$ (h) $\dfrac{1}{8}$ (j) $\dfrac{111}{100}$ (l) $\dfrac{273}{100}$ (n) $\dfrac{3}{25}$ (p) $\dfrac{2,469}{20,000}$

4.32. Express as a common fraction: **(9.2)**

(a) $.3\frac{1}{3}$ (c) $.2\frac{1}{4}$ (e) $.09\frac{1}{11}$ (g) $1.6\frac{2}{3}$ (i) $.37\frac{1}{2}$ (k) $.83\frac{1}{3}$ (m) $.15\frac{5}{8}$ (o) $.156\frac{1}{2}$

(b) $.5\frac{1}{2}$ (d) $.14\frac{2}{7}$ (f) $.16\frac{2}{3}$ (h) $.06\frac{2}{3}$ (j) $3.7\frac{1}{2}$ (l) $8.3\frac{1}{3}$ (n) $.156\frac{1}{4}$ (p) $.0333333\frac{1}{3}$

Ans. (a) $\dfrac{1}{3}$ (c) $\dfrac{9}{40}$ (e) $\dfrac{1}{11}$ (g) $\dfrac{5}{3}$ (i) $\dfrac{3}{8}$ (k) $\dfrac{5}{6}$ (m) $\dfrac{5}{32}$ (o) $\dfrac{5}{32}$

(b) $\dfrac{11}{20}$ (d) $\dfrac{1}{7}$ (f) $\dfrac{1}{6}$ (h) $\dfrac{1}{15}$ (j) $\dfrac{15}{4}$ (l) $\dfrac{25}{3}$ (n) $\dfrac{5}{32}$ (p) $\dfrac{1}{30}$

4.33. Perform the indicated operation using a calculator: **(10.1)**

(a) $431.2 + 612.8 + 1001.07$

(b) $50.003 - 4.05$

(c) 49.1×27.8 (d) $.07\overline{)1.0076}$

Ans. (a) 2045.07   (b) 45.953

(c) 1364.98   (d) 14.394285

4.34. Express as a repeating decimal using a calculator: **(10.2)**

(a) 2/9   (b) 7/15   (c) 8/99

Ans. (a) $.22\ldots = .\bar{2}$ (b) $.4\bar{6}$ (c) $\overline{.08}$

Chapter 5

Per Cents

1. MEANINGS OF PER CENT

Per cent means hundredths. The symbol for per cent, %, is an excellent one since it combines a 1 and two 0s the digits of 100. Thus, 25% means 25 hundredths, 25/100, or .25.

25% may also mean *25 out of a hundred*. Figure 5-1 shows a square containing 100 small squares. Observe that out of the 100 squares, 25 of them are shaded and the remaining 75 are unshaded.

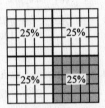

Fig. 5-1

When we say that 25% of the square is shaded, we mean that 25 out of 100 squares are shaded. Also, 75%, or 75 out of 100 squares, are unshaded. Since the entire square contains 100 out of the 100 squares, the entire square is 100% of itself.

The statement "The entire square is the sum of the shaded and the unshaded parts" in terms of per cent is: $100\% = 25\% + 75\%$.

To summarize: Per cent has meaning (1) as hundredths, (2) as a decimal fraction whose denominator is 100, (3) as a 2-place decimal, and (4) as a number out of 100.

Rule: A given number of per cent is equivalent to a fraction whose numerator is the given number and whose denominator is 100.

$$\text{Thus, } 75\% = \frac{75}{100}$$

1.1 PER CENTS OF A GEOMETRIC FIGURE

In each large square of Fig. 5-2, find the per cent that is occupied by the shaded letter, and the per cent that is unshaded.

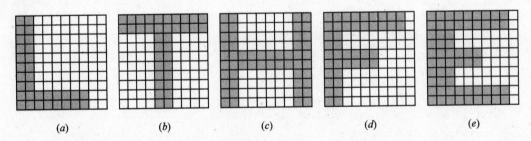

(a)　　　　　(b)　　　　　(c)　　　　　(d)　　　　　(e)

Fig. 5-2

Illustrative Solution (b) Since the letter T occupies 36 of the 100 small squares, the shaded part is 36%. The unshaded part, 64 squares, is 64%.

Ans. (a) 32%, 68%　　(c) 52%, 48%　　(d) 42%, 58%　　(e) 60%, 40%

1.2 PROBLEMS INVOLVING PER CENTS WHOSE SUM IS 100%

(a) What per cent of a class is absent if the per cent present is (1) 75%, (2) $87\frac{1}{2}$%, (3) 100%?

(b) What per cent of a mixture of water and alcohol is water if the per cent of alcohol is (1) 10%, (2) $5\frac{1}{2}$%, (3) $2\frac{1}{4}$%?

(c) What per cent of 100 radio batteries is good if it is found that the number of defective batteries is (1) 5, (2) 21, (3) 39?

(d) One of three partners receives 50% of the profits. What per cent of the profits does the second partner get if the per cent of profit of the remaining partner is (1) 20%, (2) $12\frac{1}{2}$%, (3) $16\frac{2}{3}$%?

(*e*) What per cent of the original price do you pay at a sale advertised at (1) 10% off, (2) 5% off, (3) $37\frac{1}{2}$% off?

Illustrative Solution   (*d*) (1) Since the other two partners receive 50% + 20%, of the profits, the remaining partner gets 100% − 70%, or 30%, of the profits.

Ans. (*a*) (1) 25%   (2) $12\frac{1}{2}$%   (3) 0%
       (*b*) (1) 90%   (2) $94\frac{1}{2}$%   (3) $97\frac{3}{4}$%
       (*c*) (1) 95%   (2) 79%   (3) 61%
       (*d*) (2) $37\frac{1}{2}$%   (3) $33\frac{1}{3}$%
       (*e*) (1) 90%   (2) 95%   (3) $62\frac{1}{2}$%

1.3   MEANINGS OF PER CENT

Express each of the following as per cents:
(*a*) (1) 23 hundredths   (2) 6.5 hundredths   (3) $\frac{1}{2}$ hundredth
(*b*) (1) 3 out of a hundred   (2) $12\frac{1}{4}$ out of a hundred   (3) 4.15 out of a hundred
(*c*) (1) .65   (2) $.04\frac{1}{2}$ (3) .015

(*d*) (1) $\dfrac{4}{100}$ (2) $\dfrac{40}{100}$ (3) $\dfrac{7}{10}$

Illustrative Solution

(*c*) (3) $.015 = \dfrac{15}{1000} = \dfrac{15 \div 10}{1000 \div 10} = \dfrac{1.5}{100} = 1.5\%$ or $1\frac{1}{2}\%$ *Ans.*

Ans. (*a*) (1) 23%   (2) 6.5% or $6\frac{1}{2}$%   (3) $\frac{1}{2}$% or .5%
       (*b*) (1) 3%   (2) $12\frac{1}{4}$% or 12.25%   (3) 4.15%
       (*c*) (1) 65%   (2) $4\frac{1}{2}$%
       (*d*) (1) 4%   (2) 40%   (3) 70%

2. PER CENTS AND DECIMALS

Expressing a Per Cent as a Decimal

Two steps are needed to convert 7.5% into the decimal .075: (1) Omit the % sign. (2) Move the decimal point of the per cent two places to the *left*.

Note in the following procedure the use of the arrow to indicate the new position of the decimal point.

To Express a Per Cent as a Decimal

Express as a decimal: (*a*) 7%   (*b*) 8.8%   (*c*) 15.5%   (*d*) 275%

PROCEDURE	SOLUTIONS			
Omit % and move the decimal point two places to the *left*:	.07% .07 *Ans.*	.08.8% .088 *Ans.*	.15.25% .1525 *Ans.*	2.75% 2.75 *Ans.*

Expressing a Decimal as a Per Cent

Two steps are needed to convert .075 into the per cent 7.5%: (1) Add the % sign. (2) Move the decimal point of the decimal two places to the *right*.

To Express a Decimal as a Per Cent

Express as a per cent: (a) .37   (b) .475   (c) 5.75   (d) 9.4

PROCEDURE   SOLUTIONS

Add % and move the   .37.%   .47.5%   5.75.%   9.40.%
decimal point two places
to the *right*:   37% *Ans.*   47.5% *Ans.*   575% *Ans.*   940% *Ans.*

Since 100% = 1, any per cent less than 100% is less than 1; and any per cent greater than 100% is greater than 1.

Rule: A per cent less than 100% is less than 1, and a per cent greater than 100% is greater than 1.

Thus, 200% = 2, 500% = 5, and 750% = 7.5 or $7\frac{1}{2}$.

2.1   **EXPRESSING A PER CENT AS A DECIMAL**

Express each as a decimal: (a) 35%,   (b) 5%,   (c) 45.2%,   (d) 345.2%,   (e) .2%,   (f) 1.050%.

Illustrative Solution   (c) Express 45.2% as a decimal by omitting % and moving the decimal point two places to the left; thus, .45.2% = .452   *Ans.*

Ans.   (a) .35   (b) .05   (d) 3.452   (e) .002   (f) 10.50 or 10.5

2.2   **EXPRESSING A DECIMAL AS A PER CENT**

Express each as a per cent: (a) .55,   (b) .05,   (c) .657,   (d) 6.048,   (e) .006,   (f) $.00\frac{1}{4}$,   (g) 16.508.

Illustrative Solution   (d) Express 6.048 as a per cent by adding % and moving the decimal point two places to the right; thus, 6.04.8% = 604.8%   *Ans.*

Ans.   (a) 55%   (b) 5%   (c) 65.7%   (e) .6%   (f) $\frac{1}{4}$%   (g) 1,650.8%

2.3   **PROBLEM SOLVING INVOLVING FINDING OF PER CENTS**

(a) What per cent of the original price is the new price if the original price has been (1) halved, (2) doubled, (3) quadrupled?

(b) What per cent of the games played has a team won if its winning average is (1) .600, (2) .570, (3) .525?

(c) What per cent of times at bat did a baseball player get a hit if his batting average is (1) .300, (2) .275, (3) .254?

(d) On a test of 100 questions, what per cent were answered correctly if the number of wrong answers was (1) 4, (2) 20, (3) 39?

(e) In 100 quarts of saltwater, what per cent of the mixture is water if the number of quarts of salt is (1) 24, (2) $3\frac{1}{2}$, (3) 15.5?

Illustrative Solution   (c) (2) Convert .275 to per cent by adding % and moving the decimal point two places to the right; thus, .27.5% = 27.5% or $27\frac{1}{2}$%.

Ans.   (a) (1) 50%   (2) 200%   (3) 400%
         (b) (1) 60%   (2) 57%   (3) 52.5% or $52\frac{1}{2}$%
         (c) (1) 30%   (3) 25.4% or $25\frac{2}{5}$%
         (d) (1) 96%   (2) 80%   (3) 61%
         (e) (1) 76%   (2) $96\frac{1}{2}$% or 96.5%   (3) 84.5% or $84\frac{1}{2}$%

3. PER CENTS AND FRACTIONS

Expressing a Per Cent as a Common Fraction

Three steps are needed to convert 15% into the common fraction 3/20: (1) Omit the % sign. (2) Divide by 100. (3) Reduce to lowest terms, or simpify as a common fraction, or both.

To Express a Per Cent as a Common Fraction

Express as a common fraction: (a) 35% (b) 140% (c) $3\frac{1}{3}$%

PROCEDURE	SOLUTIONS		
1. Omit % and divide by 100:	$\dfrac{35}{100}$	$\dfrac{140}{100}$	$\dfrac{3\frac{1}{3}\%}{100}$
2. Reduce to lowest terms, or simplify as a common fraction, or both:	$\dfrac{\overset{7}{\cancel{35}}}{\underset{20}{\cancel{100}}}=\dfrac{7}{20}$ _Ans._	$\dfrac{\overset{7}{\cancel{140}}}{\underset{5}{\cancel{100}}}=\dfrac{7}{5}$ _Ans._	$\dfrac{3\frac{1}{3}\times 3}{100\times 3}=\dfrac{10}{300}=\dfrac{1}{30}$ _Ans._

Expressing a Common Fraction as a Per Cent

Since per cent means hundredths, a common fraction can be converted to a per cent by dividing the numerator by the denominator and carrying the division to hundredths.

To Express a Common Fraction as a Per Cent

Express as a per cent: (a) $1\frac{1}{8}=9/8$ (b) 5/6 (c) 3/7

PROCEDURE

SOLUTIONS

1. Divide the numerator by the denominator and carry the division to two places—that is, to hundredths:

1. $\begin{array}{r} 1.12 \text{ R }4 \\ 8\overline{)9.^{1}0^{2}0} \end{array}$

$\begin{array}{r} .83 \\ 6\overline{)5.00} \\ \underline{4\,8} \\ 20 \\ \underline{18} \\ R=2 \end{array}$

$\begin{array}{r} .42 \\ 7\overline{)3.00} \\ \underline{2\,8} \\ 20 \\ \underline{14} \\ R=6 \end{array}$

2. For any nonzero remainder, annex the fraction remainder/divisor reduced to lowest terms:

2. $1.12\frac{4}{8}=1.12\frac{1}{2}$

$.83\frac{2}{6}=.83\frac{1}{3}$

$.42\frac{6}{7}$

3. Express the decimal obtained in step 2 as a per cent:

3. $112\frac{1}{2}\%$ _Ans._

$83\frac{1}{3}\%$ _Ans._

$42\frac{6}{7}\%$ _Ans._

Expressing a Common Fraction as a Per Cent if the Denominator Is a Factor of 100

If the denominator of a common fraction is a factor of 100, converting the fraction to per cent is readily accomplished by multiplying by 100%. Keep in mind that the factors of 100 are 2, 4, 5, 10, 20, 25, and 50. Thus,

$$\frac{11}{20}=\frac{11}{\underset{}{\cancel{20}}}\times \overset{5\%}{\cancel{100\%}}=55\% \qquad \frac{7}{25}=\frac{7}{\underset{}{\cancel{25}}}\times \overset{4\%}{\cancel{100\%}}=28\%$$

Rounding to the Nearest Per Cent

Round to the nearest per cent as follows:

1. Drop the end fraction of a per cent if it is less than $\frac{1}{2}$.
   Thus, round $83\frac{1}{3}\%$ to 83% since $\frac{1}{3}$ is less than $\frac{1}{2}$.

2. Add 1% after dropping the end fraction if it is equal to or more than $\frac{1}{2}$.
   Thus, round $16\frac{2}{3}\%$ to 17% since $\frac{2}{3}$ is more than $\frac{1}{2}$. Also, round $12\frac{1}{2}\%$ to 13%.

Equivalent Fractions, Decimals, and Per Cents

Use Table 5-1 for ready reference. Use it also to develop an understanding of the interrelationships amongst the most frequently used fractions, decimals, and per cents. You can verify this data using a calculator.

Table 5-1

Fraction	Per Cent	Decimal	Fraction	Per Cent	Decimal
$\frac{1}{2}$	50%	.50, .5	$\frac{1}{8}$	$12\frac{1}{2}\%$	$.12\frac{1}{2}$, .125
$\frac{1}{3}$	$33\frac{1}{3}\%$	$.33\frac{1}{3}$	$\frac{3}{8}$	$37\frac{1}{2}\%$	$.37\frac{1}{2}$, .375
$\frac{2}{3}$	$66\frac{2}{3}\%$	$.66\frac{2}{3}$	$\frac{5}{8}$	$62\frac{1}{2}\%$	$.62\frac{1}{2}$, .625
$\frac{1}{4}$	25%	.25	$\frac{7}{8}$	$87\frac{1}{2}\%$	$.87\frac{1}{2}$, .875
$\frac{3}{4}$	75%	.75	$\frac{1}{10}$	10%	.10, .1
$\frac{1}{5}$	20%	.20, .2	$\frac{3}{10}$	30%	.30, .3
$\frac{2}{5}$	40%	.40, .4	$\frac{1}{9}$	$11\frac{1}{9}\%$	$.11\frac{1}{9}$
$\frac{2}{5}$	60%	.60, .6	$\frac{1}{11}$	$9\frac{1}{11}\%$	$.09\frac{1}{11}$
$\frac{4}{5}$	80%	.80, .8	$\frac{1}{12}$	$8\frac{1}{3}\%$	$.08\frac{1}{3}$
$\frac{1}{6}$	$16\frac{2}{3}\%$	$.16\frac{2}{3}$	$\frac{1}{20}$	5%	.05
$\frac{5}{6}$	$83\frac{1}{3}\%$	$.83\frac{1}{3}$	$\frac{1}{25}$	4%	.04
$\frac{1}{7}$	$14\frac{2}{7}\%$	$.14\frac{2}{7}$	$\frac{1}{50}$	2%	.02
$\frac{2}{7}$	$28\frac{4}{7}\%$	$.28\frac{4}{7}$	$\frac{1}{100}$	1%	.01

3.1   EXPRESSING A PER CENT AS A COMMON FRACTION

Express each per cent as a common fraction: (a) 40%,   (b) 375%,   (c) $12\frac{1}{2}\%$,   (d) $4\frac{1}{4}\%$,   (e) .7%.

Solutions

(a) $\dfrac{40 \div 20}{100 \div 20}$ (b) $\dfrac{375 \div 25}{100 \div 25}$ (c) $\dfrac{12\frac{1}{2} \times 2}{100 \times 2}$ (d) $\dfrac{4\frac{1}{4} \times 4}{100 \times 4}$ (e) $\dfrac{.7 \times 10}{100 \times 10}$

$\dfrac{2}{5}$ *Ans.* $\dfrac{15}{4}$ or $3\frac{3}{4}$ *Ans.* $\dfrac{25}{200} = \dfrac{1}{8}$ *Ans.* $\dfrac{17}{400}$ *Ans.* $\dfrac{7}{1,000}$ *Ans.*

3.2   EXPRESSING A COMMON FRACTION BOTH AS A PER CENT AND AS A ROUNDED PER CENT

Express each fraction as a per cent and rounded to the nearest per cent: (a) 4/7, (b) 13/16, (c) 13/12, (d) 7/15.

Solutions

(a)
```
    .5 7 R 1
7)4.0⁵0
```
$57\frac{1}{2}\%$, 57% *Ans.*

(b)
```
      .81
16)13.00
   12 8
      20
      16
   R = 4
```
$81\frac{1}{4}\%$, 81% *Ans.*

(c)
```
      1.08
12)13.00
   12
    1 00
      96
   R = 4
```
$108\frac{1}{3}\%$, 108% *Ans.*

(d)
```
      .46
15)7.00
    6 0
    1 00
      90
   R = 10
```
$46\frac{2}{3}\%$, 47% *Ans.*

3.3   EXPRESSING A COMMON FRACTION AS A PER CENT: MULTIPLYING BY 100%

Express each fraction as a per cent: (a) 11/5,   (b) 9/20,   (c) 13/25,   (d) 153/50,   (e) 35/200.

Solutions   Multiply each fraction by 100% or 1.

(a) $\dfrac{11}{5} \times \overset{20\%}{\cancel{100\%}}$ (b) $\dfrac{9}{20} \times \overset{5\%}{\cancel{100\%}}$ (c) $\dfrac{13}{25} \times \overset{4\%}{\cancel{100\%}}$ (d) $\dfrac{153}{50} \times \overset{2\%}{\cancel{100\%}}$ (e) $\dfrac{35}{200} \times \overset{1\%}{\cancel{100\%}}$

220% *Ans.*   45% *Ans.*   52% *Ans.*   306% *Ans.* $\dfrac{35}{2}\%$ or $17\frac{1}{2}\%$ *Ans.*

3.4   PROBLEM SOLVING INVOLVING PER CENTS AND EQUIVALENT FRACTIONS

(a) What fractional part of a test is answered correctly if the test score is (1) 90%, (2) 75%, (3) 65%?

(b) What fractional part of the games that a team played was won if its winning average is (1) 50%, (2) 80%, (3) 84%?

(c) What fractional part of the original price is the new price if the reduction at a sale is (1) 5% off, (2) 40% off, (3) $7\frac{1}{2}\%$ off?

(d) What per cent of the length of a segment is the length of a greater segment if the greater segment is (1) twice as long, (2) $2\frac{1}{2}$ times as long, (3) $5\frac{1}{4}$ times as long?

(e) What per cent of a class of 20 students is the number of girls if the number of boys in the class is (1) 7, (2) 12, (3) 18?

Illustrative Solution   (e) (1) In a class of 20, if 7 are boys, then 13 are girls. Hence, the girls are 13/20 of the class. As a per cent,

$$\frac{13}{20} = \frac{13}{20} \times 100\% = 65\% \ Ans.$$

Ans.   (a) (1) $\dfrac{9}{10}$ (2) $\dfrac{3}{4}$ (3) $\dfrac{13}{20}$ (c) (1) $\dfrac{19}{20}$ (2) $\dfrac{3}{5}$ (3) $\dfrac{37}{40}$ (e) (2) 40%   (3) 10%

       (b) (1) $\dfrac{1}{2}$ (2) $\dfrac{4}{5}$ (3) $\dfrac{21}{25}$ (d) (1) 200%   (2) 250%   (3) 525%

128 PER CENTS [CHAP. 5

4. FINDING A PER CENT OF A GIVEN NUMBER

Rule: To find a per cent of a number, *multiply* the per cent and the number.

Thus, to find $33\frac{1}{3}\%$ of 27, multiply $33\frac{1}{3}\%$ and 27; that is, $\frac{1}{3} \times 27 = 9$ *Ans.*
Also, to find 12.3% of 200, multiply 12.3% and 200; that is, $.123 \times 200 = 24.6$ *Ans.*
As in the illustrations above, either one of two methods may be used:

Method 1: Express the given per cent as a *fraction*, using Table 5-1 when the terms of the fraction are small numbers.
Method 2: Express the given per cent as a *decimal*.

To Find a Per Cent of a Given Number: Method 1

Find: (a) $62\frac{1}{2}\%$ of 248 (b) $16\frac{2}{3}\%$ of 24.36

PROCEDURE SOLUTIONS

1. Express the per cent as a fraction: (See Table 5-1.)

1. $62\frac{1}{2}\% = \frac{5}{8}$ $16\frac{2}{3}\% = \frac{1}{6}$

2. Multiply the given number by the fraction found in step 1:

2. $\frac{5}{8} \times \overset{31}{248} = 155$ *Ans.* $\frac{1}{6} \times \overset{4.06}{24.36} = 4.06$ *Ans.*

To Find a Per Cent of a Given Number: Method 2

Find: (a) 22% of 350 (b) $9\frac{1}{2}\%$ of 4,000

PROCEDURE SOLUTIONS

1. Express the per cent as a decimal: 1. 22% = .22 $9\frac{1}{2}\% = .09\frac{1}{2} = .095$

2. Multiply the given number by the decimal found in step 1:

2.
```
  350
× .22
  700
 700
77.00
```
Ans. 77
```
   4000
×  .095
  20000
 36000
380.000
```
Ans. 380

Important Note: The above calculations can be easily performed using a calculator. You should verify many of the results in this chapter using your calculator.

4.1 FINDING A PER CENT OF A GIVEN NUMBER USING FRACTIONS

Find (a) $66\frac{2}{3}\%$ of 35.1, (b) $28\frac{4}{7}\%$ of 20.37, (c) $11\frac{1}{9}\%$ of 6.147, (d) $8\frac{1}{3}\%$ of .4056, (e) $87\frac{1}{2}\%$ of 42.8.

Solutions Use Table 5-1.

(a) $\frac{2}{3} \times \overset{11.7}{35.1}$ (b) $\frac{2}{7} \times \overset{2.91}{20.37}$ (c) $\frac{1}{9} \times \overset{.683}{6.147}$ (d) $\frac{1}{12} \times \overset{.0338}{.4056}$ (e) $\frac{7}{8} \times \overset{10.7}{42.8}$

23.4 *Ans.* 5.82 *Ans.* .683 *Ans.* .0338 *Ans.* 37.45 *Ans.*

4.2 FINDING A PER CENT OF A NUMBER USING DECIMALS

Find (a) 4% of 5,000, (b) $3\frac{1}{2}\%$ of 484, (c) 72% of 465, (d) .25% of 20,000, (e) 225% of 48.

Solutions

(a)	5,000	(b)	484	(c)	465	(d)	20,000	(e)	2.25

(a) 5,000
    × .04
    200.00

    200 *Ans.*

(b) 484
    × .035
    2420
    1452
    16.940

    16.94 *Ans.*

(c) 465
    × .72
    930
    3255
    334.80

    334.8 *Ans.*

(d) 20,000
    × .0025
    50.0000

    50 *Ans.*

(e) 2.25
    × 48
    1800
    900
    108.00

    108 *Ans.*

4.3   PROBLEM SOLVING INVOLVING PER CENTS OF A NUMBER

(a) A team won $66\frac{2}{3}$% of its games. How many games did it win if the number of games played was (1) 54, (2) 117, (3) 144?

(b) If the sales tax rate is 7%, how much is the sales tax to the nearest cent on a purchase of (1) $40, (2) $57.90, (3) $452.50?

(c) A square is divided into 400 small squares. How many of these squares are left unshaded if the per cent of the squares shaded is (1) 20%, (2) $37\frac{1}{2}$%, (3) 98%?

(d) How many pupils in a school of 1,200 pupils are absent if the per cent of those attending school is (1) 97%, (2) 92.5%, (3) $89\frac{2}{3}$%?

(e) How many miles, to the nearest mile, remain to be traveled on a trip of 2,500 miles if the per cent of the trip already traveled is (1) 88%, (2) 78.5%, (3) 42.75%?

(f) A family budget plan allows 30% for food, 25% for rent, 20% for savings, 10% for clothing, 7% for recreation, and the rest for miscellaneous. According to the budget plan, how much will be spent for each item if the annual earnings is (1) $30,000, (2) $25,000, (3) $27,260?

Illustrative Solution   (c) (3) If 98% of the squares are shaded, then 2% are unshaded. Hence, the number of unshaded squares is 2% of 400, which equals .02 × 400, or 8 *Ans.*

Ans.   (a) (1) 36   (2) 78   (3) 96
         (b) (1) $2.80   (2) $4.05   (3) $31.68
         (c) (1) 320   (2) 250
         (d) (1) 36   (2) 90   (3) 124
         (e) (1) 300   (2) 538   (3) 1,431

(f)

	Food	Rent	Savings	Clothing	Recreation	Miscellaneous
(1)	$9,000	$7,500	$6,000	$3,000	$2,100	$2,400
(2)	$7,500	$6,250	$5,000	$2,500	$1,750	$2,000
(3)	$8,178	$6,815	$5,452	$2,726	$1,908.20	$2,180.80

5. FINDING WHAT PER CENT ONE NUMBER IS OF A SECOND NUMBER

Rule: To find what per cent one number is of a second number, *divide* the first number by the second number, carrying the quotient to two decimal places; then express the result as a per cent.

Thus, to find what per cent 3 is of 12, divide 3 by 12:

$$\begin{array}{r} .25 \\ 12\overline{)3.00} \end{array} \quad \textit{Ans. } 25\%$$

The same result is found by reducing the fraction 3/12 to 1/4 or 25%.

To Find What Per Cent One Number Is of a Second Number

Find to the nearest per cent: (a) What % of 16 is 11?   (b) What % is 16 of 11?

PROCEDURE   SOLUTIONS

1. State the problem in the form "One   1. "11 is what % of 16?"   "16 is what % of 11?"
   number is what per cent of a second
   number?":

2. Divide the first number by the second   2.   .68   1.45
   number, carrying the quotient to two   16)11.00   11)16.00
   decimal places: 9 6   11
                                                   ────   ────
                                                   1 40   5 0
                                                   1 28   4 4
                                                   ────   ────
                                                   R = 12   60
                                                                                  55
                                                                                ────
                                                                               R = 5

3. Express the result obtained in step 2 as a 3. $68\frac{3}{4}\% = 69\%$, $145\frac{5}{11}\% = 145\%$,
   rounded per cent: to nearest % Ans.   to nearest % Ans.

5.1   FINDING WHAT PER CENT ONE NUMBER IS OF ANOTHER NUMBER

Find each result as a per cent and rounded to the nearest per cent: (a) What per cent of 7 is 8?
(b) What per cent is 7 of 6?   (c) 13 is what per cent of 27?   (d) ?% of 12 is 31?

Solutions   State each as "One number is what per cent of a second number?"

(a) 8 is what % of 7?   (b) 7 is what % of 6?   (c) 13 is what % of 27?   (d) 31 is what % of 12?

```
   1.14 R 2            1.16 R 4              .48                   2.58
 7)8.1030           6)7.1040            27)13.00              12)31.00
114 2/7 %, 114% Ans.  116 2/3 %, 117% Ans.    10 8                  24
                                             ────                  ────
                                             2 20                   7 0
                                             2 16                   6 0
                                             ────                  ────
                                                4               1 00
                                    48 4/27 %, 48% Ans.             96
                                                                  ────
                                                                    4
                                                        258 1/3 %, 258% Ans.
```

5.2   PROBLEM SOLVING INVOLVING FINDING WHAT PER CENT ONE NUMBER IS OF A SECOND NUMBER

(a) If a woman's earnings is $500, what per cent of her earnings is saved if (1) she saves $18, (2) she saves $145?

(b) What per cent of the games played did a school team win if it played a total of 30 games and (1) won 24 of them, (2) won 12 of them, (3) lost 13 of them?

(c) What per cent of test questions were correctly answered if (1) 18 were right and 7 were wrong, (2) 18 were right out of total of 30, (3) 18 were wrong out of a total of 40?

(d) What per cent of the games played did a team win if (1) 12 were won and 8 were lost, (2) 23 were won out of 25 games played, (3) the team won 3 of every 4 games, (4) the team lost 5 of every 8 games played?

(e) A worker making $150 a week had his pay raised to $200 a week. (1) What per cent of his original salary is his new salary? (2) What per cent of his original salary is the increase? (3) What per cent of the new salary is his original salary?

(f) A price of $9 is increased by $2.70. (1) What per cent of the original price is the increase? (2) What per cent of the original price is the new price? (3) What per cent of the new price is the original price?

Solutions

(a) (1) $\dfrac{180}{500} = 36\%$ (2) $\dfrac{145}{500} = 29\%$

(b) (1) $\dfrac{24}{30} = 80\%$ (2) $\dfrac{12}{30} = 40\%$ (3) $\dfrac{17}{30} = 56\frac{2}{3}\%$

(c) (1) $\dfrac{18}{25} = 72\%$ (2) $\dfrac{18}{30} = 60\%$ (3) $\dfrac{22}{40} = 55\%$

(d) (1) $\dfrac{12}{20} = 60\%$ (2) $\dfrac{23}{25} = 92\%$ (3) $\dfrac{3}{4} = 75\%$ (4) $\dfrac{3}{8} = 37\frac{1}{2}\%$ or 37.5%

(e) (1) $\dfrac{200}{150} = 133\frac{1}{3}\%$ (2) $\dfrac{50}{150} = 33\frac{1}{3}\%$ (3) $\dfrac{150}{200} = 75\%$

(f) (1) $\dfrac{2.70}{9.00} = 30\%$ (2) $\dfrac{11.70}{9.00} = 130\%$ (3) $\dfrac{9.00}{11.70} = 76\frac{12}{13}\%$

6. FINDING A NUMBER WHEN A PER CENT OF IT IS GIVEN

Rule: To find a number when a per cent of it is a given value, *divide* the given value by the per cent.

Thus, to find the length of a line when we are given that 40% of it is 12 feet, divide 12 by 40%; that is, $12 \div 40\% = 12 \div .4 = 30$. *Ans.* 30 ft. Using a second method to find $12 \div 40\%$, we may change 40% to 2/5. Hence,

$$12 \div \frac{2}{5} = 12 \times \frac{5}{2} = 30 \quad Ans. \ 30\text{ft}.$$

Method 1: Express the per cent as a *fraction*, using Table 5-1 when the terms of the fraction are small numbers.

Method 2: Express the given per cent as a *decimal*.

To Find a Number When a Per Cent of It Is Given: Method 1

Find a number if: (a) $12\frac{1}{2}\%$ of the number is 11.   (b) $28\frac{4}{7}\%$ of the number is 18.

PROCEDURE   SOLUTIONS

1. Express the per cent as a fraction: 1. $12\frac{1}{2}\% = \dfrac{1}{8}$ $28\frac{4}{7}\% = \dfrac{2}{7}$
   (See Table 6-1).

2. Divide the given value by the fraction: 2. $11 \div \dfrac{1}{8} = 11 \times 8$ $18 \div \dfrac{2}{7} = \overset{9}{\cancel{18}} \times \dfrac{7}{\underset{1}{\cancel{2}}}$
                                 $= 88 \ Ans.$ $= 63 \ Ans.$

To Find a Number When a Per Cent of It Is Given: Method 2

Find a number if: (a) 24% of the number is 8.4.   (b) $5\frac{1}{4}\%$ of the number is 210.

PROCEDURE   SOLUTIONS

1. Express the per cent as a decimal: 1. $24\% = .24$ $5\frac{1}{4}\% = .0525$

2. Divide the given value by the decimal: 2.   35.   4000.

$$.24\,\overline{)8.40}$$
$$^{\wedge}\ \underline{7\ 2}\ {}^{\wedge}$$
$$1\ 20$$
$$\underline{1\ 20}$$

$$.0525\,\overline{)210.0000}$$
$$^{\wedge}\ \underline{210\ 0}\ {}^{\wedge}$$

Ans. 35   *Ans.* 4,000

6.1   FINDING A NUMBER WHEN A PER CENT OF IT IS GIVEN, USING FRACTIONS

Find each number.

(a) 230 is 50% of what number?   (c) 45 is 125% of what number?

(b) 100 is 40% of what number?   (d) $83\frac{1}{3}$% of ? is 1.625?

Solutions   Use Table 6-1.

(a) $230 \div \frac{1}{2}$ (b) $100 \div \frac{2}{5}$ (c) $45 \div \frac{5}{4}$ (d) $1.625 \div \frac{5}{6}$

    230×2 50 9 .325

    460 *Ans.* $100 \times \frac{5}{2}$ $\cancel{45} \times \frac{4}{\cancel{5}}$ $\cancel{1.625} \times \frac{6}{\cancel{5}}$

                250 *Ans.*   36 *Ans.*   1.95 *Ans.*

6.2   FINDING A NUMBER WHEN A PER CENT OF IT IS GIVEN, USING DECIMALS

Find each number.

(a) 80% of what number is 96?   (c) 4.2% of what number is 22.4?

(b) 115% of what number is 414?   (d) 2.25 is $\frac{3}{4}$% of what number?

Solutions

(a) $80\% = .80 = .8$ (b) $115\% = 1.15$ (c) $4.2\% = .042$ (d) $\frac{3}{4}\% = .00\frac{3}{4} = .0075$

$$.8 \overline{)96.0} \quad 120 \ Ans.$$

$$1.15 \overline{)414.00} \quad 360 \ Ans.$$

$$.042 \overline{)22.400} \quad 533\frac{1}{3} \ Ans.$$

$$.0075 \overline{)2.2500} \quad 300 \ Ans.$$

6.3   PROBLEM SOLVING INVOLVING FINDING A NUMBER WHEN A PER CENT OF IT IS GIVEN

(a) How many kilometers is a distance if $12\frac{1}{2}$% of the distance is (1) 40 km, (2) 3.5 km, (3) $4\frac{1}{4}$ km?

(b) A team lost 15 games in a season. How many games did it play if it won (1) 40%, (2) 90%, (3) $37\frac{1}{2}$% of all games played?

(c) A down payment of $10,800 was made when a house was bought. What was the price of the house if the down payment was (1) 18%, (2) $16\frac{2}{3}$%, (3) 4.5% of the price?

(d) The reduction on a suit at a sale was 14% of the original price of the suit. What was the original price if the reduction was (1) $7, (2) $5.67, (3) $45.50?

(e) The sale price of a hat was $24. What was the regular price if the reduction was (1) 20%, (2) 25%, (3) 40%?

Illustrative Solutions

(b) (3) If the winning per cent is $37\frac{1}{2}$%, the losing per cent is $62\frac{1}{2}$%. To find the games played, divide 15 by $62\frac{1}{2}$% or 5/8. Thus,

$$15 \div \frac{5}{8} = 15 \times \frac{8}{5} = 24 \ Ans.$$

(e) (1) The $24 sale price was 80% of the regular price. Hence, the regular price was $24/.80 = $30 *Ans.*

Ans. (a) (1) 320   (2) 28   (3) 34

       (b) (1) 25   (2) 150

       (c) (1) $60,000   (2) $64,000   (3) $240,000

       (d) (1) $50   (2) $40.50   (3) $325

       (e) (2) $32   (3) $40

7. FINDING A NEW VALUE, GIVEN THE OLD VALUE AND THE PER CENT CHANGE

If shoes priced at $50 are increased 10% of the price, the new price is found by adding the increase to the old price as follows:

1. Find the price increase: 10% of $50 = .10 × $50 = $5.00
2. Add the increase to the old price: $50 + $5.00 = $55.00 *Ans.*

If shoes priced at $50 are reduced 10% of the price, the new price is found by subtracting the reduction from the old price as follows:

1. Find the price reduction: 10% of $50 = .10 × $50 = $5.00
2. Subtract the reduction from the old price: $50 − $5.00 = $45.00 *Ans.*

To Find a New Value, Given the Old Value and the Per Cent Change

(*a*) Find the new salary if the old salary of $180 per day is increased 25%.
(*b*) Find the new cost if the old cost of $180 is decreased by $11\frac{1}{9}$%.

PROCEDURE	SOLUTIONS	
	(*a*)	(*b*)
1. Find the increase or decrease by multiplying the per cent change and the old value:	1. 25% × 180 $= \frac{1}{4} \times 180 = 45$	$11\frac{1}{9} \times 180$ $= \frac{1}{9} \times 180 = 20$
2. Find the new value by adding the increase to the old value, or by subtracting the decrease from the old value:	2. 180 + 45 = 225 *Ans.* $225 per day	180 − 20 = 160 *Ans.* $160

Alternate Method:

In (*a*), an increase of 25% makes the new salary 125% of the old salary; that is 1.25 × 18 = 22.5. *Ans.* $25.50 per day.

7.1 FINDING A NEW VALUE, GIVEN THE OLD VALUE AND THE PER CENT CHANGE

Find the new value if (*a*) 10 is increased 45%, (*b*) 45 is increased 10%, (*c*) 84 is reduced 25%, (*d*) 25 is reduced 84%.

Solutions

(*a*) 10 × .45 = 4.5
    10 + 4.5 = 14.5 *Ans.*

(*c*) 84 × .25 = 21
    84 − 21 = 63 *Ans.*

(*b*) 45 × .1 = 4.5
    45 + 4.5 = 49.5 *Ans.*

(*d*) 25 × .84 = 21
    25 − 21 = 4 *Ans.*

7.2 PROBLEM SOLVING INVOLVING FINDING NEW VALUES GIVEN OLD VALUES AND PER CENT CHANGES

(*a*) What is the new price of a radio if its original price of $45.60 is (1) increased 100%, (2) increased 7.5%, (3) reduced 75%?
(*b*) What is the new salary if the old salary of $120 per week is increased (1) 5%, (2) 50%, (3) 500%?
(*c*) What is the new length of a metal rod if the old length of 12.4 ft is (1) reduced 25%, (2) reduced 2.5%, (3) increased 250%?

(*d*) What is the new batting average of a baseball player if his average of .320 is (1) increased 10%, (2) decreased 15%, (3) decreased 6.25%?

(*e*)

Name a segment having an endpoint at *A* whose length is equal to (1) $\overline{AB}$ increased 100%, (2) $\overline{AG}$ decreased 50%, (3) $\overline{AF}$ decreased 20%.

Illustrative Solution

(*b*) (1) The increase is .05 × $120 = $6. Hence, the new salary is $120 + $6 = $126 per week. *Ans.*
Alternate Method: The new salary is 105% of $120, or 1.05 × $120 = $126 per week.

Ans.   (*a*) (1) $91.20   (2) $49.02   (3) $11.40
        (*b*) (2) $180 per week   (3) $720 per week
        (*c*) (1) 9.3 ft   (2) 12.09 ft   (3) 43.4 ft
        (*d*) (1) .352   (2) .272   (3) .300
        (*e*) (1) $\overline{AC}$ (2) $\overline{AD}$ (3) $\overline{AE}$

8. FINDING A PER CENT CHANGE GIVEN THE OLD AND THE NEW VALUES

If John's weight has increased from 80 pounds to 100 pounds, the per cent change can be found dividing the increase by the old weight as follows:

1. Find the weight increase: $100 - 80 = 20$

2. Divide the increase by the old weight: $20 \div 80 = \dfrac{20}{80} = \dfrac{1}{4}$

3. Express the result found in step 2 as a per cent: $\dfrac{1}{4} = 25\%$ *Ans.*

If George's weight has decreased from 80 pounds to 70 pounds, the per cent change can be found by dividing the decrease by the old weight as follows:

1. Find the weight decrease: $80 - 70 = 10$

2. Divide the decrease by the old weight: $10 \div 80 = \dfrac{10}{80} = \dfrac{1}{8}$

3. Express the result found in step 2 as a per cent: $\dfrac{1}{8} = 12\frac{1}{2}\%$ *Ans.*

To Find a Per Cent Change Given the Old and New Values

(*a*) Find the per cent change if the price of meat increases from $1.25 to $1.50.
(*b*) Find the per cent change if the price of a pie decreases from $1.25 to $1.05.

PROCEDURE	SOLUTIONS	
	(*a*)	(*b*)
1. Find the increase or decrease:	1. $105¢ - 125¢ = 25¢$	$125¢ - 105¢ = 20¢$
2. Divide the change by the old value:	2. $25¢ \div 125¢ = \dfrac{25}{125} = \dfrac{1}{5}$	$20¢ \div 125¢ = \dfrac{20}{125} = \dfrac{4}{25}$
3. Express the result found in step 2 as	3. $\dfrac{1}{5} = 20\%$ *Ans.*	$\dfrac{4}{25} = 16\%$ *Ans.*

8.1   Finding a Per Cent Change Given the Old and the New Values

Find the per cent change to the nearest per cent if (*a*) 10 is increased to 18,   (*b*) 18 is reduced to 10,   (*c*) 4.2 is increased to 5.4,   (*d*) 5.4 is reduced to 4.2.

Solutions

(*a*) $18 - 10 = 8$ (*b*) $18 - 10 = 8$

$$\frac{8}{18} = 18\overline{)8.00}\quad\begin{array}{r}.44\\\hline\end{array}$$

$$\begin{array}{r}72\\\hline 80\\72\\\hline R = 8\end{array}$$

$\dfrac{8}{10} = 80\%$ *Ans.*

Ans. 45%, to nearest %

(*c*) $5.4 - 4.2 = 1.2$

$$\frac{1.2}{4.2} = 4.2\overline{)1.2\,00}\quad\begin{array}{r}.28\\\hline\end{array}$$

$$\begin{array}{r}8\,4\\\hline 3\,60\\3\,36\\\hline R = 24\end{array}$$

Ans. 29%, to nearest %

(*d*) $5.4 - 4.2 = 1.2$

$$\frac{1.2}{5.4} = 5.4\overline{)1.2\,00}\quad\begin{array}{r}.22\\\hline\end{array}$$

$$\begin{array}{r}1\,0\,8\\\hline 1\,20\\1\,08\\\hline R = 12\end{array}$$

Ans. 22%, to nearest %.

8.2   PROBLEM SOLVING

(*a*) What is the per cent change when a service charge of $2.50 an hour is (1) increased to $3.00 an hour, (2) decreased to $2.25 an hour, (3) increased to $5 an hour?

(*b*) What is the per cent change when a group of 100 is (1) increased to 200, (2) reduced to 73, (3) increased to 315?

(*c*) What is the per cent change when a grade of 60% is changed to (1) 70%, (2) 55%, (3) 84%?

(*d*) What is the per cent change when a speed limit of 50 mph is changed to (1) 55 mph, (2) 40 mph, (3) 70 mph?

(*e*)

$$\begin{array}{ccccccc} A & B & C & D & E & F & G \\ | & | & | & | & | & | & | \\ 0 & 1 & 2 & 3 & 4 & 5 & 6 \end{array} \longrightarrow$$

What is the per cent change needed to (1) increase $\overline{AB}$ to $\overline{AC}$, (2) reduce $\overline{AF}$ to $\overline{AE}$, (3) increase $\overline{AC}$ to $\overline{AF}$?

Illustrative Solution   (*b*) (3) The increase is $315 - 100 = 215$. Thus,

$$\frac{\text{increase}}{\text{old value}} = \frac{215}{100} \text{ or } 215\% \text{ } Ans.$$

Ans.   (*a*) (1) 20%   (2) 10%   (3) 100%
        (*b*) (1) 100%   (2) 27%
        (*c*) (1) $16\frac{2}{3}\%$ (2) $8\frac{1}{3}\%$ (3) 40%
        (*d*) (1) 10%   (2) 20%   (3) 40%
        (*e*) (1) 100%   (2) 20%   (3) 150%

Supplementary Problems

5.1.   In each part of Fig. 5-3, find the per cent that is occupied by the shaded letter,   **(1.1)**
and the per cent that is unshaded.

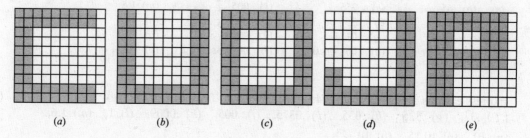

(*a*)   (*b*)   (*c*)   (*d*)   (*e*)

Fig. 5-3

Ans.   (*a*) 48%, 52%   (*b*) 52%, 48%   (*c*) 64%, 36%   (*d*) 42%, 58%   (*e*) 60%, 40%

5.2.   (*a*)   What per cent of a school is present if the per cent absent is (1) 70%,   **(1.2)**
           (2) $62\frac{1}{2}$%, (3) 5.5%, (4) $\frac{1}{2}$%?

      (*b*)   What per cent of an alloy of gold and silver is silver if the per cent gold is (1) 20%, (2) $3\frac{1}{2}$%,
           (3) 97.5%, (4) $\frac{1}{4}$%?

      (*c*)   What per cent of light bulbs are good if out of 100, the number of defective bulbs is (1) 8, (2)
           32, (3) 65, (4) 78?

      (*d*)   What is the per cent discount on a coat at a sale if the per cent of the original price that you
           pay is (1) 90%, (2) 75%, (3) $82\frac{1}{2}$%, (4) 74.5%?

      (*e*)   Three children inherited an estate. If the share of one child is 45%, what is the share of the
           second child if that of the remaining child is (1) 30%, (2) 35%, (3) $5\frac{1}{2}$%, (4) 20.5%?

  Ans.   (*a*) (1) 30%   (2) $37\frac{1}{2}$%   (3) 94.5%   (4) $99\frac{1}{2}$%

          (*b*) (1) 80%   (2) $96\frac{1}{2}$%   (3) 2.5%   (4) $99\frac{3}{4}$%

          (*c*) (1) 92%   (2) 68%   (3) 35%   (4) 22%

          (*d*) (1) 10%   (2) 25%   (3) $17\frac{1}{2}$%   (4) 25.5% or $25\frac{1}{2}$%

          (*e*) (1) 25%   (2) 20%   (3) $49\frac{1}{2}$%   (4) 34.5% or $34\frac{1}{2}$%

5.3.   Express each of the following as per cents: **(1.3)**

      (*a*)   (1) 35 hundredths (2) 3.5 hundredths (3) .5 hundredth (4) 2/5 hundredth

      (*b*)   (1) 7 out of a hundred (2) $37\frac{1}{2}$ out of a hundred (3) .25 out of a hundred

      (*c*)   (1) .01 (2) .71 (3) $.07\frac{1}{4}$ (4) .075

      (*d*)   (1) $\dfrac{9}{100}$ (2) $\dfrac{95}{100}$ (3) $\dfrac{9}{10}$ (4) $\dfrac{350}{1{,}000}$

  Ans.   (*a*) (1) 35%   (2) 3.5%   (3) .5% or $\frac{1}{2}$%   (4) $\frac{2}{5}$% or .4%

          (*b*) (1) 7%   (2) $37\frac{1}{2}$%   (3) .25% or $\frac{1}{4}$%

          (*c*) (1) 1%   (2) 71%   (3) $7\frac{1}{4}$%   (4)   7.5% or $7\frac{1}{2}$%

          (*d*) (1) 9%   (2) 95%   (3) 90%   (4) 35%

5.4.   Express each as a decimal: **(2.1)**

      (*a*) 65%   (*d*) $5\frac{1}{2}$%   (*g*) 1.25%   (*j*) $\frac{1}{4}$%   (*l*) $\frac{3}{25}$%   (*n*) 200%

      (*b*) 99%   (*e*) 25.5%   (*h*) $6\frac{1}{4}$ (*o*) 1,200%

      (*c*) 5%   (*f*) $35\frac{1}{4}$%   (*i*) .5%   (*k*) .03%   (*m*) 150%   (*p*) $787\frac{1}{2}$%

  Ans.   (*a*) .65   (*e*) .255   (*i*) .005   (*m*) 1.5

         (*b*) .99   (*f*) $.35\frac{1}{4}$ or .3525   (*j*) $.00\frac{1}{4}$ or .0025   (*n*) 2

         (*c*) .05   (*g*) .0125   (*k*) .0003   (*o*) 12

         (*d*) $.05\frac{1}{2}$ or .055   (*h*) $.06\frac{1}{4}$ or .0625   (*l*) $.00\frac{3}{25}$ or .0012   (*p*) $7.87\frac{1}{2}$ or 7.875

5.5.   Express each as a per cent: (*a*) .45,   (*b*) .70,   (*c*) .07,   (*d*) $.07\frac{1}{2}$,   (*e*) .3,   **(2.2)**
      (*f*) $.3\frac{1}{2}$,   (*g*) .325,   (*h*) .035,   (*i*) .0375,   (*j*) .005,   (*k*) .0009,   (*l*) 1,   (*m*) 1.6,
      (*n*) 10,   (*o*) 10.15,   (*p*) $10.12\frac{1}{2}$.

  Ans.   (*a*) 45%   (*e*) 30%   (*i*) 3.75% or $3\frac{3}{4}$%   (*m*) 160%

         (*b*) 70%   (*f*) 35%   (*j*) .5% or $\frac{1}{2}$%   (*n*) 1,000%

         (*c*) 7%   (*g*) 32.5%   (*k*) .09%   (*o*) 1,015%

         (*d*) $7\frac{1}{2}$% or 7.5%   (*h*) 3.5% or $3\frac{1}{2}$%   (*l*) 100%   (*p*) $1{,}012\frac{1}{2}$% or 1,012.5%

5.6.   (a)   What per cent of the original cost is the new cost if the original cost has   **(2.3)**
been (1) decreased 20%, (2) kept the same, (3) increased 20%, (4) tripled?

    (b)   What per cent of the games played has a team lost if its winning average is (1) .400, (2) .600, (3) .560, (4) .565?

    (c)   What per cent of times at bat did a baseball player not get a hit if his batting average is (1) .200, (2) .250, (3) .320, (4) .325?

    (d)   On a test of 24 questions, what per cent was answered incorrectly if the per cent of correct answers was (1) 75%, (2) $87\frac{1}{2}$%, (3) 25%, (4) 100%?

    (e)   In an alloy of 250 kilograms of lead and copper, what per cent of the alloy is lead if the per cent copper in the alloy is (1) 10%, (2) 5%, (3) 1.2%, (4) $\frac{1}{2}$%

Ans.   (a) (1) 80%   (2) 100%   (3) 120%   (4) 300%
        (b) (1) 60%   (2) 40%   (3) 44%   (4) 43.5% or $43\frac{1}{2}$%
        (c) (1) 80%   (2) 75%   (3) 68%   (4) 67.5% or $67\frac{1}{2}$%
        (d) (1) 25%   (2) $12\frac{1}{2}$%   (3) 75%   (4) 0%
        (e) (1) 90%   (2) 95%   (3) 98.8%   (4) $99\frac{1}{2}$% or 99.5%

5.7.   Express each per cent as a common fraction: (a) 30%,   (b) 75%,   (c) 80%,   **(3.1)**
(d) 2%,   (e) 5%,   (f) 7%,   (g) 16%,   (h) 72%,   (i) 27%
(j) 12.5%,   (k) $37\frac{1}{2}$%,   (l) $22\frac{2}{9}$%,   (m) $6\frac{2}{3}$%,   (n) .75%,   (o) $6\frac{1}{4}$%,
(p) $33\frac{1}{3}$%,   (q) 150%,   (r) 225%,   (s) $512\frac{1}{2}$%,   (t) $666\frac{2}{3}$%,   (u) $702\frac{1}{2}$%

Ans.   (a) $\dfrac{3}{10}$ (d) $\dfrac{1}{50}$ (g) $\dfrac{4}{25}$ (j) $\dfrac{1}{8}$ (m) $\dfrac{1}{15}$ (p) $\dfrac{1}{3}$ (s) $\dfrac{41}{8}$

      (b) $\dfrac{3}{4}$ (e) $\dfrac{1}{20}$ (h) $\dfrac{18}{25}$ (k) $\dfrac{3}{8}$ (n) $\dfrac{3}{400}$ (q) $\dfrac{3}{2}$ (t) $\dfrac{20}{3}$

      (c) $\dfrac{4}{5}$ (f) $\dfrac{7}{100}$ (i) $\dfrac{27}{100}$ (l) $\dfrac{2}{9}$ (o) $\dfrac{1}{16}$ (r) $\dfrac{9}{4}$ (u) $\dfrac{281}{40}$

5.8.   Express each fraction as a per cent and rounded to the nearest per cent: (a) 1/3,   **(3.2)**
(b) 4/3,   (c) 10/3,   (d) 1/6,   (e) 13/6,   (f) 7/8,   (g) 15/8,   (h) 2/7,   (i) 16/7,
(j) 10/9,   (k) 3/16,   (l) 35/16,   (m) 5/12,   (n) 125/12.

Ans.   (a) $33\frac{1}{3}$%, 33%   (e) $216\frac{2}{3}$%, 217%   (i) $228\frac{4}{7}$%, 229%   (m) $41\frac{2}{3}$%, 42%

      (b) $133\frac{1}{3}$%, 133%   (f) $87\frac{1}{2}$%, 88%   (j) $111\frac{1}{9}$%, 111%   (n) $1{,}041\frac{2}{3}$%, 1,042%

      (c) $333\frac{1}{3}$%, 333%   (g) $187\frac{1}{2}$%, 188%   (k) $18\frac{3}{4}$%, 19%

      (d) $16\frac{2}{3}$, 17%   (h) $28\frac{4}{7}$%, 29%   (l) $218\frac{3}{4}$%, 219%

5.9.   Express each fraction as a per cent: (a) 17/100,   (b) 9/100,   (c) 137/100,   **(3.3)**
(d) 3/10,   (e) 13/10,   (f) 1/2,   (g) 1/4,   (h) 5/2,   (i) 11/5,   (j) 9/20,
(k) 21/20,   (l) 3/25,   (m) 13/25,   (n) 33/25,   (o) 29/50,   (p) 101/50.

Ans.   (a) 17% (c) 137% (e) 130% (g) 25% (i) 220% (k) 105% (m) 52% (o) 58%
      (b) 9%   (d) 30%   (f) 50%   (h) 250% (j) 45%   (l) 12%   (n) 132% (p) 202%

5.10.   (a)   What fractional part of a test is answered incorrectly if the test score is   **(3.4)**
(1) 75%, (2) 80%, (3) 85%, (4) 98%?

    (b)   What fractional part of the games that a team played was lost if its winning average is (1) 50%, (2) 60%, (3) 90%, (4) 95%?

    (c)   What per cent of a chorus of 40 men and women are women if the number of men is (1) 30, (2) 25, (3) 16, (4) 14?

(d) A segment is increased in length. What per cent of the length of the original segment is the new length if the original segment is made (1) $1\frac{1}{2}$ times as long, (2) three times as long, (3) $4\frac{1}{8}$ times as long, (4) $5\frac{5}{6}$ times as long?

(e) What per cent of the original price is a new price if the old price was (1) reduced one-fifth, (2) increased one-third, (3) reduced three-quarters, (4) increased one and two-sevenths times?

Ans. (a) (1) $\frac{1}{4}$ (2) $\frac{1}{5}$ (3) $\frac{3}{20}$ (4) $\frac{1}{50}$

(b) (1) $\frac{1}{2}$ (2) $\frac{2}{5}$ (3) $\frac{1}{10}$ (4) $\frac{1}{20}$

(c) (1) 25%   (2) $37\frac{1}{2}$%   (3) 60%   (4) 65%

(d) (1) 150%   (2) 300%   (3) $412\frac{1}{2}$%   (4) $583\frac{1}{3}$%

(e) (1) 80%   (2) $133\frac{1}{3}$%   (3) 25%   (4) $228\frac{4}{7}$%

5.11. Find: (a) 50% of 24,   (b) $33\frac{1}{3}$% of 390,   (c) $66\frac{2}{3}$% of 1.5,   **(4.1)**
(d) 25% of 280,   (e) 75% of 3.2,   (f) 20% of 70,   (g) 40% of 3.5,   (h) 60% of 1,500,   (i) 80% of .65,   (j) $16\frac{2}{3}$% of 420,   (k) $83\frac{1}{3}$% of 5.43,
(l) $14\frac{2}{7}$% of .497,   (m) $28\frac{4}{7}$% of 63,   (n) $12\frac{1}{2}$% of 1.6,   (o) $37\frac{1}{2}$% of 4,000,
(p) $62\frac{1}{2}$% of 7.28,   (q) $87\frac{1}{2}$% of 8.8,   (r) 10% of 43.56,   (s) 30% of 50.1,
(t) $11\frac{1}{9}$% of 54.9,   (u) $8\frac{1}{3}$% of 1.224,   (v) 5% of 12.5,   (w) 4% of 76.825,
(x) 2% of 7,500,   (y) 1% of 843,000,   (z) 250% of 45.68.
Verify your answers using a calculator.

Ans. (a) 12   (e) 2.4   (i) .52   (m) 18   (q) 7.7   (u) .102   (y) 8,430
(b) 130   (f) 14   (j) 70   (n) .2   (r) 4.356   (v) .625   (z) 114.2
(c) 1   (g) 1.4   (k) 4.525   (o) 1,500   (s) 15.03   (w) 3.073
(d) 70   (h) 900   (l) .071   (p) 4.55   (t) 6.1   (x) 150

5.12. Find: (a) 30% of 50,   (b) 3% of 5,000,   (c) 15% of 450,   (d) 22% of 150,   **(4.2)**
(e) 76% of 125,   (f) 85% of 14.62,   (g) 3.5% of 600,   (h) 6.25% of 1,500,
(i) $2\frac{1}{2}$% of 360,   (j) $3\frac{1}{3}$% of 6,000,   (k) $\frac{1}{2}$% of 500,   (l) .05% of 1,000,
(m) 100% of 4.36,   (n) 300% of 12.5,   (o) 520% of 75,   (p) 72.25% of 45.6.
Verify your answers using a calculator.

Ans. (a) 15   (c) 67.5   (e) 95   (g) 21   (i) 9   (k) 2.5   (m) 4.36   (o) 390
(b) 150   (d) 33   (f) 12.427   (h) 93.75   (j) 200   (l) .5   (n) 37.5   (p) 32.946

5.13. (a) If the sales tax rate is 6%, how much is the sales tax to the nearest cent   **(4.3)**
on a purchase of (1) $50, (2) $64, (3) $75.90, (4) $125.63?

(b) A team lost $12\frac{1}{2}$% of its games. How many games did it win if the number of games played was (1) 120, (2) 96, (3) 72, (4) 48?

(c) A square contains 1,000 equal squares. How many of these 1,000 squares are shaded if the part occupied by the shaded squares is (1) 40%, (2) 54%, (3) 49.6%, (4) $11\frac{1}{2}$%?

(d) How many kilometers, to the nearest kilometer, have been traveled on a trip of 5,000 km if the part of the trip already traveled is (1) 70%, (2) 65%, (3) 51.8%, (4) 0.48%?

(e) How many pupils in a school are absent if the attendance is 85% of the school's enrollment, which is (1) 1,000 pupils, (2) 2,500 pupils, (3) 4,500 pupils, (4) 5,480 pupils?

(f) A family budget plan allows 35% for food, 20% for rent, 15% for savings, 10% for clothing, 8% for recreation, and the rest for miscellaneous. According to the budget plan, how much will be spent for each item if the annual earnings is (1) $22,500, (2) $12,000, (3) $35,000?

Ans.   (*a*) (1) $3   (2) $3.84   (3) $4.55   (4) $7.54
      (*b*) (1) 105   (2) 84   (3) 63   (4) 42
      (*c*) (1) 400   (2) 540   (3) 496   (4) 115
      (*d*) (1) 3,500 (2) 3,250   (3) 2,590   (4) 24
      (*e*) (1) 150   (2) 375   (3) 675   (4) 822

(f)

	Food	Rent	Savings	Clothing	Recreation	Miscellaneous
(1)	$ 7,875	$4,500	$3,375	$2,250	$1,800	$2,700
(2)	$ 4,200	$2,400	$1,800	$1,200	$ 960	$1,440
(3)	$12,250	$7,000	$5,250	$3,500	£2,800	$4,200

5.14.   Find each result as a per cent and rounded to the nearest per cent.   **(5.1)**

(*a*) What per cent of 2 is 1?   (*b*) What per cent of 1 is 2?
(*c*) 25 is what per cent of 100?   (*d*) 100 is what per cent of 25?
(*e*) What per cent is 3 of 10?   (*f*) What per cent is 10 of 3?
(*g*) ?% of 60 is 10?   (*h*) ?% is 60 of 10?
(*i*) 12 is ? % of 32?   (*j*) 32 is ? % of 12?
(*k*) What per cent of 700 is 300?   (*l*) 700 is what per cent of 300?

Ans.   (When two answers are given, the second one is the answer to the nearest per cent.)

(*a*) 50%   (*d*) 400%   (*g*) $16\frac{2}{3}$%, 17%   (*j*) $266\frac{2}{3}$%, 267%
(*b*) 200%   (*e*) 30%   (*h*) 600%   (*k*) $42\frac{6}{7}$%, 43%
(*c*) 25%   (*f*) $333\frac{1}{3}$%, 333%   (*i*) $37\frac{1}{2}$%, 38%   (*l*) $233\frac{1}{3}$%, 233%

5.15.   (*a*) If a man's earnings is $200, what per cent of his earnings is spent (1) if he   **(5.2)**
           spends $50, (2) if he spends $36, (3) if he saves $158.40, (4) if he saves $46.36?

(*b*) What per cent of the games played did a school team lose if it played a total of 25 games and (1) lost 10 of them, (2) lost 12 of them, (3) won 10 of them, (4) won 24 of them?

(*c*) A worker making $300 a week had his pay raised to $375 a week. (1) What per cent of his original salary is his new salary? (2) What per cent of the new salary is his original salary? (3) What per cent of his original salary is the increase? (4) What per cent of his new salary is the increase?

(*d*) A price of $10 is increased $2. (1) What per cent of the original price is the increase? (2) What per cent of the original price is the new price? (3) What per cent of the new price is the original price? (4) What per cent of the new price is the increase?

Ans.   (*a*) (1) 25%   (2) 18%   (3) 20.8%   (4) 76.82%
      (*b*) (1) 40%   (2) 48%   (3) 60%   (4) 4%
      (*c*) (1) 125%   (2) 80%   (3) 25%   (4) 20%
      (*d*) (1) 20%   (2) 120%   (3) $83\frac{1}{3}$%   (4) $16\frac{2}{3}$%

5.16.   Find each number.   **(6.1)**

(*a*) 10 is 50% of what number?   (*b*) 50 is 10% of what number?
(*c*) 25% of what number is 20?   (*d*) 20% of what number is 25?
(*e*) 40 is 200% of what number?   (*f*) 200 is 40% of what number?
(*g*) $12\frac{1}{2}$ is 5% of what number?   (*h*) $12\frac{1}{2}$% of what number is 5?
(*i*) $33\frac{1}{3}$% of what number is $16\frac{2}{3}$%?   (*j*) $16\frac{2}{3}$% of what number is $33\frac{1}{3}$?

Ans.   (*a*) 20   (*c*) 80   (*e*) 20   (*g*) 250   (*i*) 50
         (*b*) 500   (*d*) 125   (*f*) 500   (*h*) 40   (*j*) 200

5.17.   Find each number.   **(6.2)**

(*a*)   60% of what number is 36?   (*b*)   60 is 36% of what number?
(*c*)   84 is 42% of what number?   (*d*)   42 is 84% of what number?
(*e*)   7.75% of what number is 9.3?   (*f*)   9.3% of what number is 7.75?
(*g*)   .096 is .24% of?   (*h*)   .096% of what number is .24?

Ans.   (*a*) 60   (*c*) 200   (*e*) 120   (*g*) 40
         (*b*) $166\frac{2}{3}$ (*d*) 50   (*f*) $83\frac{1}{3}$ (*h*) 250

5.18.   (*a*)   How many meters is a length if $16\frac{2}{3}\%$ of the length is (1) 5 m, (2) 4.3 m,   **(6.3)**
                   (3) 56.5 m, (4) $24\frac{2}{3}$ m?
           (*b*)   A team won 42 games in a season. How many games did it play if it lost (1) 25%, (2) 16%, (3)
                   40%, (4) 65% of the total games played?
           (*c*)   A sale price is 85% of the original price of a coat. What was the original price if the reduction
                   was (1) $3, (2) $5.25, (3) $9.84, (4) $21.75?
           (*d*)   A down payment of $480 was made when a used car was bought. What was the price of the
                   car if the down payment was (1) 20%, (2) 16%, (3) 12%, (4) $6\frac{1}{4}\%$ of the price?

Ans.   (*a*) (1) 30   (2) 25.8   (3) 339   (4) 148
         (*b*) (1) 56   (2) 50   (3) 70   (4) 120
         (*c*) (1) $20   (2) $35   (3) $65.60   (4) $145
         (*d*) (1) $2,400 (2) $3,000 (3) $4,000   (4) $7,680

5.19.   Find the new value if (*a*) 25 is increased 20%, (*b*) 20 is increased 25%, (*c*) 30 is   **(7.1)**
           decreased 10%, (*d*) 10 is decreased 30%, (*e*) 100 is increased $12\frac{1}{2}\%$, (*f*) $12\frac{1}{2}$ is increased 100%, (*g*)
           60 is reduced 25%, (*h*) 25 is reduced 60%, (*i*) 64 is increased 12.8%, (*j*) 12.8 is increased 64%, (*k*)
           17.5 is reduced 4%, (*l*) 4 is reduced 17.5%, (*m*) 720 is increased 15%, (*n*) 15 is increased 720%, (*o*)
           24.8 is reduced 35%.

Ans.   (*a*) $25 + (25 \times 20\%) = 30$ (*i*) $64 + (64 \times 12.8\%) = 72.192$
         (*b*) $20 + (25\% \times 20) = 25$ (*j*) $12.8 + (12.8 \times 64\%) = 20.992$
         (*c*) $30 - (30 \times 10\%) = 27$ (*k*) $17.5 - (17.5 \times 4\%) = 16.8$
         (*d*) $10 - (10 \times 30\%) = 7$ (*l*) $4 - (4 \times 17.5\%) = 3.3$
         (*e*) $100 + (100 \times 12\frac{1}{2}\%) = 112\frac{1}{2}$ (*m*) $720 + (720 \times 15\%) = 828$
         (*f*) $12\frac{1}{2} + (12\frac{1}{2} \times 100\%) = 25$ (*n*) $15 + (15 \times 720\%) = 123$
         (*g*) $60 - (60 \times 25\%) = 45$ (*o*) $24.8 - (24.8 \times 35\%) = 16.12$
         (*h*) $25 - (25 \times 60\%) = 10$

5.20.   (*a*)   What is the new price of a television set if its original price of $450 is   **(7.2)**
                   (1) increased 20%, (2) increased 2.4%, (3) decreased 24%?
           (*b*)   What is the new speed of a car if the old speed of 60 miles per hour is (1) increased 10%, (2)
                   increased 5.8%, (3) decreased $12\frac{1}{2}\%$?
           (*c*)   What is the new winning average of a team if the old average of .540 is (1) increased 15%, (2)
                   increased $16\frac{2}{3}\%$, (3) decreased 35%?
           (*d*)   What is the new salary if the old salary of $50,480 is (1), increased 50%, (2) decreased 10%,
                   (3) decreased 6.5%?

(e)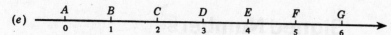

Name a segment having an endpoint at A whose length is (1) $\overline{AG}$ decreased $33\frac{1}{3}\%$, (2) $\overline{AE}$ decreased 75%, (3) $\overline{AC}$ increased 200%.

Ans.   (a) (1) $540   (2) $460.80   (3) $342
        (b) (1) 66 mph   (2) 63.48 mph   (3) $52\frac{1}{2}$ or 52.5 mph
        (c) (1) .621   (2) .630   (3) .351
        (d) (1) $75,720   (2) $45,432   (3) $47,198.50
        (e) (1) $\overline{AE}$ (2) $\overline{AB}$ (3) $\overline{AG}$

5.21. Find the per cent change if (a) 5 is increased to 10, (b) 10 is decreased to 5,   **(8.1)** (c) 80 is increased to 100, (d) 100 is decreased to 80, (e) 25 is increased to 40, (f) 40 is reduced to 25, (g) 666 is increased to 888, (h) 888 is decreased to 666, (i) 5.25 is decreased to 3.25, (j) 3.25 is increased to 5.25.

Ans.   (a) $\dfrac{10-5}{5}=100\%$ (e) $\dfrac{40-25}{25}=60\%$ (i) $\dfrac{5.25-3.25}{5.25}=38\frac{2}{21}\%$

      (b) $\dfrac{10-5}{10}=50\%$ (f) $\dfrac{40-25}{40}=37\frac{1}{2}\%$ (j) $\dfrac{5.25-3.25}{3.25}=61\frac{7}{13}\%$

      (c) $\dfrac{100-80}{80}=25\%$ (g) $\dfrac{888-666}{666}=33\frac{1}{3}\%$

      (d) $\dfrac{100-80}{100}=20\%$ (h) $\dfrac{888-666}{888}=25\%$

5.22. (a) What is the per cent change when a price of $3.60 is (1) increased to $4.80,   **(8.2)** (2) decreased to $2.40, (3) increased to $5.22, (4) decreased to $1.62?

    (b) What is the per cent change when a travel time of 3 hours is (1) increased to 5 hours, (2) increased to 3.5 hours, (3) decreased to 2.1 hours, (4) decreased to 2.79 hours?

    (c) What is the per cent change when a grade of 75% is changed to (1) 90%, (2) 72%, (3) 80%, (4) 83%?

    (d) What is the per cent change when a group of 20 is (1) increased to 21, (2) increased to 50, (3) reduced to 5, (4) reduced to 17?

(e)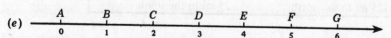

What is the per cent change needed to (1) increase $\overline{AC}$ to $\overline{AD}$, (2) increase $\overline{AD}$ to $\overline{AF}$, (3) reduce $\overline{AG}$ to $\overline{AF}$, (4) reduce $\overline{AF}$ to $\overline{AD}$?

Ans.   (a) (1) $33\frac{1}{3}\%$ (2) $33\frac{1}{3}\%$ (3) 45%   (4) 55%
        (b) (1) $66\frac{2}{3}\%$ (2) $16\frac{2}{3}\%$ (3) 30%   (4) 7%
        (c) (1) 20%   (2) 4%   (3) $6\frac{2}{3}\%$ (4) $10\frac{2}{3}\%$
        (d) (1) 5%   (2) 150%   (3) 75%   (4) 15%
        (e) (1) 50%   (2) $66\frac{2}{3}\%$ (3) $16\frac{2}{3}\%$ (4) 40%

Chapter 6

Signed Numbers

1. POSITIVE AND NEGATIVE NUMBERS AND INTEGERS

The temperatures listed in Fig. 6-1 are the Fahrenheit temperatures for five cities on a winter's day. Note how these temperatures are shown on a *number line* or *number scale*. On this number line, plus and minus signs are used to distinguish between temperatures above zero and those below zero. Such numbers are *positive* or *negative numbers*. Since positive numbers have plus signs and negative numbers have minus signs, they are called *signed numbers*. Zero, 0, is an unsigned number since 0 is neither positive nor negative.

Positive and negative numbers are used to represent quantities that are opposites of each other. Thus, if +40, read "plus 40," represents 40° above zero, then −40, read "minus 40," represents 40° below zero.

A negative number can be indicated only by means of the minus sign (−). However, a positive number may be shown by using a plus sign (+) or by using no sign at all. Thus, 35° above zero may be indicated either by +35 or by 35.

Table 6-1 illustrates pairs of opposites which may be represented by +25 and −25.

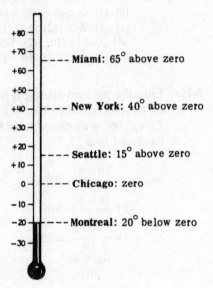

Fig. 6-1

Table 6-1

+25	−25
$25 deposited	$25 withdrawn
25 mph faster	25 mph slower
25 lb gained	25 lb lost
25 mi to the north	25 mi to the south

Integers

You have already observed several types of sets of numbers in this text. In particular, we have described whole numbers and natural numbers. The collection of all whole numbers together with the negative whole numbers is the set of integers.

Absolute Value of a Signed Number

In any pair of opposites, the positive of the pair is the *absolute value* of each of the numbers. The absolute value of a signed number a is written $|a|$. Thus, 40 is the absolute value of each of the opposites +40 and −40. Hence, using the absolute value symbol, we write $40 = |+40|$ and $40 = |-40|$.

The absolute value of 0 is defined as 0; that is $|0| = 0$.

Note that the absolute value of each of a pair of opposites may be written by simply removing the sign of either number.

The Number Line for Signed Numbers

Figure 6-2 is a horizontal number line; the number line may also be drawn vertically, as in Fig. 6-1. Each point on the number line is the *graph* of a signed number, and each signed number is the *coordinate* of a point on the number line.

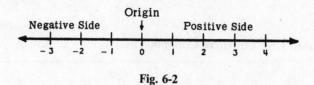

Fig. 6-2

Procedure to Construct the Number Line

1. Draw a line and choose any point of the line as the origin, with 0 associated with it. The number 0 is the coordinate of the point chosen as the origin, and the origin is the graph of 0.
2. Choose a point to the right of the origin and associate this point with the number 1. The distance from the origin to the unit point is the *unit length*. The unit point is the graph of 1, and 1 is the coordinate of the unit point.
3. Using the unit length, mark off successive points to the right and label these points $+1, +2, +3, +4, \ldots$.
4. Using the unit length, mark off successive points to the left and label these points $-1, -2, -3, -4, \ldots$.

1.1   WORDS OPPOSITE IN MEANING

State the words that are opposite in meaning to the following:

(a) gain   (e) north   (i) deposit   (m) A.D.
(b) rise   (f) east   (j) asset   (n) expand
(c) above   (g) right-hand   (k) earnings   (o) accelerate
(d) up   (h) forward   (l) receipt   (p) clockwise

Solutions

(a) loss   (e) south   (i) withdrawal   (m) B.C.
(b) fall   (f) west   (j) liability   (n) contract
(c) below   (g) left-hand   (k) spendings   (o) decelerate
(d) down   (h) backward   (l) payment   (p) counterclockwise

1.2   EXPRESSING QUANTITIES AS SIGNED NUMBERS

What quantity is represented by

(a) -10, if $+10$ means 10 yd gained?   (e) -100, if $+100$ means 100 ft east?
(b) -5, if $+5$ means \$5 earned?   (f) -3, if $+3$ means 3 steps right?
(c) $+15$, if -15 means 15 mi south?   (g) 20, if -20 means 20 oz underweight?
(d) $+8$, if -8 means 8 hr earlier?   (h) 5, if -5 means 5 flights down?

Solutions

(a) 10 yd lost   (c) 15 mi north   (e) 100 ft west   (g) 20 oz overweight
(b) \$5 spent   (d) 8 hr later   (f) 3 steps left   (h) 5 flights up

1.3   ABSOLUTE VALUE OF A SIGNED NUMBER

Evaluate:

(a) $|25| + |-25|$ (c) $\left|\frac{1}{2}\right| + \left|-\frac{1}{2}\right| + \left|\frac{1}{4}\right| - \left|-\frac{1}{4}\right|$ (e) $(3|-1|)(5|-2|)$ (g) $(|-10|)^2$

(b) $|36| - |-36|$ (d) $2|3.5| - 4|-.25|$ (f) $\dfrac{10| + 5|}{10| - 5|}$

Solutions

(a) $25 + 25 = 50$ (c) $\dfrac{1}{2} + \dfrac{1}{2} + \dfrac{1}{4} - \dfrac{1}{4} = 1$ (e) $(3 \times 1)(5 \times 2) = 3(10) = 30$ (g) $(10)^2 = 100$

(b) $36 - 36 = 0$ (d) $2(3.5) - 4(.25) = 7 - 1 = 6$ (f) $\dfrac{10(5)}{10(5)} = 1$

1.4 POINTS ON A NUMBER LINE AND THEIR COORDINATES

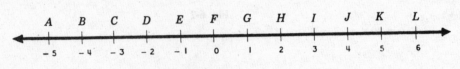

Fig. 6-3

Using the number line, Fig. 6-3, with reference to the graphs and coordinates shown, name (a) the coordinates of points A, D, and the origin; (b) the graphs of the natural numbers; (c) the coordinates of points midway between (1) A and C, (2) C and G, (3) A and G; (d) the graphs of coordinates that are greater than -5 and less than or equal to 0; (e) the graphs of coordinates whose absolute value is less than 3.

Solutions

(a) $-5, -2, 0$ (c) (1) -4, (2) -1, (3) -2 (e) D, E, F, G, H
(b) G, H, I, J, K, L (d) B, C, D, E, F

1.5 WHOLE NUMBERS, INTEGERS, AND NATURAL NUMBERS

Tell whether each of the following is a whole number, natural number, or integer:
(a) 5 (b) 0 (c) -10 (d) $|-10|$

Solutions

(a) All three (b) Whole, integer (c) Integer (d) All three

2. COMPARING SIGNED NUMBERS

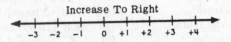

Fig. 6-4 Horizontal Number Line

On a horizontal number line, signed numbers increase to the right. Hence, the graph of the greater of two numbers is to the right of the graph of the smaller. Because of the positions of the graphs of -2 and 1, we conclude that -2 is less than 1. Using the inequality symbol, $<$, which is read "is less than," we write $-2 < 1$. Using the inequality symbol, $>$, which is read "is greater than," we write $1 > -2$.

The inequality symbols, $>$ and $<$, need not be confused if you remember each symbol as an arrowhead pointing toward the smaller number (in a true statement). From Fig. 6-4, note that 0 is *between* -1 and 2. It follows that -1 is less than 0 and 0 is less than 2; that is, $-1 < 0$ and $0 < 2$. In compact form, these facts may be combined into $-1 < 0 < 2$, read "-1 is less than 0, which is less than 2."

Rules for Comparing Signed Numbers

Rule 1: Any positive number is greater than 0.

Thus, $+5 > 0$. Note on the number line that each positive number is to the right of 0, to the right signifying "greater than."

Rule 2: Any negative number is less than 0.

Thus, $-5 < 0$. Note on the number line that each negative number is to the left of 0, to the left signifying "less than."

Rule 3: Any positive number is greater than any negative number.

Thus, $+\frac{1}{4} > -1,000$. Note on the number line that each positive number is to the right of each negative number.

Rule 4: The greater of two positive numbers has the greater absolute value.

Thus, since $+10 > +5$, then $|+10| > |+5|$. This is true because the absolute value of a positive number is the positive number.

Rule 5: The greater of two negative numbers has the smaller absolute value.

Thus, since $-5 > 10$, then $|-5| < |-10|$. This is true since $|-5| = 5$ while $|-10| = 10$.

2.1 COMPARING SIGNED NUMBERS

Which is greater?

(a) $+\frac{1}{4}$ or 0   (b) $+50$ or $+30$ (c) -30 or 0   (d) -30 or -10 (e) $+10$ or -100

Solutions

(a) $+\frac{1}{4}$ (Rule 1)   (b) $+50$ (Rule 4)   (c) 0 (Rule 2)   (d) -10 (Rule 5)   (e) $+10$ (Rule 3)

2.2 DETERMINING THE CORRECT EQUALITY OR INEQUALITY SYMBOL

In each, obtain a true statement by replacing the question mark with either $=$, $<$, or $>$:

(a) $-2 \; ? \; 0$ (d) $-5 \times 1 \; ? \; -5$ (g) $25 \times 1 \; ? \; 25 + 1$

(b) $0 \; ? \; -25$ (e) $\frac{1}{2} \; ? \; \frac{1}{4}$ (h) $|-25| \; ? \; |25|$

(c) $20 \times 0 \; ? \; 0$ (f) $\frac{1}{2} \times \frac{1}{2} \; ? \; \frac{1}{4}$ (i) $|-5| \; ? \; |-10|$

Solutions

(a) $<$ (b) $>$ (c) $=$ (d) $=$ (e) $>$ (f) $=$ (g) $<$ (h) $=$ (i) $<$

2.3 USING INEQUALITY SYMBOLS

Express each relation verbally and state whether each statement is true or false:

(a) $-3 > -20$ (d) $5 > 0 > -5$ (g) $|-25| \neq |+25|$

(b) $2 < 6 < 8$ (e) $-2 < 0 < -3$

(c) $-10 < -\frac{1}{2} < 0$ (f) $2 + 2 \neq 2 \times 2$

Solutions

(a) "-3 is greater than -20." True.

(b) "2 is less than 6, which is less than 8." True.

(c) "-10 is less than $-\frac{1}{2}$, which is less than 0." True.

(d) "5 is greater than 0, which is greater than −5." True.

(e) "−2 is less than 0, which is less than −3." False, since 0 is greater than −3.

(f) "2 plus 2 is not equal to 2 times 2." False, since 4 = 4.

(g) "Absolute value of −25 does not equal absolute value of +25." False, since 25 = 25.

3. ADDING SIGNED NUMBERS

In algebra, adding signed numbers means combining them to obtain a single number which represents the total or combined effect. Thus, if Mary first gains 10 lb and then loses 15 lb, the total effect of the two changes is a loss of 5 lb. This is shown by adding signed numbers: $(+10) + (−15) = −5$.

The symbol +, used in connection with signed numbers, has two meanings: (1) + may mean "add" or (2) + may mean "positive number." Thus $(−8) + (+15)$ means *add positive* 15 to negative 8.

Rules for Adding Signed Numbers

Rule 1: To add two signed numbers with **like** signs, **add** their absolute values. The sign of the sum is the common sign of the numbers.

Thus, to add $(+7)$ and $(+3)$ or to add $(−7)$ and $(−3)$, add the absolute values 7 and 3. The sign of the sum is the common sign. Hence, $(+7) + (+3) = +10$ and $(−7) + (−3) = −10$. The sum $(+7\frac{1}{2}) + (+3)$ $= 10\frac{1}{2}$; the sum $(−7\frac{1}{2}) + (−3) = −10\frac{1}{2}$.

Rule 2: To add two signed numbers with **unlike** signs, **subtract** the smaller absolute value from the other. The sign of the sum is the sign of the number having the greater absolute value.

Thus, to add $(+7)$ and $(−3)$ or $(−7)$ and $(+3)$, subtract the absolute value 3 from the absolute value 7. The sign of the sum is the sign of the number having the larger absolute value. Hence, $(+7) + (−3) = +4$ and $(−7) + (+3) = −4$. The sum $(+7\frac{1}{2}) + (−3) = 4\frac{1}{2}$.

Rule 3: Zero is the sum of two signed numbers with unlike signs and the same absolute value. Such numbers are opposites or *additive inverses* of each other.

Thus, $(+27) + (−27) = 0$. The numbers +27 and −27 are opposites or additive inverses of each other.

3.1   COMBINING BY MEANS OF SIGNED NUMBERS

Using signed numbers, find each sum: (a) 10 yd gained plus 5 yd lost, (b) 8 steps right plus 10 steps left, (c) $5 earned plus $5 spent, (d) 80 mi south plus 100 mi south.

Solutions

(a) 10 yd gained plus 5 yd lost
$\qquad (+10) + (−5)$
    Ans. +5 or 5 yd gained

(b) 8 steps right plus 10 steps left
$\qquad (+8) + (−10)$
    Ans. −2 or 2 steps left

(c) $5 earned plus $5 spent
$\qquad (+5) + (−5)$
    Ans. 0 or no change

(d) 80 mi south plus 100 mi south
$\qquad (−80) + (−100)$
    Ans. −180 or 180 mi south

3.2   ADDING SIGNED NUMBERS WITH LIKE SIGNS

Add: (a) $(+8), (+2)$;   (b) $(−8), (−2)$;   (c) $(−30), (−14\frac{1}{2})$.

Solutions Apply Rule 1.

	(a) $(+8)+(+2)$	(b) $(-8)+(-2)$	(c) $(-30)+(-14\frac{1}{2})$
1. Add absolute values:	$8+2=10$	$8+2=10$	$30+14\frac{1}{4}=44\frac{1}{2}$
2. Use the common sign:	*Ans.* $+10$	*Ans.* -10	*Ans.* $-44\frac{1}{2}$

3.3   ADDING SIGNED NUMBERS WITH UNLIKE SIGNS

Add: (a) $(+7), (-5)$;   (b) $(-17), (+10)$.

Solutions   Apply Rule 2.

	(a) $(+7)+(-5)$	(b) $(-17)+(+10)$
1. Subtract absolute values:	$7-5=2$	$17-10=7$
2. Use the sign of number having greater absolute value:	Sign of $+7$ is $+$.	Sign of -17 is $-$.
	Ans. $+2$	*Ans.* -7

3.4   ADDING SIGNED NUMBERS WHICH ARE OPPOSITES OF EACH OTHER

Add: (a) $(-18), (+18)$;   (b) $(+30\frac{1}{2}), (-30\frac{1}{2})$;   (c) $(-1.75), (+1.75)$.

Solutions   Sum is always zero, by Rule 3.

3.5   ADDING SIGNED NUMBERS

Add $+25$ to   (a) $+30$,   (b) -30.   (c) -25.   To -20, add   (d) -30,   (e) $+10$,   (f) $+20$,   (g) (-30.5).

Solutions

(a) $(+30)+(+25)=+55$ (Rule 1)   (d) $(-20)+(-30)=-50$ (Rule 1)
(b) $(-30)+(+25)=-5$ (Rule 2)   (e) $(-20)+(+10)=-10$ (Rule 2)
(c) $(-25)+(+25)=\ \ 0$ (Rule 3)   (f) $(-20)+(+20)=\ \ 0$ (Rule 3)
                                              (g) $(-20)+(-30.5)=-50.5$ (Rule 1)

4.   SIMPLIFYING THE ADDITION OF SIGNED NUMBERS

1. Parentheses may be omitted.
2. The symbol $+$ may be omitted when it means "add."
3. If the first signed number is positive, its $+$ sign may be omitted.

Thus, $8+9-10$ may be written instead of $(+8)+(+9)+(-10)$.

To Simplify Adding Signed Numbers

Add: $(+23), (-12), (-8), (+10)$.

PROCEDURE	SOLUTION
1. Add all the positive numbers: *(Their sum is positive.)*	1. $\begin{array}{r} 23 \\ \underline{10} \\ 33 \end{array}$
2. Add all the negative numbers: *(Their sum is negative.)*	2. $\begin{array}{r} -12 \\ \underline{-8} \\ -20 \end{array}$
3. Add the resulting sums:	3. $33-20=13$ *Ans.*

4.1   SIMPLIFYING THE ADDITION OF SIGNED NUMBERS

Express in simplified form, then add:
(a) $(+27)+(-15)+(+3)+(-5)$ (c) $(+11.2)+(+13.5)+(-6.7)+(+20.9)$
(b) $(-8)+(+13)+(-20)+(+9)$

Solutions

(a) $27 - 15 + 3 - 5 = 10$ (b) $-8 + 13 - 20 + 9 = -6$ (c) $11.2 + 13.5 - 6.7 + 20.9 = 38.9$

4.2 ADDING POSITIVES AND NEGATIVES SEPARATELY

Add: (a) $(+27)$, (-15), $(+3)$, (-5);   (b) (-8), $(+13)$, (-20), $(+9)$;   (c) $(+11.2)$, $(+13.5)$, (-6.7), $(+20.9)$, (-3.1).

Solutions

(a) Simplify:

$27 - 15 + 3 - 5$

Add +'s	Add −'s
27	−15
3	−5
30	−20

Add results: $30 - 20$
Ans. 10

(b) Simplify:

$-8 + 13 - 20 + 9$

Add +'s	Add −'s
13	− 8
9	−20
22	−28

Add results: $22 - 28$
Ans. −6

(c) Simplify:

$11.2 + 13.5 - 6.7 + 20.9 - 3.1$

Add +'s	Add −'s
11.2	−6.7
13.5	−3.1
20.9	−9.8
45.6	

Add results: $45.6 - 9.8$
Ans. 35.8

4.3 USING SIGNED NUMBERS AND NUMBER SCALES TO SOLVE PROBLEMS

In May, Tom made deposits of $30 and $20. Later, he withdrew $40 and $30. Find the change in his balance due to these changes, using   (a) signed numbers and   (b) number line.

Solutions

(a) Simplify: $30 + 20 - 40 - 30$

Add +'s	Add −'s
30	−40
20	−30
50	−70

$50 - 70 = -20$
Ans. $20 less

(b) Represent the successive changes by arrows parallel to the number line, as in Fig. 6-5. The last of the 4 arrows ends at −20. This means $20 less than the original balance. *Ans.*

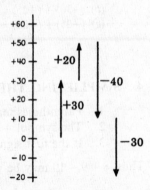

Fig. 6-5

5. SUBTRACTING SIGNED NUMBERS

The symbol −, used in connection with signed numbers, has two meanings: (1) − may mean "subtract" or (2) − may mean "negative number." Thus, $(+8) - (-15)$ means *subtract negative* 15 from positive 8.

Rule: **To subtract** a signed number, **add** its opposite.

Thus, to subtract $(+10)$, add (-10). For example,

$$(+18) - (+\mathbf{10}) = (+18) + (-\mathbf{10}) = +8 \quad Ans.$$

Also, to subtract (-10), add $(+10)$. For example,

$$(+30) - (-\mathbf{10}) = (+30) + (+\mathbf{10}) = +40 \quad Ans.$$

And, to subtract $(-10\frac{1}{2})$, add $+10\frac{1}{2}$. For example $(+30) - (-10\frac{1}{2}) = 40\frac{1}{2}$.

5.1   SUBTRACTING A POSITIVE NUMBER

Subtract: (a) (+8) from (+29),   (b) (+80) from (+80),   (c) (+18) from (−40),
          (d) (+20) from (−50.5).

Solutions   To subtract a positive, add its opposite negative.

(a) $(+29) - (+8)$ (b) $(+80) - (+80)$ (c) $(-40) - (+18)$ (d) $(-50.5) - (+20)$
     $(+29) + (-8)$ $(+80) + (-80)$ $(-40) + (-18)$ $(-50.5) + (-20)$

Ans. 21   *Ans.* 0   *Ans.* −58   *Ans.* −70.5

5.2   SUBTRACTING A NEGATIVE NUMBER

Subtract: (a) (−7) from (+20),   (b) (−67) from (−67),   (c) (−27) from (−87).

Solutions   To subtract a negative, add its opposite positive.

(a) $(+20) - (-7)$ (b) $(-67) - (-67)$ (c) $(-87) - (-27)$
     $(+20) + (+7)$ $(-67) + (+67)$ $(-87) + (+27)$
         $20 + 7$ $-67 + 67$ $-87 + 27$

Ans. 27   *Ans.* 0   *Ans.* −60

5.3   SUBTRACTING VERTICALLY

Subtract the lower number from the upper:
(a) +40   (b) +40   (c) −40   (d) −40.5
     +10   +55   −75   −81.7

Solutions   To subtract a signed number, add its opposite.

(a) $+40$ $+40$ (b) $+40$ $+40$ (c) -40 -40
     $\underline{-(+10)} \rightarrow \underline{+(-10)}$ $\underline{-(+55)} \rightarrow \underline{+(-55)}$ $\underline{-(-75)} \rightarrow \underline{+(+75)}$
                      $+30$ *Ans.* -15 *Ans.* $+35$ *Ans.*

(d) -40.5 -40.5
     $\underline{-(-81.7)} \rightarrow \underline{+(+81.7)}$
                     $+41.2$ *Ans.*

5.4   SUBTRACTING SIGNED NUMBERS

Subtract +3 from   (a) +11,   (b) −15.   From −8, subtract   (c) −15,   (d) −5.

Solutions

(a) $(+11) - (+3)$ (b) $(-15) - (+3)$ (c) $(-8) - (-15)$ (d) $(-8) - (-5)$
     $(+11) + (-3)$ $(-15) + (-3)$ $(-8) + (+15)$ $(-8) + (+5)$
        $11 - 3$ $-15 - 3$ $-8 + 15$ $-8 + 5$

Ans. 8   *Ans.* −18   *Ans.* +17   *Ans.* −3

5.5   COMBINING ADDITION AND SUBTRACTION OF SIGNED NUMBERS

Evaluate: (a) $(+10) + (+6) - (-2)$,   (b) $(+8) - (-12) - (+5) + (-3)$,   (c) $(-11) - (+5) + (-7)$

Solutions

(a) $(+10) + (+6) - (-2)$ (b) $(+8) - (-12) - (+5) + (-3)$ (c) $(-11) - (+5) + (-7)$
     $(+10) + (+6) + (+2)$ $(+8) + (+12) + (-5) + (-3)$ $(-11) + (-5) + (-7)$
        $10 + 6 + 2$ $8 + 12 - 5 - 3$ $-11 - 5 - 7$

Ans. 18   *Ans.* 12   *Ans.* −23

5.6   FINDING THE CHANGE BETWEEN TWO SIGNED NUMBERS

Find the change from +20 to −60, using   (a) number line and   (b) signed numbers.

Solutions

(a) See Fig. 6-6.

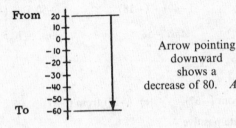

Arrow pointing
downward
shows a
decrease of 80.   *Ans.*

Fig. 6-6

(b) Subtract 20 from −60:

$$-60 - (+20)$$
$$-60 + (-20)$$
$$-80 \ Ans.$$

5.7   FINDING THE DISTANCE BETWEEN TWO LEVELS

Using   (a) signed numbers and   (b) a number line, find the distance from 300 ft below sea level to (1) 800 ft below sea level, (2) sea level, (3) 100 ft above sea level.

Solutions

(a) (1) $-800 - (-300)$ (2) $0 - (-300)$ (3) $100 - (-300)$
        $-800 + (+300)$ $0 + (+300)$ $100 + (+300)$
              -500 $+300$ $+400$

   Ans. 500 ft down   *Ans.* 300 ft up   *Ans.* 400 ft up

(b) See Fig. 6-7.

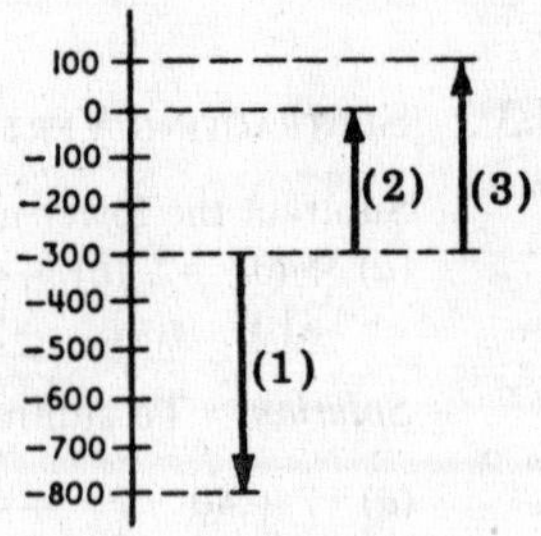

Fig. 6-7

5.8   FINDING A TEMPERATURE CHANGE

On Monday, the temperature changed from −10° to 20°. On Tuesday, the change was from 20° to −20°. Find each change, using   (a) signed numbers and   (b) a number line.

Solutions

(a)   Monday Change   Tuesday Change
                   $20 - (-10)$ $-20 - (+20)$
                   $20 + (+10)$ $-20 + (-20)$
                        $+30$ -40

              Ans. Rise of 30°   *Ans.* Drop of 40°

(b) See Fig. 6-8.

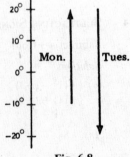

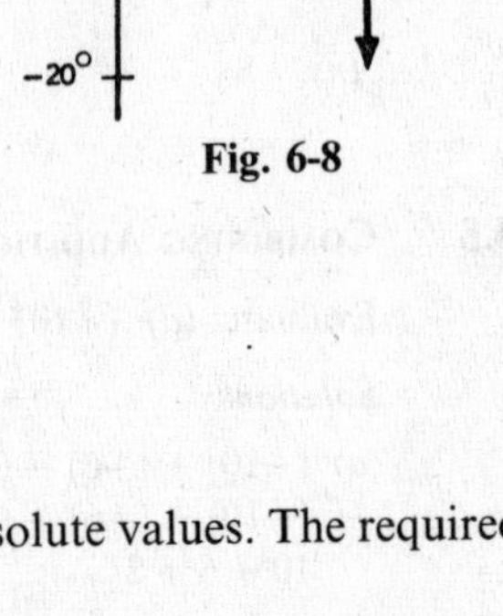

Fig. 6-8

6. MULTIPLYING SIGNED NUMBERS

Multiplying Two Signed Numbers

Rule 1: To multiply two signed numbers with **like** signs, multiply their absolute values. The required product is **positive**.

Thus, $(+5)(+4) = +20$ and $(-5)(-4) = +20$.

Rule 2: To multiply two signed numbers with **unlike** signs, multiply their absolute values. The required product is **negative**.
Thus, $(-7)(+2) = -14$ and $(+7)(-2) = -14$.

Rule 3: Zero times any signed number equals zero (multiplicative property of zero).
Thus, $0(-8\frac{1}{2}) = 0$ and $(44.7)(0) = 0$.

Multiplying More Than Two Signed Numbers

Rule 4: The product is positive if all the signed numbers are positive or if there is an even number of negatives.
Thus, $(+10)(+4)(-3)(-5) = +600$.

Rule 5: The product is negative if there is an odd number of negatives.
Thus, $(+10)(-4)(-3)(-5) = -600$.

Rule 6: The product is zero if any number is zero.
Thus, $(-5)(+82)(0)(316\frac{1}{2}) = 0$.

6.1 **MULTIPLYING NUMBERS WITH LIKE SIGNS**

Multiply: (a) $(+5)(+9)$, (b) $(-5)(-11)$, (c) $(-3)(-2.4)$.

Solutions Apply Rule 1.

	(a) $(+5)(+9)$	(b) $(-5)(-11)$	(c) $(-3)(-2.4)$
1. Multiply absolute values:	$5(9) = 45$	$5(11) = 55$	$3(2.4) = 7.2$
2. The product is positive:	*Ans.* $+45$	*Ans.* $+55$	*Ans.* $+7.2$

6.2 **MULTIPLYING NUMBERS WITH UNLIKE SIGNS**

Multiply: (a) $(+8)(-9)$, (b) $(-20)(+5)$, (c) $(+3\frac{1}{2})(-10)$.

Solutions Apply Rule 2.

	(a) $(+8)(-9)$	(b) $(-20)(+5)$	(c) $(+3\frac{1}{2})(-10)$
1. Multiply absolute values:	$8(9) = 72$	$20(5) = 100$	$(3\frac{1}{2})(10) = 35$
2. The product is negative:	*Ans.* -72	*Ans.* -100	*Ans.* -35

6.3 **MULTIPLYING SIGNED NUMBERS**

Multiply $+10$ by (a) $+18$, (b) 0, (c) -13. Multiply -20 by (d) $+2.5$, (e) 0, (f) $-2\frac{1}{4}$.

Solutions

(a) $(+10)(+18) = +180$ (Rule 1)   (d) $(-20)(+2.5) = -50$ (Rule 2)
(b) $(+10)(0) = 0$ (Rule 3)   (e) $(-20)0 = 0$ (Rule 3)
(c) $(+10)(-13) = -130$ (Rule 2)   (f) $(-20)(-2\frac{1}{4}) = +45$ (Rule 1)

6.4 **MULTIPLYING MORE THAN TWO SIGNED NUMBERS**

Multiply:
(a) $(+2)(+3)(+4)$ (d) $(-2)(-3)(-4)$ (g) $(-1)(-2)(+5)(+10)$
(b) $(+2)(+3)(-4)$ (e) $(+2)(+3)(-4)(0)$ (h) $(-1)(+2)(+5)(+10)$
(c) $(+2)(-3)(-4)$ (f) $(-1)(-2)(-5)(-10)$ (i) $(-1)(-2)(-5)(+10)$

Solutions

(a) $+24$ (Rule 4)   (d) -24 (Rule 5)   (g) $+100$ (Rule 4)
(b) -24 (Rule 5)   (e) 0 (Rule 6)   (h) -100 (Rule 5)
(c) $+24$ (Rule 4)   (f) $+100$ (Rule 4)   (i) -100 (Rule 5)

6.5   USING SIGNED NUMBERS TO SOLVE PROBLEMS

Complete each statement, using signed numbers to obtain the answer. (*a*) If George deposits $50 each week, then after 3 weeks his bank balance will be ( ). (*b*) If Henry has been spending $5 a day, then 4 days ago he had ( ). (*c*) If the temperature falls 6° each day, then after 3 days it will be ( ). (*d*) If a team has been gaining 10 yd on each play, then 3 plays ago it was ( ). (*e*) If a car has been traveling west at 30 mph, then 3 hr ago it was ( ).

Solutions

(*a*) Let +50 = $50 weekly deposit
       +3 = 3 weeks later
    Then (+50)(+3) = 150

Ans. $150 more

(*b*) Let −5 = $5 daily spending
       −4 = 4 days earlier
    Then, (−5)(−4) = +20

Ans. $20 more

(*c*) Let −6 = 6° daily fall
      +3 = 3 days later
   Then, (−6)(+3) = −18

Ans. 18° lower

(*d*) Let +10 = 10 yd gain per play
      −3 = 3 plays earlier
   Then, (+10)(−3) = −30

Ans. 30 yd farther back

(*e*) Let −30 = 30 mph westward
      −3 = 3 hr earlier
   Then, (−30)(−3) = +90

Ans. 90 mi farther east

7. FINDING POWERS OF SIGNED NUMBERS

The following rules apply to powers whose exponent is a positive integer.

Rules for Powers of Signed Numbers

Rule 1: For positive base, power is always positive.

Thus $(+2)^3 = +8$ since $(+2)^3 = (+2)(+2)(+2)$. For positive bases, + signs are omitted. Thus, $2^3 = 8$.

Rule 2: For negative base, power is positive when exponent is even, and negative when exponent is odd.

Thus, $(-2)^4 = +16$ since $(-2)^4 = (-2)(-2)(-2)(-2)$; and $(-2)^5 = -32$ since $(-2)^5 = (-2)(-2)(-2)(-2)(-2)$.

7.1   FINDING POWERS WHEN BASE IS POSITIVE

Find each power:

(*a*) 3^2 (*d*) $(+5)^3$ (*g*) $\left(\dfrac{2}{3}\right)^2$

(*b*) 2^4 (*e*) $(+7)^2$ (*h*) $\left(\dfrac{1}{2}\right)^3$

(*c*) 10^3 (*f*) $(+1)^{10}$ (*i*) $\left(\dfrac{1}{10}\right)^4$

Solutions

(*a*) 9   (*b*) 16   (*c*) 1000   (*d*) 125   (*e*) 49   (*f*) 1   (*g*) $\dfrac{4}{9}$ (*h*) $\dfrac{1}{8}$ (*i*) $\dfrac{1}{10,000}$

7.2   FINDING POWERS WHEN BASE IS NEGATIVE

Find each power:

(*a*) $(-3)^2$ (*c*) $(-1)^{100}$ (*e*) $(-.5)^2$

(*b*) $(-3)^3$ (*d*) $(-1)^{105}$ (*f*) $(-.2)^3$

(g) $\left(-\frac{1}{2}\right)^2$ (h) $\left(-\frac{1}{3}\right)^3$ (i) $\left(-\frac{2}{5}\right)^3$

Solutions

(a) 9   (b) −27   (c) 1   (d) −1   (e) .25   (f) −.008   (g) $\frac{1}{4}$ (h) $-\frac{1}{27}$ (i) $-\frac{8}{125}$

7.3 FINDING POWERS OF SIGNED NUMBERS

Find each power:

(a) 3^2 (d) $\left(\frac{1}{2}\right)^3$ (g) $(-1)^{100}$

(b) -3^2 (e) $\left(-\frac{1}{2}\right)^3$ (h) -1^{100}

(c) $(-3)^2$ (f) $-\left(\frac{1}{2}\right)^3$ (i) $(-1)^{111}$

Solutions

(a) 9   (b) −9   (c) 9   (d) $\frac{1}{8}$ (e) $-\frac{1}{8}$ (f) $-\frac{1}{8}$ (g) 1   (h) −1   (i) −1

7.4 FINDING BASES, EXPONENTS, OR POWERS

Complete each:

(a) $(-3)^? = 81$ (c) $(?)^5 = -32$ (e) $\left(-\frac{3}{4}\right)^3 = ?$

(b) $2^? = 128$ (d) $(?)^{121} = -1$ (f) $-.2^4 = ?$

Solutions

(a) 4   (b) 7   (c) −2   (d) −1   (e) $-\frac{27}{64}$ (f) −.0016

8. DIVIDING SIGNED NUMBERS

Rules for Dividing Signed Numbers

   Rule 1: To divide two signed numbers with **like** signs, divide the absolute value of the first by that of the second. The required quotient is **positive**.

Thus, $\frac{+8}{+2} = +4$ and $\frac{-8}{-2} = +4$.

   Rule 2: To divide two signed numbers with **unlike** signs, divide the absolute value of the first by that of the second. The required quotient is **negative**.

Thus, $\frac{+12}{-4} = -3$ and $\frac{-12}{+4} = -3$.

   Rule 3: Zero divided by any signed number is zero.

Thus, $\frac{0}{+17} = 0$ and $\frac{0}{-17} = 0$.

Rule 4: Dividing a signed number by zero is an impossible operation.

Thus, $\dfrac{18}{0}$, $\dfrac{+18}{0}$, and $\dfrac{-18}{0}$ are meaningless.

Combining Multiplication and Division of Signed Numbers

To multiply and divide signed numbers at the same time, (1) isolate their absolute values to obtain the absolute value of the answer; (2) find the sign of the answer from the following rules:

Rule 5: The quotient is positive if all the numbers are positive or if there is an even number of negatives.

Thus, $\dfrac{(+12)(+5)}{(+3)(+2)}$, $\dfrac{(+12)(+5)}{(-3)(-2)}$, and $\dfrac{(-12)(-5)}{(-3)(-2)}$ equal +10.

Rule 6: The quotient is negative if there is an odd number of negatives.

Thus, $\dfrac{(+12)((+5)}{(+3)(-2)}$, $\dfrac{(+12)(-5)}{(-3)(-2)}$, and $\dfrac{(-12)(-5)}{(+3)(-2)}$ equal −10.

Rule 7: The quotient is zero if one of the numbers in the dividend is 9.

Thus, $\dfrac{(+53)(0)}{(-17)(-84)} = 0$.

Rule 8: A zero divisor makes the operation impossible.

Thus, $\dfrac{(+53)(-17)}{(+84)0}$ is meaningless.

8.1   DIVIDING SIGNED NUMBERS WITH LIKE SIGNS

Divide: (a) (+12) by (+6),   (b) (−12) by (−4),   (c) (−24.6) by (−3).

Solutions   Apply Rule 1.

$$(a)\ \frac{+12}{+6}\qquad\qquad (b)\ \frac{-12}{-4}\qquad\qquad (c)\ \frac{-24.6}{-3}$$

1. Divide absolute values: $\dfrac{12}{6} = 2$ $\dfrac{12}{4} = 3$ $\dfrac{24.6}{3} = 8.2$

2. Make quotient positive: *Ans.* +2   *Ans.* +3   *Ans.* +8.2

8.2   DIVIDING SIGNED NUMBERS WITH UNLIKE SIGNS

Divide: (a) (+20) by (−5),   (b) −24 by +8,   (c) +8 by −16.

Solutions   Apply Rule 2.

$$(a)\ \frac{+20}{-5}\qquad\qquad (b)\ \frac{-24}{+8}\qquad\qquad (c)\ \frac{+8}{-16}$$

1. Divide absolute values: $\dfrac{20}{5} = 4$ $\dfrac{24}{8} = 3$ $\dfrac{8}{16} = \dfrac{1}{2}$

2. Make quotient negative: *Ans.* −4   *Ans.* −3   *Ans.* $-\dfrac{1}{2}$

8.3   DIVIDING SIGNED NUMBERS

Divide +9 by   (a) +3,   (b) 0,   (c) −10. Divide each by −12:   (d) +36,   (e) 0,   (f) −3.

Solutions

(a) $\dfrac{+9}{+3} = +3$ (Rule 1)   (d) $\dfrac{+36}{-12} = -3$ (Rule 2)

(b) $\dfrac{+9}{0}$ is meaningless   (Rule 4)   (e) $\dfrac{0}{-12} = 0$ (Rule 3)

(c) $\dfrac{+9}{-10} = -\dfrac{9}{10}$ (Rule 2)   (f) $\dfrac{-3}{-12} = +\dfrac{1}{4}$ (Rule 1)

8.4   COMBINING MULTIPLICATION AND DIVISION OF SIGNED NUMBERS

Evaluate: (a) $\dfrac{(+12)(+8)}{(-3)(-4)}$ (b) $\dfrac{(-4)(-2)(-3)}{(-5)(-10)}$

Solutions

1. Isolate absolute values: (a) $\dfrac{(12)(8)}{(3)(4)} = 8$ (b) $\dfrac{(4)(2)(3)}{(5)(10)} = \dfrac{12}{25}$

2. Find the sign of the answer: $\dfrac{(+)(+)}{(-)(-)} = +$ $\dfrac{(-)(-)(-)}{(-)(-)} = -$

$Ans.$ $+8$ $Ans.$ $-\dfrac{12}{25}$

8.5   ZERO IN DIVIDEND OR DIVISOR

Evaluate: (a) $\dfrac{(+27)0}{(-3)(+15)}$ (b) $\dfrac{(+27)(-3)}{(+15)0}$

Solutions   (a) 0   (Rule 7)   (b) meaningless   (Rule 8)

9. EVALUATING EXPRESSIONS HAVING SIGNED NUMBERS

To evaluate an expression whose variables take on signed-number values: (1) substitute the values given for the variables, enclosing them in parentheses; (2) perform the operations in the correct order, doing the power operations first.

9.1   EVALUATING EXPRESSIONS HAVING ONE VARIABLE

Evaluate if $y = -2$:

(a) $2y + 5$ (b) $20 - 3y$ (c) $4y^2$ (d) $2y^3$ (e) $20 - y^5$

Solutions

(a) $2(-2) + 5$ (b) $20 - 3(-2)$ (c) $4(-2)^2$ (d) $2(-2)^3$ (e) $20 - (-2)^5$
$\quad\;\; -4 + 5$ $\quad\;\; 20 + 6$ $\quad\;\; 4(4)$ $\quad\;\; 2(-8)$ $\quad\;\; 20 - (-32)$

$Ans.$ 1 $Ans.$ 26 $Ans.$ 16 $Ans.$ -16 $Ans.$ 52

9.2   EVALUATING EXPRESSIONS HAVING TWO VARIABLES

Evaluate if $x = -3$ and $y = +2$:

(a) $2x - 5y$ (b) $20 - 2xy$ (c) $3xy^2$ (d) $2x^2 - y^2$

Solutions

(a) $2(-3) - 5(+2)$ (b) $20 - 2(-3)(+2)$ (c) $3(-3)(+2)^2$ (d) $2(-3)^2 - (+2)^2$
$\quad\;\; 6 - 10$ $\quad\;\; 20 + 12$ $\quad\;\; 3(-3)(4)$ $\quad\;\; 2(+9) - (+4)$

$Ans.$ -16 $Ans.$ 32 $Ans.$ -36 $Ans.$ 14

9.3 EVALUATING EXPRESSIONS HAVING THREE VARIABLES

Evaluate if $x = +1$, $y = -2$, $z = -3$:

(a) $4xy^2 + z$ (b) $x^2 + y^2 - z^2$ (c) $\dfrac{3x - 5y}{2z}$ (d) $\dfrac{2y^2}{x - z}$

Solutions

(a) $4(+1)(-2)^2 + (-3)$ (b) $(+1)^2 + (-2)^2 - (-3)^2$ (c) $\dfrac{3(+1) - 5(-2)}{2(-3)}$ (d) $\dfrac{2(-2)^2}{(+1) - (-3)}$

     $4(1)(4) - 3$ $1 + 4 - 9$

Ans. 13   *Ans.* -4 $\dfrac{3 + 10}{-6}$ $\dfrac{2 \cdot 4}{1 + 3}$

                                                  Ans. $-2\frac{1}{6}$ *Ans.* 2

Supplementary Problems

6.1.   State the words that are opposite in meaning to the following: **(1.1)**

    (a)   farther   (d)   longer   (g) wider   (j) stronger
    (b)   slower   (e)   cheaper   (h) credit
    (c)   hotter   (f)   younger   (i) heavier

    Ans.   (a)   nearer   (d)   shorter   (g) narrower   (j) weaker
             (b)   faster   (e)   costlier   (h) debit
             (c)   colder   (f)   older   (i) lighter

6.2.   State the quantity represented by   **(1.2)**

    (a) $+12$, if -12 means 12 yd lost   (d) -32, if $+32$ means \$32 earned
    (b) -7, if $+7$ means 7 flights up   (e) $+15$, if -15 means \$15 withdrawn
    (c) $+25$, if -25 means 25 mi westward   (f) -13, if $+13$ means 13 steps forward

    Ans.   (a)   12 yd gained   (c)   25 mi eastward   (e)   \$15 deposited
             (b)   7 flights down   (d)   \$32 spent   (f)   13 steps backward

6.3.   Evaluate: **(1.3)**

    (a) $|3\frac{1}{2}| + |-3\frac{1}{2}|$ (c) $-5|-3||-2||+4|$ (e) $|-5|^2$

    (b) $|97| - |-97|$ (d) $\dfrac{18(|+9|)}{6(|-9|)}$ (f) $\dfrac{2|+2| + 3|-3|}{4|-4| - 5|+3|}$

    Ans.   (a) $3\frac{1}{2} + 3\frac{1}{2} = 7$ (c) $-5(3)(2)(4) = -120$ (e) $5^2 = 25$

                 (b) $97 - 97 = 0$ (d) $\dfrac{18(9)}{6(9)} = 3$ (f) $\dfrac{2(2) + 3(3)}{4(4) - 5(3)} = \dfrac{13}{1} = 13$

6.4.   With reference to the graphs and coordinates shown in Fig. 6-9, name   **(1.4)**

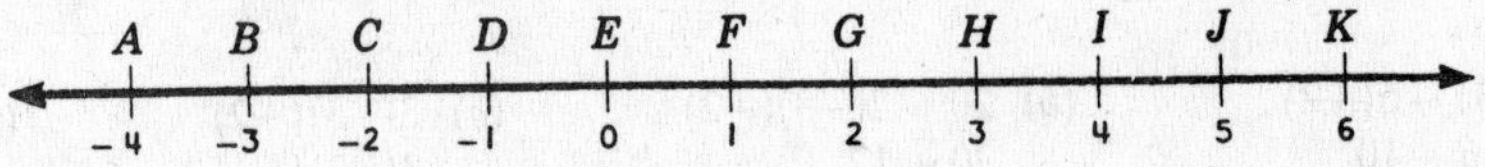

Fig. 6-9

(a) the graphs of the whole numbers; (b) the coordinates of points midway between (1) B and D, (2) E and K, (3) D and E, (4) G and J; (c) the graphs of coordinates that are at least -1 and no more than 3; (d) the graphs of coordinates whose absolute value is equal to or less than 2; (e) the graphs of the coordinates that differ from -1 by 3.

Ans. (a) E, F, G, H, I, J, K

(b) (1) -2, (2) 3, (3) $-\frac{1}{2}$, (4) $3\frac{1}{2}$

(c) D, E, F, G, H

(d) C, D, E, F, G

(e) A, G

6.5. Tell whether each is a natural number, whole number, or integer: **(1.5)**
(a) 17 (b) -17 (c) $|-17|$ (d) 0

Ans. (a) All three (b) Integer (c) All three (d) Whole, integer

6.6. Which is greater? **(2.1)**

(a) $-\frac{1}{2}$ or 0   (b) $-\frac{1}{2}$ or -30 (c) $+\frac{1}{2}$ or $-2\frac{1}{2}$ (d) -110 or -120

Ans. (a) 0   (b) $-\frac{1}{2}$ (c) $+\frac{1}{2}$ (d) -110

6.7. In each, obtain a true statement by replacing the question mark with either $=$, $<$, or $>$: **(2.2)**
(a) $-2 ? -1$ (e) $50 + 0 ? 50 \times 1$ (h) $|-7| ? |-8|$

(b) $-100 ? 0$ (f) $\frac{1}{2} ? \frac{1}{4} \times \frac{1}{4}$ (i) $2(-7) ? 0$

(c) $0 ? -\frac{1}{4}$ (g) $-7 ? -8$ (j) $2|-7| ? 0$

(d) $50 \times 0 ? 0$

Ans. (a) $<$ (b) $<$ (c) $>$ (d) $=$ (e) $=$ (f) $>$ (g) $>$ (h) $<$ (i) $<$ (j) $>$

6.8. Express each relation verbally and state whether each statement is true or false: **(2.3)**

(a) $-\frac{1}{2} < -\frac{1}{4}$ (c) $-3 < -4 < -5$ (e) $2 \times 1 \neq 2 + \frac{1}{2}$

(b) $-3 < 0 < 3$ (d) $-100 > -1000 > 1,000,000$ (f) $|-3| \neq |+3|$

Ans. (a) "$-\frac{1}{2}$ is less than $-\frac{1}{4}$." True.
   (b) "-3 is less than 0, which is less than 3." True.
   (c) "-3 is less than -4, which is less than -5." False, since $-3 > -4 > -5$.
   (d) "-100 is less than -1000, which is less than 1,000,000." False, since $-1000 < -100$.
   (e) "2×1 does not equal $2 + \frac{1}{2}$." True.
   (f) "Absolute value of -3 does not equal the absolute value of $+3$." False, since $3 = 3$.

6.9. Using signed numbers, find each sum: **(3.1)**

(a) 20 lb gained plus 7 lb lost   (d) $5 gained plus $15 spent
(b) 17 lb lost plus 3 lb gained   (e) 8° rise plus 20° rise
(c) $5 spent plus $15 spent   (f) 8° drop plus 8° rise

Ans. (a) $(+20) + (-7) = +13$ or 13 lb gained   (d) $(+5) + (-15) = -10$ or $10 spent
   (b) $(-17) + (+3) = -14$ or 14 lb lost   (e) $(+8) + (+20) = +28$ or 28° rise
   (c) $(-5) + (-15) = -20$ or $20 spent   (f) $(-8) + (+8) = 0$ or no change

6.10. Add: (*a*) $(+10\frac{1}{2})$, $(+7\frac{1}{2})$;   (*b*) (-1.4), (-2.5);   (*c*) $(+5)$, $(+3)$, $(+12)$;   **(3.2)**
(*d*) (-1), (-2), (-3).

Ans. (*a*) $+18$ (*b*) -3.9 (*c*) $+20$ (*d*) -6

6.11. Add: (*a*) $(+6\frac{1}{4})$, $(-3\frac{1}{2})$;   (*b*) $(-.23)$, $(+.18)$;   (*c*) $(+23)$, (-13), $(+12)$;   **(3.3)**
(*d*) $(+5)$, (-8), $(+17)$.

Ans. (*a*) $+2\frac{3}{4}$ (*b*) $-.05$ (*c*) $+22$ (*d*) $+14$

6.12. Add: (*a*) $(+25.7)$, (-25.7);   (*b*) $(+120)$, (-120), $(+19)$;   (*c*) $(+28)$,   **(3.4)**
(-16), (-28) $(+16)$.

Ans. (*a*) 0 (*b*) $+19$ (*c*) 0

6.13. Add -32 to (*a*) -25,   (*b*) $+25$,   (*c*) $+32$,   (*d*) $+100$. To $+11$,   **(3.5)**
add (*e*) -29,   (*f*) -11,   (*g*) $+117$,   (*h*) -108.

Ans. (*a*) -57 (*b*) -7 (*c*) 0 (*d*) $+68$ (*e*) -18 (*f*) 0 (*g*) $+128$ (*h*) -97

6.14. Express in simpified form and add: **(4.1, 4.2)**
(*a*) (-14). (-7), $(+22)$, (-35) (*b*) $(+3\frac{1}{4})$, $(+8\frac{1}{2})$, $(-40\frac{3}{4})$ (*c*) (-1.78), (-3.22), $(+16)$

Ans. (*a*) $-14-7+22-35$ (*b*) $3\frac{1}{4}+8\frac{1}{2}-40\frac{3}{4}$ (*c*) $-1.78-3.22+16$
        Sum is -34 Sum is -29 Sum is 11

6.15. In a football game, a team gained 8 yd on the first play, gained 1 yd on the second   **(4.3)**
play, lost 12 yd on the third play, and lost 6 yd on the fourth play. Find the change in position due
to these changes, using (*a*) signed numbers and (*b*) a number scale.

Ans. (*a*) $(+8)+(+1)+(-12)+(-6)$ (*b*)
        $8+1-12-6$
        -9

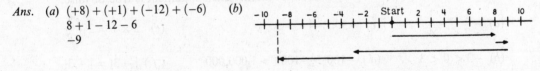

Ans. Total change is 9 yd loss.

6.16. Subtract: **(5.1)**

(*a*) $(+11)$ from $(+16)$ (*c*) $(+3.5)$ from $(+7.2)$ (*e*) $(+3\frac{1}{3})$ from $(+11\frac{2}{3})$
(*b*) $(+47)$ from (-47) (*d*) $(+7.2)$ from (-3.5) (*f*) $(+17\frac{2}{3})$ from (-8)

Ans. (*a*) $+5$ (*b*) -94 (*c*) $+3.7$ (*d*) -10.7 (*e*) $+8\frac{1}{3}$ (*f*) $-25\frac{2}{3}$

6.17. Subtract: **(5.2)**

(*a*) (-11) from $(+16)$ (*c*) $(-.81)$ from $(+.92)$ (*e*) $(-7\frac{1}{5})$ from $(+2\frac{4}{5})$
(*b*) (-47) from $(+47)$ (*d*) $(-.23)$ from $(-.27)$ (*f*) $(-3\frac{5}{6})$ from (-4)

Ans. (*a*) $+27$ (*b*) $+94$ (*c*) $+1.73$ (*d*) $-.04$ (*e*) $+10$ (*f*) $-\frac{1}{6}$

6.18. Subtract the lower number from the upper: (5.3)

(a) $+22$ (b) $+17$ (c) -30.7 (d) $-.123$ (e) $-13\frac{2}{3}$ (f) $-27\frac{1}{2}$
$\underline{+17}$ $\underline{+22}$ $\underline{+30.7}$ $\underline{+.265}$ $\underline{-10\frac{1}{3}}$ $\underline{+3\frac{1}{4}}$

Ans. (a) $+5$ (b) -5 (c) -61.4 (d) $-.388$ (e) $-3\frac{1}{3}$ (f) $-31\frac{1}{4}$

6.19. Subtract $+4$ from   (a) $+8\frac{1}{2}$,   (b) $+4$,   (c) $+2.3$, and   (d) -25. From -10 (5.4)
subtract   (e) -23,   (f) -8,   (g) -3.9, and   (h) $+20\frac{1}{4}$.

Ans. (a) $+4\frac{1}{2}$ (b) 0 (c) -1.7 (d) -29 (e) $+13$ (f) -2 (g) -6.1 (h) $-30\frac{1}{4}$

6.20. Combine: (5.5)

(a) $(+11)+(+7)-(+24)$ (d) $(+25)-(-6)-(-22)+(+40)$
(b) $(-11)-(-5)-(-14)$ (e) $(-3.7)+(-2.4)-(+7.8)+(-11.4)$
(c) $(+.13)-(+.07)-(+.32)$ (f) $(+2\frac{1}{2})-(-1\frac{1}{4})-(+5\frac{3}{4})-(-7)$

Ans. (a) -6 (b) $+8$ (c) $-.26$ (d) $+93$ (e) -25.3 (f) $+5$

6.21. Find the change from: (5.6)

(a) $+10$ to -20 (d) -45 to 0 (g) $-\frac{1}{2}$ to $+5\frac{1}{2}$
(b) -10 to $+20$ (e) -45 to $+45$ (h) $+5\frac{1}{2}$ to $+4\frac{1}{2}$
(c) $+45$ to 0 (f) $+45$ to -45 (i) $+62\frac{1}{4}$ to $+112\frac{1}{2}$

Ans. (a) -30, a decrease of 30   (d) $+45$, an increase of 45   (g) $+6$, an increase of 6
(b) $+30$, an increase of 30   (e) $+90$, an increase of 90   (h) -1, a decrease of 1
(c) -45, a decrease of 45   (f) -90, a decrease of 90   (i) $+50\frac{1}{4}$, an increase of $50\frac{1}{4}$

6.22. Find the distance from 500 m above sea level to   (a) 1200 m above sea level, (5.7)
(b) sea level,   (c) 2000 m below sea level.

Ans. (a) 700 m up   (b) 500 m down   (c) 2500 m down

6.23. On a Monday, the hourly temperatures from 1 P.M. to 6 P.M. were (5.8)

1 P.M.	2 P.M.	3 P.M.	4 P.M.	5 P.M.	6 P.M.
$-8°$	$-5°$	$0°$	$4°$	$-4°$	$-20°$

Find each hourly temperature change.

Ans. (a) 3° rise from 1 to 2 P.M.   (c) 4° rise from 3 to 4 P.M.   (e) 16° drop from 5 to 6 P.M.
(b) 5° rise from 2 to 3 P.M.   (d) 8° drop from 4 to 5 P.M.

6.24. Multiply: (6.1)

(a) $(+3)(+22)$ (c) $(-5)(-4)$ (e) $(+8)(+2\frac{1}{2})$ (g) $(-2\frac{1}{2})(-1\frac{1}{2})$

(b) $(+.3)(+2.2)$ (d) $(-.05)(-.04)$ (f) $(-35)(-2\frac{1}{5})$ (h) $\left(+\frac{5}{4}\right)\left(+\frac{15}{2}\right)$

Ans. (a) $+66$ (b) $+.66$ (c) $+20$ (d) $+.002$ (e) $+20$ (f) $+77$ (g) $+3\frac{3}{4}$ (h) $+9\frac{3}{8}$

6.25. Multiply: **(6.2)**

(a) $(-20)(+6)$ (c) $(-.8)(+.11)$ (e) $(+6)(-7\frac{2}{3})$ (g) $\left(-\frac{7}{2}\right)\left(+\frac{11}{4}\right)$

(b) $(+13)(-8)$ (d) $(+3.4)(-21)$ (f) $(-2\frac{1}{7})(+21)$ (h) $(+2\frac{1}{5})(-3\frac{2}{5})$

Ans. (a) -120 (b) -104 (c) $-.088$ (d) -71.4 (e) -46 (f) -45 (g) $-\frac{77}{8}$ (h) $-7\frac{12}{25}$

6.26. Multiply $+8$ by (a) $+7$,   (b) 0,   (c) -4. Multiply -12 by (d) $+.9$,   (e) $-1\frac{1}{2}$,   **(6.3)**
(f) -8.21.

Ans. (a) $+56$ (b) 0 (c) -32 (d) -10.8 (e) $+18$ (f) $+98.52$

6.27. Multiply: **(6.4)**

(a) $(+3)(+4)(+12)$ (d) $(-1)(-1)(+1)(-1)$ (g) $\left(+\frac{1}{2}\right)\left(+\frac{1}{2}\right)\left(+\frac{1}{2}\right)\left(-\frac{1}{2}\right)$

(b) $(+.3)(+.4)(+1.2)$ (e) $(-2)(-2)(-2)(-2)$ (h) $\left(-\frac{1}{2}\right)(+8)\left(-\frac{1}{4}\right)(+16)$

(c) $(+.3)(-4)(-.12)$ (f) $(-2)(+5)(-5)(+4)$ (i) $(-1\frac{1}{2})(+2\frac{1}{2})(+3\frac{1}{2})$

Ans. (a) $+144$ (c) $+.144$ (e) $+16$ (g) $-\frac{1}{16}$ (i) $-13\frac{1}{8}$
       (b) $+.144$ (d) -1 (f) $+200$ (h) $+16$

6.28. Complete each statement. In each case, indicate how signed numbers may be used   **(6.5)**
to obtain the answer.

(a) If George withdraws \$10 each week, then after 5 weeks, his bank balance will be (   ).
(b) If Tom has been earning \$150 a day, then in 3 days, he will have (   ).
(c) If the temperature has risen 8° each day, then five days ago, it was (   ).
(d) If a car has been traveling east at 40 mph, then 3 hours ago, it was (   ).
(e) If a school decreases in register 20 pupils per day, then 12 days ago, the register was (   ).

Ans. (a) \$50 less   (c) 40° lower   (e) 240 pupils more
           $(-10)(+5)=-50$ $(+8)(-5)=-40$ $(-20)(-12)=+240$
       (b) \$450 more   (d) 120 mi farther west
           $(+150)(+3)=+450$ $(+40)(-3)=-120$

6.29. Find each power: (a) 2^3,   (b) $(+4)^2$,   (c) 10^4,   (d) $.3^2$ (e) $(+.2)^3$,   **(7.1)**
(f) $(1/4)^2$,   (g) $(1/10)^3$ (h) $(2/5)^2$.

Ans. (a) 8   (b) 16   (c) 10,000   (d) .09   (e) .008   (f) 1/16   (g) 1/1000   (h) 4/25

6.30. Find each power: (a) $(-1)^5$,   (b) $(-1)^{82}$,   (c) $(-.3)^3$,   (d) $(-.1)^4$,   **(7.2)**

(e) $(-.12)^2$,   (f) $\left(-\frac{2}{5}\right)^3$,   (g) $\left(-\frac{1}{3}\right)^4$.

Ans. (a) -1 (b) $+1$ (c) $-.027$ (d) $+.0001$ (e) $+.0144$ (f) $-\frac{8}{125}$ (g) $+\frac{1}{81}$

6.31. Find each power: **(7.3)**

(a) 5^2 (d) $\left(\dfrac{1}{3}\right)^2$ (g) -1^{100} (j) $-(-1)^{102}$ (m) $-\left(-\dfrac{2}{3}\right)^2$

(b) $(-5)^2$ (e) $\left(-\dfrac{1}{3}\right)^2$ (h) -1^{101} (k) $-(-1)^{103}$

(c) -5^2 (f) $(-1)^{100}$ (i) $(-1)^{101}$ (l) $-\left(\dfrac{2}{3}\right)^3$

Ans.   (a) 25   (d) $\dfrac{1}{9}$ (g) -1 (j) -1 (m) $-\dfrac{4}{9}$

(b) 25   (e) $\dfrac{1}{9}$ (h) -1 (k) 1

(c) -25 (f) 1   (i) -1 (l) $-\dfrac{8}{27}$

6.32. Complete each: **(7.4)**

(a) $(-5)^? = -125$ (c) $(-?)^4 = +.0001$ (e) $\left(+\dfrac{2}{3}\right)^4 = ?$ (g) $(?)^{171} = -1$

(b) $(+10)^? = 10{,}000$ (d) $(-?)^3 = -.343$ (f) $\left(-\dfrac{1}{2}\right)^5 = ?$ (h) $(?)^{242} = +1$

Ans.   (a) 3   (b) 4   (c) $-.1$ (d) $-.7$ (e) $\dfrac{16}{81}$ (f) $-\dfrac{1}{32}$ (g) -1 (h) $+1$ or -1

6.33. Divide: **(8.1)**

(a) $(+24)$ by $(+3)$ (c) (-49) by (-7) (e) $(+4.8)$ by $(+.2)$ (g) (-18) by (-4)
(b) $(+88)$ by $(+8)$ (d) (-78) by (-6) (f) (-95) by $(-.5)$ (h) $(+8)$ by $(+12)$

Ans.   (a) $+8$ (b) $+11$ (c) $+7$ (d) $+13$ (e) $+24$ (f) $+190$ (g) $+4\dfrac{1}{2}$ (h) $+\dfrac{2}{3}$

6.34. Divide: **(8.2)**

(a) $(+30)$ by (-5) (c) $(+13)$ by (-2) (e) $(+.2)$ by $(-.04)$ (g) (-100) by $(+500)$
(b) (-30) by $(+5)$ (d) (-36) by $(+8)$ (f) (-30) by $(+.1)$ (h) $(+100)$ by (-3)

Ans.   (a) -6 (b) -6 (c) $-6\dfrac{1}{2}$ (d) $-4\dfrac{1}{2}$ (e) -5 (f) -300 (g) $-\dfrac{1}{5}$ (h) $-33\dfrac{1}{3}$

6.35. Divide $+12$ by   (a) $+24$,   (b) $+12$,   (c) $+4$,   (d) -1,   (e) -3,   (f) -48.   **(8.3)**
Divide each by -20:   (g) $+60$,   (h) $+20$,   (i) $+5$,   (j) -255,   (k) -10,   (l) -100.

Ans.   (a) $+\dfrac{1}{2}$ (c) $+3$ (e) -4 (g) -3 (i) $-\dfrac{1}{4}$ (k) $+\dfrac{1}{2}$

(b) $+1$ (d) -12 (f) $-\dfrac{1}{4}$ (h) -1 (j) $+11\dfrac{1}{4}$ (l) $+5$

6.36. Divide: **(8.3)**

(a) $\dfrac{+25}{+5}$ (b) $\dfrac{-25}{+5}$ (c) $\dfrac{-2.5}{-.5}$ (d) $\dfrac{-.25}{+.5}$ (e) $\dfrac{-25}{-.5}$ (f) $\dfrac{-.025}{+.5}$

Ans.   (a) $+5$ (b) -5 (c) $+5$ (d) $-\dfrac{1}{2}$ (e) $+50$ (f) $-.05$

6.37. Multiply and divide as indicated: **(8.4, 8.5)**

$(a)\ \dfrac{(+20)(+12)}{(+3)(+5)}$ $(c)\ \dfrac{(+3)(+6)(+10)}{(+12)(-3)}$ $(e)\ \dfrac{0}{(-17)(-24)}$ $(g)\ \dfrac{(+1)(+2)(+3)}{(-4)(-6)}$

$(b)\ \dfrac{(+20)(-12)}{(-3)(-5)}$ $(d)\ \dfrac{(-3)(-6)(+18)}{(+12)(+3)}$ $(f)\ \dfrac{(-120)(+31)0}{(-5)(-8)(-9)}$ $(h)\ \dfrac{(+.1)(+.2)(-30)}{(-.4)(-.1)}$

Ans. $(a)\ +16$ $(b)\ -16$ $(c)\ -5$ $(d)\ +9$ $(e)\ 0$ $(f)\ 0$ $(g)\ +\dfrac{1}{4}$
$(h)\ -15$

6.38. Evaluate if $y = -3$:   **(9.1)**

$(a)\ 3y + 1$ $(c)\ 2y^2$ $(e)\ (-y)^2$ $(g)\ \dfrac{7y}{3}$ $(i)\ \dfrac{3y + 15}{y}$

$(b)\ 20 - 4y$ $(d)\ 3y^3$ $(f)\ 2 - y^3$ $(h)\ y(y^2 - 2)$ $(j)\ 2y^2 - 5y + 27$

Ans. $(a)\ -8$ $(b)\ 32$ $(c)\ 18$ $(d)\ -81$ $(e)\ 9$ $(f)\ 29$ $(g)\ -7$ $(h)\ -21$ $(i)\ -2$
$(j)\ 60$

6.39. Evaluate if $x = -1$ and $y = +3$:   **(9.2)**

$(a)\ x + y$ $(c)\ x^2 + y^2$ $(e)\ 4xy - x^2$ $(g)\ 3xy^2$ $(i)\ \dfrac{y^2}{6x}$

$(b)\ y - 2x$ $(d)\ 3xy$ $(f)\ x^2y + 10$ $(h)\ x^3 + 10y$ $(j)\ \dfrac{y + x}{y - x}$

Ans. $(a)\ 2$ $(b)\ 5$ $(c)\ 10$ $(d)\ -9$ $(e)\ -13$ $(f)\ 13$ $(g)\ -27$ $(h)\ 29$

$(i)\ -\dfrac{3}{2}$ $(j)\ \dfrac{1}{2}$

6.40. Evaluate if $x = -2$, $y = -1$, and $z = +3$:   **(9.3)**

$(a)\ x + y + z$ $(c)\ x^2 + y^2 + z^2$ $(e)\ 2xyz$ $(g)\ xy + z^2$ $(i)\ \dfrac{2x - 3y}{4z}$

$(b)\ 2x + 2y - 2z$ $(d)\ x^3 - y^2 + z$ $(f)\ xy + yz$ $(h)\ y^2 - 5xz$ $(j)\ \dfrac{x^2 - y^2}{z^2}$

Ans. $(a)\ 0$ $(b)\ -12$ $(c)\ 14$ $(d)\ -6$ $(e)\ 12$ $(f)\ -1$ $(g)\ 11$ $(h)\ 31$

$(i)\ -\dfrac{1}{12}$ $(j)\ \dfrac{1}{3}$

Chapter 7

Fundamentals of Algebra: Laws and Operations

1. RELATING FUNDAMENTALS OF ARITHMETIC AND ALGEBRA

In this first chapter in algebra, we are going to lead you from arithmetic to algebra. Underlying algebra as well as arithmetic are the fundamental ideas that you learned in earlier chapters. As we recall each of the fundamental ideas, we will develop each of them in greater depth and scope. You are now ready to move up to a higher level of mathematics, the level of algebra.

The Four Fundamental Operations

Underlying both arithmetic and algebra are the four fundamental operations:

1. ADDITION (sum)
2. SUBTRACTION (difference)

3. MULTIPLICATION (product)
4. DIVISION (quotient)

The result of each operation is named in parenthesis. The result in subtraction may also be called the *remainder*.

Variables and Constants

A *variable* is a letter or other symbol which holds a place for, or represents, any number in a specified set of numbers. The set whose numbers may replace a variable is called the *replacement set* or the *domain* of the variable.

Thus, if x represents, 1, 10, or any other whole number, then x is a variable and the set of whole numbers is the replacement set or domain of x. Numbers such as 1 and 10 that may be replaced are the *values* of the variable.

A *constant* is a letter or other symbol that represents only one number. For example, 5 and π are constants. If a variable represents a set having a single member, such as $\{5\}$, then the variable may be regarded as a constant.

A *variable expression* is an expression that contains a variable. Thus, if n is a variable, then

$$7n, \quad n+7, \quad \frac{1}{2}n - 5, \quad \frac{3n+1}{5n+1}$$

are variable expressions.

To *evaluate* a variable expression is to find its value for given values of the variable. Thus, using the replacement set $\{1, 2, 5, 10\}$ as the set of given values of x, the set of values of the expression $2x + 1$ is $\{3, 5, 11, 21\}$.

A *formula* is an equation in which a variable is expressed in terms of other variables. Thus, $p = 4s$ is a formula in which the variable p is expressed in terms of the variable s.

Replacing Verbal Statements with Algebraic Equations

In the following examples, note how variable expressions may be used to replace lengthy verbal statements:

VERBAL STATEMENTS	ALGEBRAIC EQUATIONS
1. Seven times a number reduced by the same number equals six times the number.	1. $7n - n = 6n$
2. The sum of twice a number and three times the same number equals five times that number.	2. $2n + 3n = 5n$
3. The perimeter of a square equals four times the length of one of its sides.	3. $p = 4s$

Omitting the multiplication sign, as in $7n$, is the preferred method of indicating multiplication. However, the multiplication sign may not be omitted when showing the multiplication of two numbers. See Chapter 2, Section 1.

Properties of 0 and 1: Additive Identity and Multiplicative Identity

If the domain of the variable n is the set of integers, then each of the following important properties of 0 and 1 is true for each integer. In later work, these properties will be extended to apply to any "real number".

Properties of 0 and 1

VERBAL STATEMENTS	ALGEBRAIC EQUATIONS
Multiplicative Property of Zero 1. If any number is multiplied by 0, the product is 0. Thus, $1357 \times 0 = 0$.	1. $n \cdot 0 = 0$
Additive Property of Zero 2. If 0 is added to any number, the sum is the number. Thus, $1357 + 0 = 1357$. Because identically the same number remains when 0 is added to it, 0 is called the *additive identity*.	2. $n + 0 = n$
Multiplicative Property of One 3. If any number is multiplied by 1, the product is the number. Thus, $-1357 \times 1 = -1357$. Because identically the same number remains when it is multiplied by 1, 1 is called the *multiplicative identity*.	3. $n \cdot 1 = n$

1.1   REPLACING A VERBAL STATEMENT BY AN ALGEBRAIC EQUATION

Using variable expressions, replace each verbal statement by an algebraic equation: (*a*) If six times a number is reduced by the same number, the result must be five times the number. (*b*) The sum of twice a number, three times the same number and four times the same number is equivalent to nine times the number. (*c*) Increasing a number by itself and 20 is the same as doubling the number and adding 20.

Illustrative Solution

(c) $\underbrace{\text{Increasing a number by itself and 20}}_{n + n + 20}$ $\underbrace{\text{is the same as}}_{=}$ $\underbrace{\text{doubling the number}}_{2n}$ $\underbrace{\text{and adding 20.}}_{+20}$

Ans. $n + n + 20 = 2n + 20$

Ans. (a) $6n - n = 5n$ (b) $2n + 3n + 4n = 9n$

1.2   REPLACING A VERBAL RULE BY A FORMULA

Using the initial letters of words as variables, replace each verbal rule by a formula: (a) The perimeter of a square is four times the length of a side. (b) The perimeter of a rectangle is twice the length added to twice the width. (c) The area of a rectangle is the product of its length and width. (d) The selling price of an article is the sum of its cost and its profit.

Illustrative Solution (a) Using p for the perimeter of a square and s for the length of a side, the formula is $p = 4s$.

Ans. (b) $p = 2l + 2w$ (c) $a = lw$ (d) $s = c + p$

1.3   PROPERTIES OF 0 AND 1

Evaluate each numerical expression and state the property of 0 or 1 that applies:

(a) $123{,}000 \times 0$ (d) $1 + 0$ (g) $3 \times 4 \times 0$
(b) $123{,}000 \times 1$ (e) 1×0 (h) $3 \times 4 \times 1$
(c) $123{,}000 + 0$ (f) 1×1

Illustrative Solution

(a) $123{,}000 \times 1 = 123{,}000$ since the product of a number and 1 is the number.
(g) $3 \times 4 \times 0 = 0$ since the product of any number and 0 is 0.

Ans. (a) 0, multiplicative property of 0   (e) 0, multiplicative property of 0 or of 1
       (c) 123,000, additive property of 0   (f) 1, multiplicative property of 1
       (d) 1, additive property of 0   (h) 12, multiplicative property of 1

1.4   EVALUATING VARIABLE EXPRESSIONS USING A REPLACEMENT SET

Using the replacement set $\{1, 2, 5, -10\}$ as the set of values of n, evaluate the following expressions and list these values in a set:

(a) $7n$ (b) $n + 7$ (c) $5 - \dfrac{1}{2}n$ (d) $\dfrac{3n + 1}{5n - 1}$

Illustrative Solution (d) If $n = 1$, $3n + 1 = 3 \cdot 1 + 1 = 4$ and $5n - 1 = 5 \cdot 1 - 1 = 4$. Hence,

$$\frac{3n + 1}{5n - 1} = \frac{4}{4} = 1$$

If $n = 2, 5,$ and -10, then

$$\frac{3n + 1}{5n - 1} = \frac{7}{9} \qquad \frac{3n + 1}{5n - 1} = \frac{16}{24} = \frac{2}{3} \qquad \frac{3n + 1}{5n - 1} = \frac{29}{51}$$

respectively. List these values in a set: $\{1, 7/9, 2/3, 31/49\}$ *Ans.*

Ans. (a) $\{7, 14, 35, 70\}$ (b) $\{8, 9, 12, 17\}$ (c) $\{4\frac{1}{2}, 4, 2\frac{1}{2}, 0\}$

2.   COMMUTATIVE LAW OF ADDITION

Addends are numbers being added. Their *sum* is the answer obtained. Thus, in $5 + 3 = 8$, the addends are 5 and 3. Their sum is 8.

Numerical addends are numbers used as addends. Thus, in $3 + 4 + 6 = 13$, 3, 4 and 6 are numerical addends.

Literal addends are variables which represent numbers being added. Thus, in $a + b = 8$, a and b are literal addends.

Commutative Law of Addition

Interchanging addends does not change their sum.

$$a + b = b + a$$
$$a + b + c = b + c + a$$

Thus, $2 + 3 = 3 + 2$ and $3 + 4 + 6 = 4 + 6 + 3$.

Applications of the Commutative Law

1. **To simplify addition.**   Thus, $25 + 82 + 75$ by interchanging becomes $25 + 75 + 82$. The sum is $100 + 82 = 182$.
2. **To check addition.**   Thus, numbers may be added downwards and checked upwards:

Add Down	Check Up
148	148
357	357
762	762
1267	1267

3. **To rearrange addends in a preferred order.**   Thus, $b + c + a$ becomes $a + b + c$ if the literal addends are to be arranged alphabetically. Also, $3 + x$ becomes $x + 3$ if the literal addend is to precede the numerical addend.

2.1   INTERCHANGING ADDENDS TO SIMPLIFY ADDITION

Simplify each addition by interchanging addends:

(a) $20 + 73 + 280$ (c) $\frac{3}{4} + 2\frac{1}{2} + 1\frac{1}{4}$ (e) $1.95 + 2.65 + .05 + .35$

(b) $141 + 127 + (-41)$ (d) $1\frac{1}{2} + 2\frac{2}{7} + \frac{1}{2} + \frac{1}{7}$ (f) $9.4 + 18.7 + 1.3 + .6$

Solutions

(a) $20 + 280 + 73$

$\quad 300 + 73 = 373$

(b) $141 + 127 + (-41)$
$= 141 + (-41) + 127$
$= 100 + 127 = 227$

(c) $\frac{3}{4} + 1\frac{1}{4} + 2\frac{1}{2}$

$\quad 2 + 2\frac{1}{2} = 4\frac{1}{2}$

(d) $1\frac{1}{2} + \frac{1}{2} + 2\frac{2}{7} + \frac{1}{7}$
$\quad 2 + 2\frac{3}{7} = 4\frac{3}{7}$

(e) $1.95 + .05 + 2.65 + .35$

$\quad 2 + 3 = 5$

(f) $9.4 + .6 + 18.7 + 1.3$
$\quad 10 + 20 = 30$

2.2   REARRANGING ADDENDS TO OBTAIN A PREFERRED ORDER

Rearrange the addends so that literal addends are arranged alphabetically and precede numerical addends:

(a) $3 + b$ (c) $d + 10 + e$ (e) $15 + x + 10$ (g) $w + y + x$

(b) $c + a$ (d) $c + 12 + b$ (f) $20 + s + r$ (h) $b + 8 + c + a$

Illustrative Solution   (d) Arrange the literal addends alphabetically; thus, $b + c$. Since the literal addends precede the numerical addend 12, the required result is $b + c + 12$.

Ans.   (a) $b + 3$ (c) $d + e + 10$ (f) $r + s + 20$ (h) $a + b + c + 8$

(b) $a + c$ (e) $x + 25$ (g) $w + x + y$

3. COMMUTATIVE LAW OF MULTIPLICATION

Factors are numbers being multiplied. Their *product* is the answer obtained. Thus, in $5 \times 3 = 15$, the factors are 5 and 3. Their product is 15.

Numerical factors are numbers used as factors. Thus, in $2 \times 3 \times 5 = 30$, 2, 3 and 5 are numerical factors.

Literal factors are variables which represent numbers being multiplied. Thus, in $ab = 20$, a and b are literal factors.

Commutative Law of Multiplication

Interchanging factors does not change their product.

$$ab = ba$$

$$cba = abc$$

Thus, $2 \times 5 = 5 \times 2$ and $2 \times 4 \times (-5) = (-5)2 \times 4 \times 2$.

Applications of the Commutative Law

1. **To simplify multiplication**. Thus, $4 \times 13 \times 25$ by interchanging becomes $4 \times 25 \times 13$. The product is $100 \times 13 = 1300$.
2. **To check multiplication**. Thus, since $24 \times 75 = 75 \times 24$,

	24	Check:	75
	×75		×24
	120		300
	168		150
	1800		1800

3. **To rearrange factors in a preferred order**. Thus, bca becomes abc if the literal factors are arranged alphabetically. Also, $x3$ becomes $3x$ if the numerical factor is to precede the literal factor.

3.1 INTERCHANGING FACTORS TO SIMPLIFY MULTIPLICATION

Simplify each multiplication by interchanging factors:

(*a*) $2 \times 17 \times 5$ (*c*) $7\frac{1}{2} \times 7 \times 4$ (*e*) $1.25 \times 4.4 \times 4 \times 5$

(*b*) $25 \times 19 \times (-4) \times 2$ (*d*) $33\frac{1}{3} \times 23 \times 3$ (*f*) $.33 \times 225 \times 3\frac{1}{3} \times 4$

Solutions

(*a*) $2 \times 5 \times 17$ (*c*) $7\frac{1}{2} \times 4 \times 7$ (*e*) $1.25 \times 4 \times 4.4 \times 5$

    $10 \times 17 = 170$ $30 \times 7 = 210$ $5 \times 22 = 110$

(*b*) $25 \times (-4) \times 19 \times 2$ (*d*) $33\frac{1}{3} \times 3 \times 23$ (*f*) $.33 \times 3\frac{1}{3} \times 225 \times 4$

    $-100 \times 38 = -3800$ $100 \times 23 = 2300$ $1.1 \times 900 = 990$

3.2 REARRANGING FACTORS TO OBTAIN A PREFERRED ORDER

Rearrange the factors so that literal factors are arranged alphabetically and follow numerical factors:

(*a*) $b3$ (*b*) ca (*c*) $d10e$ (*d*) $c12b$ (*e*) $-15x10$ (*f*) $20sr$ (*g*) wyx (*h*) $b35ca$

Illustrative Solution   (*e*) The numerical factors -15 and 10 should precede the literal factor x. Hence, the result is $-15 \cdot 10 \cdot x$ or $-150x$.

Ans.   (*a*) $3b$ (*b*) ac (*c*) $10de$ (*d*) $12bc$ (*f*) $20rs$ (*g*) wxy (*h*) $35abc$

4. SYMBOLIZING THE FUNDAMENTAL OPERATIONS IN ALGEBRA: DIVISION BY ZERO

The symbols for the fundamental operations are as follows:

1. ADDITION: +   3. MULTIPLICATION: $\times$, ( ), $\cdot$, no sign
2. SUBTRACTION: −   4. DIVISION: $\div$; :, fraction bar

Thus,

$n + 4$ means "add n and 4" $4 \times n$, $4(n)$, $4 \cdot n$ mean "multiply n and 4"

$n - 4$ means "subtract 4 from n" $n \div 4$, $n : 4$, $\dfrac{n}{4}$ $n/4$ mean "divide n by 4"

Rule: Division by zero is an impossible operation.

Thus, $4 \div 0$ or $x \div 0$ is meaningless. Also, $4/n$ is meaningless if $n = 0$.

4.1 SYMBOLS FOR MULTIPLICATION AND DIVISION

Symbolize each, avoiding multiplication signs where possible:

(a) 8 times −11   (c) b times c (e) 5 multiplied by a and the result divided by b
(b) 8 times x (d) 8 divided by x (f) d divided by the product of 7 and e

Ans. (a) $8 \times (-11)$, $8 \cdot (-11)$, $8(-11)$ or $(8)(-11)$ (c) $b \cdot c$ or bc (avoid $b \times c$)   (e) $\dfrac{5a}{b}$ or $5a \div b$

   (b) $8 \cdot x$ or $8x$ (avoid $8 \times x$)   (d) $\dfrac{8}{x}$ or $8 \div x$ $\left(\dfrac{8}{x}\text{ is preferred}\right)$ (f) $\dfrac{d}{7e}$ or $d \div (7e)$

4.2 DIVISION BY ZERO: FRACTIONS WHOSE DENOMINATOR HAS A SINGLE VARIABLE

When is each division impossible? Give a reason for your answer.

(a) $\dfrac{5}{a}$ (b) $\dfrac{7}{2b}$ (c) $\dfrac{2}{c-6}$ (d) $\dfrac{10}{3d-9}$ (e) $\dfrac{3x}{35-7e}$

Illustrative Solution   (c) Division is impossible if $c - 6$, the denominator, equals 0. If $c - 6 = 0$, $c = 6$ *Ans.*

Ans. (a) $a = 0$ (b) $b = 0$ (d) $d = 3$ (e) $e = 5$

4.3 DIVISION BY ZERO: FRACTIONS WHOSE DENOMINATOR HAS MORE THAN ONE VARIABLE

When is each division impossible? Give a reason for your answer.

(a) $\dfrac{a}{xy}$ (b) $\dfrac{3x}{2ab}$ (c) $\dfrac{10}{x-y}$ (d) $\dfrac{x}{y-2z}$

Illustrative Solution   (b) Division is impossible if $2ab$, the denominator, equals 0. If $2ab = 0$, $a = 0$ or $b = 0$ *Ans.*

Ans. (a) $x = 0$ or $y = 0$ (c) $x = y$ (d) $y = 2z$

5. EXPRESSING ADDITION AND SUBTRACTION ALGEBRAICALLY

In algebra, changing verbal expressions into algebraic expressions is of major importance. The operations of addition and subtraction are denoted by words such as the following:

WORDS DENOTING ADDITION		WORDS DENOTING SUBTRACTION	
sum	more than	difference	less than
plus	greater than	minus	smaller than
gain	larger than	lose	fewer than
increase	enlarge	decrease	shorten
rise	grow	drop	depreciate
expand	augment	lower	diminish

The Commutative Law applies to addition but does *not* apply to subtraction. Thus, "the sum of n and 20" may be represented by either $n + 20$ or $20 + n$. But "20 minus a number" may be represented only by $20 - n$ and *not* by $n - 20$.

5.1   EXPRESSING ADDITION ALGEBRAICALLY

If n represents a number, express algebraically:

(*a*) the sum of the number and 7   (*c*) the number increased by 9   (*e*) 20 enlarged by the number
(*b*) the number plus 8   (*d*) 15 plus the number   (*f*) 25 augmented by the number

Illustrative Solution   (*e*) Use + for "enlarged" to obtain $20 + n$ or, by interchanging addends, $n + 20$.

Ans.   (*a*) $n + 7$ or $7 + n$ (*c*) $n + 9$ or $9 + n$ (*f*) $25 + n$ or $n + 25$
              (*b*) $n + 8$ or $8 + n$ (*d*) $15 + n$ or $n + 15$

5.2   EXPRESSING SUBTRACTION ALGEBRAICALLY

If n represents a number, express algebraically: (*a*) the difference if the number is subtracted from 15, (*b*) the number diminished by 20, (*c*) 25 less than the number, (*d*) 25 less the number, (*e*) the difference if 15 is subtracted from the number, (*f*) 50 subtracted from the number, (*g*) the number subtracted from 50, (*h*) the number reduced by 75.

Illustrative Solution   (*a*) Express algebraically as $15 - n$. (Do not use $n - 15$.)

Ans.   (*b*) $n - 20$ (*d*) $25 - n$ (*f*) $n - 50$ (*h*) $n - 75$
              (*c*) $n - 25$ (*e*) $n - 15$ (*g*) $50 - n$.

5.3   CHANGING VERBAL EXPRESSIONS INTO ALGEBRAIC EXPRESSIONS

Express algebraically: (*a*) the no. of km of a weight that is 10 km heavier than w km, (*b*) the no. of mi in a distance that is 40 mi farther than d mi, (*c*) the no. of degrees in a temperature 50° hotter than t degrees, (*d*) the no. of dollars in a price \$60 cheaper than p dollars, (*e*) the no. of mph in a speed 30 mph faster than r mph, (*f*) the no. of ft in a length of l ft expanded 6 ft, (*g*) the no. of oz in a weight that is 10 oz lighter than w oz.

Illustrative Solution

(*d*) Since "cheaper" means fewer dollars, use − to express it. *Ans.* $p - 60$
(*f*) Since "expanded" means greater length, use + to express it. *Ans.* $l + 6$

Ans.   (*a*) $w + 10$ (*b*) $d + 40$ (*c*) $t + 50$ (*e*) $r + 30$ (*g*) $w - 10$

6. EXPRESSING MULTIPLICATION AND DIVISION ALGEBRAICALLY

WORDS DENOTING MULTIPLICATION		WORDS DENOTING DIVISION	
multiplied by	double	divided by	ratio
times	triple or treble	quotient	half
product	quadruple		
twice	quintuple		

The Commutative Law applies to multiplication but does *not* apply to division. Thus, "the product of n and 10" may be represented by either $n10$ or $10n$ (the latter is preferred). But "a number divided by 20" may be represented only by $n/20$ and *not* by $20/n$.

6.1 EXPRESSING MULTIPLICATION OR DIVISION

State verbal expressions that may be represented by each of the following:

(a) $-5x$　　　(b) $\dfrac{y}{5}$　　　(c) $\dfrac{5w}{7}$

Ans.　(a) (1) -5 multiplied by x　　(b) (1) y divided by 5　　(c) (1) five-sevenths of w
　　　　　　(2) x multiplied by -5　　　　(2) quotient of y and 5　　　　(2) $5w$ divided by 7
　　　　　　(3) -5 times x　　　　　　　(3) ratio of y to 5　　　　　　(3) quotient of $5w$ divided by 7
　　　　　　(4) product of -5 and x　　　(4) one-fifth of y　　　　　　(4) ratio of $5w$ to 7

6.2 EXPRESSING DIVISION ALGEBRAICALLY

If n represents a number, express algebraically in the form of a fraction:
(a) the quotient of the number and 10　　　　(c) twice the number divided by 7
(b) the ratio of 10 to the number　　　　　　(d) 20 divided by the product of the number and 3

Illustrative Solution　(b) "Ratio" denotes division. Hence, express the ratio of 10 to the number in fraction form as $10/n$. (Do not use $n/10$.)

Ans.　(a) $\dfrac{n}{10}$　　(c) $\dfrac{2n}{7}$　　(d) $\dfrac{20}{3n}$

7. EXPRESSING ALGEBRAICALLY STATEMENTS INVOLVING TWO OR MORE OPERATIONS

In addition to their use in signed numbers, parentheses ( ) are used to indicate that an expression is to be treated as a single number. Thus, "double the sum of 4 and x" is written as $2(4 + x)$.

7.1 EXPRESSING ALGEBRAICALLY EXPRESSIONS INVOLVING TWO OPERATIONS

Express algebraically: (a) a increased by twice b, (b) twice the sum of a and b, (c) 30 decreased by three times c, (d) three times the difference of 30 and c, (e) 50 minus the product of 10 and p, (f) the product of 50 and the sum of p and 10, (g) 100 increased by the quotient of x and y, (h) the quotient of x and the sum of y and 100, (i) the average of s and 20.

Illustrative Solutions

(b) Express the sum of a and b as $(a + b)$, using parentheses. The correct answer is $2(a + b)$, not $2a + b$.

(i) The average of s and 20 is one-half their sum.　*Ans.* $\dfrac{1}{2}(s + 20)$ or $\dfrac{s + 20}{2}$

Ans.　(a) $a + 2b$　　　(d) $3(30 - c)$　　　(f) $50(p + 10)$　　　(h) $\dfrac{x}{7 + 100}$

　　　　(c) $30 - 3c$　　　(e) $50 - 10p$　　　(g) $100 + \dfrac{x}{y}$

7.2   EXPRESSING ALGEBRAICALLY MORE DIFFICULT EXPRESSIONS

Express algebraically: (a) half of a, increased by the product of 25 and b;   (b) four times c, decreased by one-fifth of d;   (c) half the sum of m and twice n;   (d) the average of m, r and 80;   (e) 60 diminished by one-third the product of 7 and x;   (f) twice the sum of e and 30, diminished by 40;   (g) two-thirds the sum of n and three-sevenths of p;   (h) the product of a and b, decreased by twice the difference of c and d;   (i) the quotient of x and 10, minus four times their sum.

Illustrative Solutions

(d) The average of three numbers is one-third of their sum.   *Ans.* $(m + r + 80)/3$

(h) Express the difference of c and d as $(c - d)$, using parentheses. However, the product of a and b may be expressed as ab without parentheses.   *Ans.* $ab - 2(c - d)$

Ans.   (a) $\dfrac{a}{2} + 25b$ (c) $\dfrac{m + 2n}{2}$ or $\dfrac{1}{2}(m + 2n)$ (f) $2(e + 30) - 40$ (i) $\dfrac{x}{10} - 4(x + 10)$

(b) $4c - \dfrac{d}{5}$ (e) $60 - \dfrac{7x}{3}$ (g) $\dfrac{2}{3}\left(n + \dfrac{3p}{7}\right)$

7.3   CHANGING VERBAL EXPRESSIONS INTO ALGEBRAIC EXPRESSIONS

Express algebraically: (a) a speed in mph that is 30 mph faster than twice another of r mph,   (b) a weight in lb that is 20 lb lighter than 3 times another of w lb,   (c) a temperature in degrees that is 15° colder than two-thirds another of t°,   (d) a price in cents that is 25¢ cheaper than another of D dollars,   (e) a length in inches that is 8 in. longer than another of f ft.

Illustrative Solutions

(d) D dollars equals 100D cents. "25¢ cheaper" means 25¢ less.   *Ans.* $100D - 25$

(e) f feet equals 12f inches. "8 in. longer" means 8 inches more.   *Ans.* $12f + 8$

Ans.   (a) $2r + 30$ (b) $3w - 20$ (c) $\dfrac{2t}{3} - 15$

8.   ASSOCIATIVE LAWS OF ADDITION AND MULTIPLICATION

Associative Law of Addition

The way in which quantities are added in groups of two does not change their sum.

$$a + b + c = (a + b) + c = a + (b + c)$$

Thus, the sum $2 + 3 + 5$ may be found by obtaining *partial sums* in two ways:

(1) Add 2 and 3 to obtain a partial sum of 5, then add 5 and 5:

$$2 + 3 + 5 = (2 + 3) + 5 = 5 + 5 = 10$$

(2) Add 3 and 5 to obtain a partial sum of 8, then add 2 and 8:

$$2 + 3 + 5 = 2 + (3 + 5) = 2 + 8 = 10$$

Hence, $2 + 3 + 5 = (2 + 3) + 5 = 2 + (3 + 5)$.

Associative Law of Multiplication

The way in which quantities are multiplied in groups of two does not change their product.

$$abc = (ab)c = a(bc)$$

Thus, the product $2 \cdot 3 \cdot 5$ may be found by obtaining *partial products* in two ways:

(1) Multiply 2 and 3 to obtain a partial product of 6, then multiply 6 and 5:

$$2 \cdot 3 \cdot 5 = (2 \cdot 3) \cdot 5 = 6 \cdot 5 = 30$$

(2) Multiply 3 and 5 to obtain a partial product of 15, then multiply 2 and 15:

$$2 \cdot 3 \cdot 5 = 2 \cdot (3 \cdot 5) = 2 \cdot 15 = 30$$

Hence, $2 \cdot 3 \cdot 5 = (2 \cdot 3) \cdot 5 = 2 \cdot (3 \cdot 5)$.

A sum may be simplified by using both the Commutative and Associative Laws of Addition:

$$25 + (467 + 175) = 25 + (175 + 467) \text{ Commutative Law of Addition)}$$
$$= (25 + 175) + 467 \text{ (Associative Law of Addition)}$$
$$= 200 + 467 = 667 \ Ans.$$

A product may be simplified by using both the Commutative and Associative Laws of Multiplication:

$$\left(\frac{1}{4} \times 35\right) \times 400 = \left(35 \times \frac{1}{4}\right) \times 400 \text{ (Commutative Law of Multiplication)}$$
$$= 35 \times \left(\frac{1}{4} \times 400\right) \text{ (Associative Law of Multiplication)}$$
$$= 35 \times 100 = 3500 \ Ans.$$

8.1   VERIFYING THE ASSOCIATIVE LAWS FOR THREE NUMBERS

If $a = 2$, $b = 3$, $c = 5$, and $d = 10$, show that

(a) $(a+b)+c = a+(b+c)$ (d) $(ab)c = a(bc)$
(b) $(a+c)+d = a+(c+d)$ (e) $(ac)d = a(cd)$
(c) $(d+b)+c = d+(b+c)$ (f) $(db)c = d(bc)$

Solutions

(a) $(5)+5 = 2+(8)$ and $10 = 10$ (d) $(6)5 = 2(15)$ and $30 = 30$
(b) $(7)+10 = 2+(15)$ and $17 = 17$ (e) $(10)10 = 2(50)$ and $100 = 100$
(c) $(13)+5 = 10+(8)$ and $18 = 18$ (f) $(30)5 = 10(15)$ and $150 = 150$

8.2   VERIFYING THE ASSOCIATIVE LAWS FOR FOUR NUMBERS

If $a = 1$, $b = 5$, $c = 10$, $d = 20$, and $e = 100$, show that

(a) $(a+b+c)+d = a+(b+c+d)$ (d) $(abc)d = a(bcd)$
(b) $(b+c)+(d+e) = b+(c+d+e)$ (e) $(bc)(de) = b(cde)$
(c) $(d+a+c)+e = (d+a)+(c+e)$ (f) $(dac)e = (da)(ce)$

Solutions

(a) $(16)+20 = 1+(35)$ and $36 = 36$ (d) $(50)20 = 1(1000)$ and $1000 = 1000$
(b) $(15)+(120) = 5+(130)$ and $135 = 135$ (e) $(50)(2000) = 5(20,000)$ and $100,000 = 100,000$
(c) $(31)+100 = (21)+(110)$ and $131 = 131$ (f) $(200)100 = (20)(1000)$ and $20,000 = 20,000$

9.   ORDER IN WHICH FUNDAMENTAL OPERATIONS ARE PERFORMED

In evaluating an expression, the operations involved must be performed in a certain order. Note in the following, how *multiplication and division must precede addition and subtraction*!

To Evaluate a Numerical Expression Not Containing Parentheses

Evaluate: (a) $3 + 4 \times 2$ (b) $5 \times 4 - 18 \div 6$

PROCEDURE   SOLUTIONS

1. Do **multiplications and division (M & D)** in order   1. $3 + 4 \times 2$ $5 \times 4 - 18 \div 6$
   from left to right: $3 + 8$ $20 - 3$
2. Do remaining **additions and subtractions (A & S)** in   2. $11 \ Ans.$ $17 \ Ans.$
   order from left to right:

To Evaluate an Algebraic Expression Not Containing Parentheses

Evaluate $x + 2y - \dfrac{z}{5}$ when $x = 5$, $y = 3$, $z = 20$.

PROCEDURE	SOLUTION
1. **Substitute** the value given for each variable:	1. $x + 2y - \dfrac{z}{5}$ $5 + 2(3) - \dfrac{20}{5}$
2. Do **multiplications and divisions (M & D)** in order from left to right:	2. $5 + 6 - 4$
3. Do remaining **additions and subtractions (A & S)** in order from left to right:	3. 7 *Ans.*

To Evaluate an Algebraic Expression Containing Parentheses

Evaluate: $2(a + b) + 3a - \dfrac{b}{2}$ if $a = 7$, $b = 2$.

PROCEDURE	SOLUTION
1. **Substitute** the value given for each variable:	1. $2(7 + 2) + 3 \cdot 7 - \dfrac{2}{2}$
2. **Evaluate inside parentheses**:	2. $2 \cdot 9 + 3 \cdot 7 - \dfrac{2}{2}$
3. Do **multiplications and divisions (M & D)** in order from left to right:	3. $18 + 21 - 1$
4. Do remaining **additions and subtractions (A & S)** in order from left to right:	4. 38 *Ans.*

9.1   EVALUATING NUMERICAL EXPRESSIONS

Evaluate:
(a) $24 \div 4 + 8$,   (b) $24 + 8 \div 4$,   (c) $8 \times 6 - 10 \div 5 + 12$.

Solutions

	(a)	(b)	(c)
1. Do M & D:	$24 \div 4 + 8$	$24 + 8 \div 4$	$8 \times 6 - 10 \div 5 + 12$
2. Do A & S:	$6 + 8$	$24 + 2$	$48 - 2 + 12$
	14 *Ans.*	26 *Ans.*	58 *Ans.*

9.2   EVALUATING ALGEBRAIC EXPRESSIONS

Evaluate if $a = 8$, $b = 10$, $x = 3$:

(a) $4b - \dfrac{a}{4}$ (b) $12x + ab$ (c) $\dfrac{3a}{4} + \dfrac{4b}{5} - \dfrac{2x}{3}$

Solutions

	(a) $4b - \dfrac{a}{4}$	(b) $12x + ab$	(c) $\dfrac{3a}{4} + \dfrac{4b}{5} - \dfrac{2x}{3}$
1. Substitute:	$4 \times 10 - \dfrac{8}{4}$	$12 \times 3 + 8 \times 10$	$\dfrac{3}{4} \times 8 + \dfrac{4}{5} \times 10 - \dfrac{2}{3} \times 3$
2. Do M & D:	$40 - 2$	$36 + 80$	$6 + 8 - 2$
3. Do A & S:	38 *Ans.*	116 *Ans.*	12 *Ans.*

9.3   EVALUATING NUMERICAL EXPRESSIONS CONTAINING PARENTHESES

Evaluate:
(a) $3(4 - 2) + 12$ (b) $7 - \dfrac{1}{2}(14 - 6)$ (c) $8 + \dfrac{1}{3}(4 + 2)$ (d) $20 - 5(4 - 1)$

Solutions

1. Substitute: (a) $3(4-2)+12$ (b) $7-\frac{1}{2}(14-6)$ (c) $8+\frac{1}{3}(4+2)$ (d) $20-5(4-1)$

2. Do ( ): $3 \cdot 2 + 12$ $7-\frac{1}{2} \cdot 8$ $8+\frac{1}{3} \cdot 6$ $20-5 \cdot 3$

3. Do M & D: $6+12$ $7-4$ $8+2$ $20-15$
4. Do A & S: 18 *Ans.*   3 *Ans.*   10 *Ans.*   5 *Ans.*

9.4   EVALUATING ALGEBRAIC EXPRESSIONS CONTAINING PARENTHESES

Evaluate if $a=10$, $b=2$, $x=12$:

(a) $3(x+2b)-30$ (b) $8+2\left(\frac{a}{b}+x\right)$ (c) $3x-\frac{1}{2}(a+b)$

Solutions

$\qquad$ (a) $3(x+2b)-30$ (b) $8+2\left(\frac{a}{b}+x\right)$ (c) $3x-\frac{1}{2}(a+b)$

1. Substitute: $3(12+2 \cdot 2)-30$ $8+2\left(\frac{10}{2}+12\right)$ $3 \cdot 12-\frac{1}{2}(10+2)$

2. Do ( ): $3 \cdot 16-30$ $8+2 \cdot 17$ $36-\frac{1}{2} \cdot 12$

3. Do M & D: $48-30$ $8+34$ $36-6$
4. Do A & S: 18 *Ans.*   42 *Ans.*   30 *Ans.*

9.5   EVALUATING WHEN THE VALUE OF ONE OF THE VARIABLES IS ZERO

Evaluate if $w=4$, $x=2$, and $y=0$:

(a) $wx+y$ (b) $w+xy$ (c) $\frac{w+y}{x}$ (d) $\frac{xy}{w}$ (e) $\frac{x}{w+y}$ (f) $\frac{wx}{y}$

Solutions

$\qquad$ (a) $wx+y$ (b) $w+xy$ (c) $\frac{w+y}{x}$ (d) $\frac{xy}{w}$ (e) $\frac{x}{w+y}$ (f) $\frac{wx}{y}$

1. Substitute: $4 \times 2+0$ $4+2 \times 0$ $\frac{4+0}{2}$ $\frac{2 \times 0}{4}$ $\frac{2}{4+0}$ has no meaning if $y=0$. *Ans.*

2. Do M & D: $8+0$ $4+0$ $\frac{4}{2}$ $\frac{0}{4}$ $\frac{2}{4}$

3. Do A & S: 8 *Ans.*   4 *Ans.*   2 *Ans.*   0 *Ans.* $\frac{1}{2}$ *Ans.*

10.   TERMS, FACTORS, AND COEFFICIENTS

A *term* is a number, a variable, or the product or quotient of numbers and variables. Thus, 5, $8y$, cd, $3wx$ and $2/3r$ are terms. An *expression* consists of one or more terms connected by plus or minus signs. Thus, the expression $8y+5$ consists of two terms, $8y$ and 5.

A *factor of a term* is any of the numbers or variables multiplied to form the term. Thus, 8 and y are factors of the term $8y$; 2 is also a factor, though not a *displayed* factor.

Any factor or group of factors of a term is the *coefficient* of the product of the remaining factors. Thus, in $3abc$, 3 is the *numerical coefficient* of abc, while abc is the *literal coefficient* of 3. Note that the literal coefficient is the product of the variable factors.

10.1   TERMS IN EXPRESSIONS

State the terms in each expression:

(a) $8abc$ (c) $8a + bc$ (e) $3 + bcd$

(b) $-8 + a + bc$ (d) $3b + c + d$ (f) $3(b + c) + d$

Illustrative Solutions

(b) The two plus signs separate $-8 + a + bc$ into 3 terms: -8, a, and bc.

(f) 3 and $(b + c)$ are factors of the term $3(b + c)$. There are 2 terms: $3(b + c)$ and d.

Ans. (a) $8abc$ (c) $8a, bc$ (d) $3b, c, d$ (e) $3, bcd$

10.2   FACTORS OF TERMS

State the factors of the following terms, disregarding 1 and the term itself.

(a) 21   (c) rs (e) $\frac{1}{3}m$ (g) $\frac{n+3}{5}$

(b) 121   (d) $5cd$ (f) $\frac{n}{-5}$ (h) $3(x + 2)$

Illustrative Solutions

(a) Since $21 = 3 \cdot 7$, 3 and 7 are factors.

(d) $5cd$ is the product of the factors 5, c, and d.

(g) $\frac{n+3}{5}$ is the product of the factors $\frac{1}{5}$, and $(n + 3)$

Ans. (b) 11, 11   (c) r, s (e) $1/3, m$ (f) $-1/5, n$ (h) $3, (x + 2)$

10.3   NUMERICAL AND LITERAL COEFFICIENT

State each numerical and literal coefficient:

(a) y (b) $\frac{4x}{5}$ (c) $\frac{w}{7}$ (d) $.7abc$ (e) $8(a + b)$

Solutions

	(a) y	(b) $\frac{4x}{5}$	(c) $\frac{w}{7}$	(d) $.7abc$	(e) $8(a + b)$
Numerical Coefficient:	1	$\frac{4}{5}$	$\frac{1}{7}$	.7	8
Literal Coefficient:	y	x	w	abc	$(a + b)$

11.   REPEATED MULTIPLYING OF A FACTOR: BASE, EXPONENT, AND POWER

$$\text{BASE}^{EXPONENT} = \text{POWER}$$

In $2 \cdot 2 \cdot 2 \cdot 2 \cdot 2$, the factor 2 is being multiplied repeatedly. This may be written in a shorter form as 2^5, where the repeated factor 2 is the *base* while the small 5 written above and to the right of 2 is the *exponent*. The answer 32 is called the fifth *power* of 2.

An exponent is a number which indicates how many times another number, the base, is being used as a repeated factor. The power is the answer obtained. Thus, since $3 \cdot 3 \cdot 3 \cdot 3$, or 3^4, equals 81, 3 is the base, 4 is the exponent and 81 is the fourth power of 3. Table 7-1 gives the first five powers of the most frequently used numerical bases.

Table 7-1. Powers

EXPONENT

		1	2	3	4	5
	1	1	1	1	1	1
	2	2	4	8	16	32
	3	3	9	27	81	243
BASE	4	4	16	64	256	1024
	5	5	25	12	625	3125
	10	10	100	1,000	10,000	100,000

Variable Bases: Squares and Cubes

The area of a square with side s is found by multiplying s by s. This may be written as $A = s^2$ and read "Area equals s-square." Here, A is the second power of s. See Fig. 7-1(a).

The volume of a cube with side s is found by multiplying s three times; that is, $s \cdot s \cdot s$. This may be written as $V = s^3$ and read "Volume equals s-cube." Here, V is the third power of s. See Fig. 7-1(b).

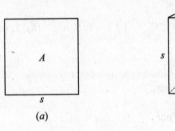
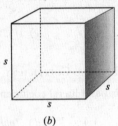

(a)   (b)

Fig. 7-1

Reading Powers

b^2 is read as "b-square," "b to the second power," "b-second" or "b to the second." x^3 is read as "x-cube," "x to the third power," "x-third" or "x to the third."

11.1 WRITING AS BASES AND EXPONENTS

Write each, using bases and exponents:

(a) $5 \cdot 5 \cdot 5$ (c) $2 \cdot 8 \cdot 8 \cdot 8 \cdot 8$ (e) $bbccc$ (g) $(3y)(3y)(3y)$ (i) $7rr(s-8)$ (k) $\dfrac{7w}{xx}$

(b) $3 \cdot 3 \cdot 7 \cdot 7$ (d) $bbbbb$ (f) $12bccd$ (h) $2(a+b)(a+b)$ (j) $\dfrac{yy}{x}$ (l) $\dfrac{2ttt}{5vvvv}$

Illustrative Solutions

(g) Since $(3y)$ is repeated as a factor 3 times, $(3y)(3y)(3y) = (3y)^3$ *Ans.*

(h) Since $(a+b)$ is repeated as a factor twice, $2(a+b)(a+b) = 2(a+b)^2$ *Ans.*

Ans.   (a) 5^3 (c) $2 \cdot 8^4$ (e) b^2c^3 (i) $7r^2(s-8)$ (k) $\dfrac{7w}{x^2}$

         (b) $3^2 7^2$ (d) b^5 (f) $12bc^2d$ (j) $\dfrac{y^2}{x}$ (l) $\dfrac{2t^3}{5v^4}$

11.2 WRITING WITHOUT EXPONENTS

Write each without exponents:

(a) 2^6 (d) x^5 (g) $(2x)^3$ (j) $\dfrac{a^5}{b^2}$

(b) $3 \cdot 4^2$ (e) $10y^4z^2$ (h) $6(5y)^2$ (k) $\dfrac{2(a+b)^2}{c^5}$

(c) $5 \cdot 7^3 \cdot 8$ (f) $8rs^2t^3$ (i) $4(a-b)^2$ (l) $\dfrac{a^2+b^2}{c^3-d^3}$

Illustrative Solutions:

(e) In $10y^4z^2$, y is repeated as a factor 4 times and z twice.   *Ans.* $10yyyyzz$

(k) In $\dfrac{2(a+b)^2}{c^5}$, $(a+b)$ is repeated as a factor twice and c 5 times.   *Ans.* $\dfrac{2(a+b)(a+b)}{ccccc}$

Ans.   (a) $2 \cdot 2 \cdot 2 \cdot 2 \cdot 2 \cdot 2$ (d) $xxxxx$ (h) $6(5y)(5y)$ (l) $\dfrac{aa+bb}{ccc-ddd}$

(b) $3 \cdot 4 \cdot 4$ (f) $8rssttt$ (i) $4(a-b)(a-b)$

(c) $5 \cdot 7 \cdot 7 \cdot 7 \cdot 8$ (g) $(2x)(2x)(2x)$ (j) $\dfrac{aaaaa}{bb}$

11.3 EVALUATING POWERS

Evaluate (Table 7-1 may be used to check values):

(a) 3^5 (d) 2^2+3^2 (g) $1^2 \cdot 1^3 \cdot 1^4$ (j) $10+3 \cdot 2^2$ (m) $(3+4^2)(3^3-5^2)$

(b) 5^4 (e) 2^3+3^3 (h) $2^3 5^2$ (k) $8 \cdot 10^2-3^3$ (n) $\dfrac{4^4}{2^5}$

(c) 10^3 (f) 10^4-4^4 (i) $\frac{1}{2} \cdot 2^4 \cdot 3^2$ (l) $\frac{1}{2} \cdot 4^2-\frac{1}{3} \cdot 3^2$ (o) $\dfrac{3^3+2^5}{10^2}$

Solutions

(a) $3 \cdot 3 \cdot 3 \cdot 3 \cdot 3 = 243$ (e) $8+27=35$ (i) $\frac{1}{2} \cdot 16 \cdot 9 = 72$ (m) $19 \cdot 2 = 38$

(b) $5 \cdot 5 \cdot 5 \cdot 5 = 625$ (f) $10{,}000-256$ (j) $10+3 \cdot 4 = 22$ (n) $\dfrac{256}{32}=8$
    $= 9{,}744$

(c) $10 \cdot 10 \cdot 10 = 1{,}000$ (g) $1 \cdot 1 \cdot 1 = 1$ (k) $8 \cdot 100-27 = 773$ (o) $\dfrac{27+32}{100}=\dfrac{59}{100}$

(d) $4+9=13$ (h) $8 \cdot 25 = 200$ (l) $\frac{1}{2} \cdot 16-\frac{1}{3} \cdot 9 = 5$

11.4 EVALUATING POWERS OF FRACTIONS AND DECIMALS

Evaluate:

(a) $.2^4$ (d) $1000(.2^3)$ (g) $\left(\dfrac{1}{2}\right)^3$ (j) $100\left(\dfrac{1}{5}\right)^2$

(b) $.5^2$ (e) $\frac{1}{3}(.3^2)$ (h) $\left(\dfrac{2}{3}\right)^2$ (k) $32\left(\dfrac{3}{2}\right)^3$

(c) $.01^3$ (f) $200(.4^4)$ (i) $\left(\dfrac{5}{3}\right)^4$ (l) $80\left(\dfrac{5}{2}\right)^3$

Solutions

(a) $(.2)(.2)(.2)(.2) = 0016$ (d) $1000(.008) = 8$ (g) $\frac{1}{2} \cdot \frac{1}{2} \cdot \frac{1}{2} = \frac{1}{8}$ (j) $100\left(\dfrac{1}{25}\right) = 4$

(b) $(.5)(.5) = .25$ (e) $\frac{1}{3}(.09) = .03$ (h) $\frac{2}{3} \cdot \frac{2}{3} = \frac{4}{9}$ (k) $32\left(\dfrac{27}{8}\right) = 108$

(c) $(.01)(.01)(.01) = .000001$ (f) $200(.0256) = 5.12$ (i) $\frac{5}{3} \cdot \frac{5}{3} \cdot \frac{5}{3} \cdot \frac{5}{3} = \frac{625}{81}$ (l) $80\left(\dfrac{125}{8}\right) = 1250$

11.5 EVALUATING POWERS OF VARIABLE BASES

Evaluate if $a = 5$, $b = 1$ and $c = 10$:

(a) a^3 (d) $2a^2$ (g) $\left(\frac{1}{2}c\right)^2$ (j) $a^2 + c^2$ (m) $c^2(a + b)$

(b) b^4 (e) $(2a)^2$ (h) $\left(\frac{b}{3}\right)^2$ (k) $(c + 3b)^2$ (n) $c(a - b^2)$

(c) c^2 (f) $(a + 2)^2$ (i) $\frac{4a^2}{c}$ (l) $5(a^2 - b^2)$ (o) $3b(a^3 - c^2)$

Solutions

(a) $5 \cdot 5 \cdot 5 = 125$ (d) $2 \cdot 25 = 50$ (g) $5^2 = 25$ (j) $25 + 100 = 125$ (m) $100 \cdot 6 = 600$

(b) $1 \cdot 1 \cdot 1 \cdot 1 = 1$ (e) $10^2 = 100$ (h) $\left(\frac{1}{3}\right)^2 = \frac{1}{9}$ (k) $13^2 = 169$ (n) $10 \cdot 4 = 40$

(c) $10 \cdot 10 = 100$ (f) $7^2 = 49$ (i) $\frac{100}{10} = 10$ (l) $5 \cdot 24 = 120$ (o) $3 \cdot 25 = 75$

Supplementary Problems

7.1. Using variable expressions, replace each verbal statement by an algebraic equation.   **(1.1)**

(a) Three times a number added to eight times the same number is equivalent to eleven times the number. (b) The difference between ten times a number and one-half of the same number is the same as nine and one-half times the number. (c) The perimeter of an equilateral triangle is equal to three times the length of one of the sides. (d) The area of a square is found by multiplying the length of a side by itself.

Ans. (a) $3n + 8n = 11n$ (b) $10n - \frac{1}{2}n = 9\frac{1}{2}n$ (c) $p = 3s$ (d) $A = ss$

7.2. Using the initial letters of words as variables, replace each verbal rule by a formula.   **(1.2)**
(a) The perimeter of a regular hexagon is six times a side. (b) The area of a triangle is one-half the product of the base and height. (c) The profit made on an article is the difference between the selling price and the cost. (d) The semiperimeter of a rectangle is the sum of the length and width. (e) The volume of a rectangular box is the product of the length, the width and the height.

Ans. (a) $p = 6s$ (b) $A = \frac{1}{2}bh$ (c) $p = s - c$ (d) $s = l + w$ (e) $V = lwh$

7.3. Evaluate each numerical expression and state the property of 0 or 1 that applies.   **(1.3)**

(a) $1,000 + 0$ *Ans.* 1,000, additive property of 0
(b) $1,000 \times 1$ *Ans.* 1,000, multiplicative property of 1
(c) $1,000 \times 0$ *Ans.* 0, multiplicative property of 0
(d) $50 \times 40 \times 0$ *Ans.* 0, multiplicative property of 0
(e) $50 \times 40 \times 1$ *Ans.* 2,000, multiplicative property of 1
(f) 0×1 *Ans.* 0, multiplicative property of 0 or of 1
(g) $0 + 1$ *Ans.* 1, additive property of 0

(h) $10 \times 0 + 10$ Ans. 10, both the multiplicative and additive properties of 0

(i) $20 \times 1 + 0$ Ans. 20, multiplicative property of 1 and additive property of 0

7.4. Using the replacement set $\{2, 3, 6, 10\}$ as the set of values of n, evaluate the following   **(1.4)**
expressions and list these values in a set:

(a) $5n$ (b) $2n + 5$ (c) $10 - \dfrac{1}{2}n$ (d) $\dfrac{2n - 1}{2n + 1}$ (e) $150\%n$

Ans. (a) $\{10, 15, 30, 50\}$ (c) $\{9, 8\frac{1}{2}, 7, 5\}$ (e) $\{3, 4\frac{1}{2}, 9, 15\}$

  (b) $\{9, 11, 17, 25\}$ (d) $\left\{\dfrac{3}{5}, \dfrac{5}{7}, \dfrac{11}{13}, \dfrac{19}{21}\right\}$

7.5. Simplify each addition by interchanging addends: **(2.1)**

(a) $64 + 138 + 36$ (c) $1\frac{1}{3} + \dfrac{3}{5} + 6\frac{2}{3}$ (e) $12\frac{1}{2}\% + 46\% + 87\frac{1}{2}\%$

(b) $15 + 78 + 15 + 170$ (d) $2\frac{1}{8} + \dfrac{13}{16} + \dfrac{3}{16} + \dfrac{3}{8}$ (f) $5.991 + 1.79 + .21 + .009$

Ans. (a) $64 + 36 + 138$ (c) $1\frac{1}{3} + 6\frac{2}{3} + \dfrac{3}{5}$ (e) $12\frac{1}{2}\% + 87\frac{1}{2} + 46\%$
  $100 + 138 = 238$ $100\% + 46\% = 146\%$

  $8 + \dfrac{3}{5} = 8\frac{3}{5}$

  (b) $15 + 15 + 170 + 78$ (d) $2\frac{1}{8} + \dfrac{3}{8} + \dfrac{13}{16} + \dfrac{3}{16}$ (f) $5.991 + .009 + 1.79 + .21$
  $200 + 78 = 278$ $6 + 2 = 8$

  $2\frac{1}{2} + 1 = 3\frac{1}{2}$

7.6. Rearrange the addends so that literal addends are arranged alphabetically and precede   **(2.2)**
numerical addends:

(a) $10 + d$ (c) $15 + g + f$ (e) $d + b + e + a$ (g) $5 + m + 4 + j + 11$

(b) $y + x$ (d) $17 + r + q + 13$ (f) $s + 12 + p + 48$ (h) $v + w + 16 + t + 50$

Ans. (a) $d + 10$ (c) $f + g + 15$ (e) $a + b + d + e$ (g) $j + m + 20$
  (b) $x + y$ (d) $q + r + 30$ (f) $p + s + 60$ (h) $t + v + w + 66$

7.7. Simplify each multiplication by interchanging factors: **(3.1)**

(a) $5 \times 26 \times 40$ (c) $10\frac{1}{2} \times 7 \times 2$ (e) $3.75 \times .15 \times 20 \times 4$

(b) $17 \times 12 \times 6$ (d) $303 \times 8 \times 1\frac{2}{3}$ (f) $66\frac{2}{3}\% \times 50 \times 27$

Ans. (a) $5 \times 40 \times 26$ (c) $10\frac{1}{2} \times 2 \times 7$ (e) $3.75 \times 4 \times .15 \times 20$
  $200 \times 26 = 5200$ $21 \times 7 = 147$ $15 \times 3 = 45$

  (b) $17 \times 6 \times 12$ (d) $303 \times 1\frac{2}{3} \times 8$ (f) $66\frac{2}{3}\% \times 27 \times 50$
  $102 \times 12 = 1224$ $505 \times 8 = 4040$ $\frac{2}{3} \times 27 \times 50 = 18 \times 50 = 900$

7.8. Rearrange the factors so that literal factors are arranged alphabetically and follow   **(3.2)**
numerical factors:

(a) $r8$ (c) $qp7$ (e) $x5y7w$ (g) $2h5k10$ (i) $cd13ab$

(b) cab (d) $4v3t$ (f) $def13c$ (h) $r11sm4$

Ans. (a) $8r$ (c) $7pq$ (e) $35wxy$ (g) $100hk$ (i) $13abcd$
  (b) abc (d) $12tv$ (f) $13cdef$ (h) $44mrs$

7.9. Symbolize each, avoiding multiplication signs where possible: **(4.1)**

    (a) the product of 8, m and n (e) f divided by 7
    (b) 5 times p times q (f) 25 divided by x
    (c) two-thirds of c (g) the product of u and v divided by 9
    (d) one-half of b multiplied by h

    Ans. (a) $8mm$ (b) $5pq$ (c) $\frac{2}{3}c$ (d) $\frac{1}{2}bh$ (e) $\frac{f}{7}$ (f) $\frac{25}{x}$ (g) $\frac{uv}{9}$

7.10. When is each division impossible?   **(4.2, 4.3)**

    (a) $\frac{7}{d}$ (c) $\frac{3}{4x}$ (e) $\frac{b}{7-c}$ (g) $\frac{50}{w-y}$ (i) $\frac{45}{x-2y}$

    (b) $\frac{r}{t}$ (d) $\frac{5}{a-8}$ (f) $\frac{10}{2x-4}$ (h) $\frac{100}{pq}$

    Ans. (a) if $d=0$ (c) if $x=0$ (e) if $c=7$ (g) if $w=y$ (i) if $x=2y$
            (b) if $t=0$ (d) if $a=8$ (f) if $x=2$ (h) if $p=0$ or $q=0$

7.11. If n represents a number, express algebraically: **(5.1, 5.2)**

    (a) 25 more than the number   (g) 30 less than the number
    (b) 30 greater than the number   (h) 35 fewer than the number
    (c) the sum of the number and 35   (i) 40 less the number
    (d) the number increased by 40   (j) 45 decreased by the number
    (e) 45 plus the number   (k) 50 minus the number
    (f) 50 added to the number   (l) 55 subtracted from the number

    Ans. (a) $n+25$ or $25+n$ (d) $n+40$ or $40+n$ (g) $n-30$ (j) $45-n$
            (b) $n+30$ or $30+n$ (e) $n+45$ or $45+n$ (h) $n-35$ (k) $50-n$
            (c) $n+35$ or $35+n$ (f) $n+50$ or $50+n$ (i) $40-n$ (l) $n-55$

7.12. Express algebraically: (a) the no. of kg of a weight that is 15 kg lighter than w kg,   **(5.3)** (b) the no. of ft in a length that is 50 ft shorter than l ft,   (c) the no. of sec in a time interval that is 1 minute less than t sec,   (d) the no. of cents in a price that is \$1 more than p cents,   (e) the no. of ft per sec (fps) in a speed that is 20 fps slower than r fps,   (f) the no. of ft in a distance that is 10 yd farther than d ft,   (g) the no. of sq ft in an area that is 30 sq ft greater than A sq ft,   (h) the no. of degrees in a temperature that is 40° colder than t°,   (i) the no. of floors in a building that is 8 floors higher than f floors,   (j) the no. of yr in an age 5 yr younger than a yr.

    Ans. (a) $w-15$ (c) $t-60$ (e) $r-20$ (g) $A+30$ (i) $f+8$
            (b) $l-50$ (d) $p+100$ (f) $d+30$ (h) $t-40$ (j) $a-5$

7.13. Express algebraically: **(6.1, 6.2)**

    (a) x times 3   (c) product of 12 and y (e) 10 divided by y
    (b) one-eighth of b (d) three-eighths of r (f) quotient of y and 10

    Ans. (a) $3x$ (b) $\frac{b}{8}$ or $\frac{1}{8}b$ (c) $12y$ (d) $\frac{3}{8}r$ or $\frac{3r}{8}$ (e) $\frac{10}{y}$ (f) $\frac{y}{10}$

7.14. Express algebraically: **(7.1, 7.2)**

    (a) b decreased by one-half c (f) twice d, less 25

    (b) one-third of g, decreased by 5   (g) 8 more than the product of 5 and x

    (c) four times r, divided by 9   (h) four times the sum of r and 9

    (d) the average of m and 60   (i) the average of 60, m, p and q

    (e) three-quarters of x, less y (j) the ratio of b to three times c

Ans. (a) $b - \dfrac{c}{2}$ (c) $\dfrac{4r}{9}$ (e) $\dfrac{3x}{4} - y$ (g) $5x + 8$ (i) $\dfrac{m + p + q + 60}{4}$

      (b) $\dfrac{g}{3} - 5$ (d) $\dfrac{m + 60}{2}$ (f) $2d - 25$ (h) $4(r + 9)$ (j) $\dfrac{b}{3c}$

7.15. Express algebraically: (a) a distance in m that is 25 m shorter than three times   **(7.3)** another of d m, (b) a weight in oz that is 5 oz more than twice another of w oz, (c) a temperature in degrees that is 8° warmer than five times another of $T°$, (d) a price in dollars that is $50 higher than one-half another of p dollars, (e) a price in cents that is 50¢ cheaper than one-third another of p cents, (f) a length in cm that is 2 cm longer than y m.

Ans. (a) $3d - 25$ (b) $2w + 5$ (c) $5T + 8$ (d) $\dfrac{p}{2} + 50$ (e) $\dfrac{p}{3} - 50$ (f) $100y + 2$

7.16. If $a = 2$, $b = 5$, $c = 10$, $d = 100$, and $e = 1,000$, show that   **(8.1, 8.2)**

    (a) $(a + b) + c = a + (b + c)$ (d) $(a + c + e) + (d + b) = (a + c) + (e + d) + b$

    (b) $(a + b) + (c + d) = (a + b + c) + d$ (e) $(ab)c = a(bc)$

    (c) $(b + d) + (a + e) = b + (d + a + e)$ (f) $(ab)(cd) = (abc)d$

Ans. (a) $(7) + 10 = 2 + (15)$ and $17 = 17$

      (b) $(7) + (110) = (17) + 100$ and $117 = 117$

      (c) $(105) + (1,002) = 5 + (1,102)$ and $1,107 = 1,107$

      (d) $(1,012) + (105) = (12) + (1,100) + 5$ and $1,117 = 1,117$

      (e) $(10)10 = 2(50)$ and $100 = 100$

      (f) $(10)(1,000) = (100)\,100$ and $10,000 = 10,000$

7.17. Evaluate: **(9.1)**

    (a) $40 - 2 \times 5$ (c) $40 \div 2 + 5$ (e) $16 \div 2 - \frac{1}{2} \cdot 10$ (g) $40 \times 2 - 40 \div 2$

    (b) $3 \times 8 - 2 \times 5$ (d) $3 + 8 - 2 \times 5$ (f) $3 + 8 \times 2 \times 5$ (h) $3 + 8 \times 2 - 5 \div 10$

Ans. (a) 30   (b) 14   (c) 25   (d) 1   (e) 3   (f) 83   (g) 60   (h) $18\frac{1}{2}$

7.18. Evaluate if $a = 5$, $b = 6$ and $c = 10$:   **(9.2)**

    (a) $a + b - c$ *Ans.* 1   (d) $\dfrac{a + b}{2}$ *Ans.* $5\frac{1}{2}$ (g) $\dfrac{4}{5}c$ or $\dfrac{4c}{5}$ *Ans.* 8

    (b) $a + 2b$ *Ans.* 17   (e) $\dfrac{3c}{a}$ *Ans.* 6   (h) $\dfrac{2}{3}b + \dfrac{3}{2}c$ *Ans.* 19

    (c) $a + \dfrac{b}{2}$ *Ans.* 8   (f) $3 + \dfrac{c}{a}$ *Ans.* 5   (i) $\dfrac{a + b}{c - 9}$ *Ans.* 11

(j) $\dfrac{ab}{c}$ *Ans.* 3   (l) $6c - 2ab$ *Ans.* 0   (n) $a + c - \dfrac{b}{2}$ *Ans.* 12

(k) $5a + 4b - 2c$ *Ans.* 29   (m) $a + \dfrac{c - b}{2}$ *Ans.* 7   (o) $\dfrac{a + c - b}{3}$ *Ans.* 3

7.19. Evaluate: **(9.3)**

  (a) $5(8 + 2)$ *Ans.* 50   (e) $8 \cdot 2(5 - 3)$ *Ans.* 32   (i) $4(4 \cdot 4 - 4)$ *Ans.* 48

  (b) $5(8 - 2)$ *Ans.* 30   (f) $3(6 + 2 \cdot 5)$ *Ans.* 48   (j) $(4 + 4)4 - 4$ *Ans.* 28

  (c) $8 + 2(5 - 3)$ *Ans.* 12   (g) $(3 \cdot 6 + 2)5$ *Ans.* 100   (k) $(4 + 4)(4 - 4)$ *Ans.* 0

  (d) $8(2 \cdot 5 - 3)$ *Ans.* 56   (h) $3(6 + 2)5$ *Ans.* 120   (l) $4 + 4(4 - 4)$ *Ans.* 4

7.20. Evaluate if $a = 4$, $b = 3$ and $c = 5$:   **(9.4)**

  (a) $a(b + c)$ *Ans.* 32   (e) $\frac{1}{2}(a + b) + c$ *Ans.* $8\frac{1}{2}$ (i) $3a + 2(c - b)$ *Ans.* 16

  (b) $b(c - a)$ *Ans.* 3   (f) $3(b + 2c)$ *Ans.* 39   (j) $3(a + 2c) - b$ *Ans.* 39

  (c) $c(a - b)$ *Ans.* 5   (g) $3(b + 2)c$ *Ans.* 75   (k) $3(a + 2c - b)$ *Ans.* 33

  (d) $\frac{1}{2}(a + b + c)$ *Ans.* 6   (h) $(3b + 2)c$ *Ans.* 55   (l) $3(a + 2)(c - b)$ *Ans.* 36

7.21. Evaluate if $x = 3$, $y = 2$ and $z = 0$:   **(9.5)**

  (a) $x + y + z$ *Ans.* 5   (f) $\dfrac{z}{x}$ *Ans.* 0   (k) $xz + yz$ *Ans.* 0

  (b) $x - y - z$ *Ans.* 1   (g) $\dfrac{x}{z}$ *Ans.* valueless   (l) $\dfrac{z}{x + y}$ *Ans.* 0

  (c) $x(y + z)$ *Ans.* 6   (h) xyz *Ans.* 0   (m) $\dfrac{x}{y + z}$ *Ans.* $1\frac{1}{2}$

  (d) $z(x + y)$ *Ans.* 0   (i) $xy + z$ *Ans.* 6   (n) $x + \dfrac{z}{y}$ *Ans.* 3

  (e) $y(x + z)$ *Ans.* 6   (j) $x + yz$ *Ans.* 3   (o) $\dfrac{y + z}{x}$ *Ans.* $\dfrac{2}{3}$

7.22. State the terms in each expression: **(10.1)**

  (a) $5xyz$ (c) $5 + x + y + z$ (e) $3ab + c$

  (b) $5 + xyz$ (d) $3a + bc$ (f) $3a(b + c)$

Ans. (a) $5xyz$ (c) $5, x, y, z$ (e) $3ab, c$

     (b) $5, xyz$ (d) $3a, bc$ (f) $3a(b + c)$ is a single term.

7.23. State the factors of the following, disregarding 1 and the term itself: **(10.2)**

  (a) 77   (b) 25   (c) pq (d) $\dfrac{3}{4}x$ (e) $\dfrac{w}{10}$ (f) $8(x - 5)$ (g) $\dfrac{y - 2}{4}$

Ans. (a) 7, 11   (b) 5, 5   (c) p, q (d) $\dfrac{3}{4}, x$ (e) $\dfrac{1}{10}, w$ (f) $8, (x - 5)$ (g) $\dfrac{1}{4}, (y - 2)$

7.24. State each numerical and literal coefficient: **(10.3)**

  (a) w (b) $\dfrac{1}{8}x$ (c) $\dfrac{n}{10}$ (d) $.03ab$ (e) $\dfrac{3y}{10}$ (f) $\dfrac{2a}{3b}$ (g) $\dfrac{3}{5}(a - b)$

Ans.

	(a)	(b)	(c)	(d)	(e)	(f)	(g)
Numerical Coefficient:	1	$\frac{1}{8}$	$\frac{1}{10}$	.03	$\frac{3}{10}$	$\frac{2}{3}$	$\frac{3}{5}$
Literal Coefficient:	w	x	n	ab	y	$\frac{a}{b}$	$(a - b)$

7.25. Write each, using bases and exponents: **(11.1)**

(a) $7 \cdot 3 \cdot 3$ (b) $7xyyy$ (c) $\dfrac{7x}{yyy}$ (d) $(7x)(7x)$ (e) $(a+5)(a+5)$ (f) $\dfrac{2rrw}{5stvv}$

Ans.   (a) $7 \cdot 3^2$ (b) $7xy^3$ (c) $\dfrac{7x}{y^3}$ (d) $(7x)^2$ (e) $(a+5)^2$ (f) $\dfrac{2r^2w}{5stv^2}$

7.26. Write each without exponents: **(11.2)**

(a) $4 \cdot 7^2$ (b) $\dfrac{1}{2}y^4$ (c) $\dfrac{5a}{b^4}$ (d) $(ab)^3$ (e) $(x+2)^2$ (f) $\dfrac{a^2 - b^3}{c + d^2}$

Ans.   (a) $4 \cdot 7 \cdot 7$ (b) $\dfrac{1}{2}yyyy$ (c) $\dfrac{5a}{bbbb}$ (d) $(ab)(ab)(ab)$ (e) $(x+2)(x+2)$ (f) $\dfrac{aa - bbb}{c + dd}$

7.27. Evaluate (Table 7-1 may be used to check values): **(11.3)**

(a) $3^3 - 2^3$ (c) $10^3 + 5^4$ (e) $1^2 + 2^2 + 3^2$ (g) $1^5 + 1^4 + 1^3 + 1^2$ (i) $\dfrac{1}{2} \cdot 2^2 + \dfrac{1}{3} \cdot 3^3$

(b) $5^2 \cdot 2^5$ (d) $10^2 \div 2$ (f) $5^3 \div 5$ (h) $2^5 - 4 \cdot 2^2$

Ans.   (a) 19   (b) 800   (c) 1625   (d) 50   (e) 14   (f) 25   (g) 4   (h) 16
   (i) 11

7.28. Evaluate: **(11.4)**

(a) $.1^2 \cdot 9^2$ (b) $.3 \cdot 4^2$ (c) $(-3)^2 \cdot 4^2$ (d) $40\left(\dfrac{1}{2}\right)^3$ (e) $\left(\dfrac{2}{5}\right)^3$ (f) $\dfrac{(-2)^3}{5^2}$ (g) $\dfrac{10}{.1^2}$

Ans.   (a) .81   (b) 4.8   (c) 144   (d) 5   (e) $\dfrac{8}{125}$ (f) $\dfrac{-8}{25}$ (g) 1000

7.29. Evaluate if $a = 3$ and $b = 2$: **(11.5)**

(a) a^2b	Ans. 18	(e) $(a+b)^2$	Ans. 25	(i) $a^3 - b^3$	Ans. 19
(b) ab^2	Ans. 12	(f) $a^2 + b^2$	Ans. 13	(j) $(a-b)^3$	Ans. 1
(c) $(ab)^2$	Ans. 36	(g) a^3b	Ans. 54	(k) a^2b^3	Ans. 72
(d) $a + b^2$	Ans. 7	(h) $(ab)^3$	Ans. 216	(l) a^3b^2	Ans. 108

7.30. Evaluate if $w = -1$, $x = 3$, $y = 4$: **(11.5)**

(a) $2w^2$	Ans. 2	(e) $(y+x)^2$	Ans. 49	(i) $w^2 + x^2 + y^2$	Ans. 26
(b) $(2w)^2$	Ans. 4	(f) $y^2 - x^2$	Ans. 7	(j) y^3x	Ans. 192
(c) $(x+2)^2$	Ans. 25	(g) $(y-x)^2$	Ans. 1	(k) yx^3	Ans. 108
(d) $y^2 + x^2$	Ans. 25	(h) $(w+x+y)^2$	Ans. 36	(l) $(yx)^3$	Ans. 1728

Chapter 8

Fundamentals of Algebra: Equations and Formulas

1. VARIABLES AND EQUATIONS

The question "What number increased by 7 equals 10?" may be answered by expressing the problem in the form of an equation, $n + 7 = 10$. In this equation, n is a variable whose replacement set is assumed to be the set of all numbers.

The equation $n + 7 = 10$ is solved by comparing it with the addition fact $3 + 7 = 10$; thus,

$$\left. \begin{array}{l} n + 7 = 10 \\ 3 + 7 = 10 \end{array} \right\} \longrightarrow \text{Conclusion: } n = 3 \ Ans.$$

A *solution* or *root* of an equation is a value of the variable for which the equation becomes a true statement. A solution or root of an equation is said to *satisfy* the equation.

Thus, 3 is a solution or root of the equation $n + 7 = 10$ and satisfies the equation.

Solving an Equation: Solution Set or Truth Set

Solving an equation is the process of finding the roots or solutions of the equation. The *solution set* or *truth set* of an equation is the set of roots of the equation.

Thus, $n + 7 = 10$ is solved when the root 3 is found. The solution set of $n + 7 = 10$ is {3}.

Identities and Equivalent Expressions

We have found that the equation $n + 7 = 10$ has one root, 3, and no other. However, there are equations that are satisfied by any number! For example, $3n - 2n = n$ is such an equation. Such an equation is an *identity* or *unconditional equation*, and both the left side and the right side are *equivalent expressions*. To show that $3n - 2n = n$ is an identity, simply combine like terms to obtain $n = n$.

Substitution Rule for Equivalent Expressions

In any process, an expression may be replaced by an equivalent expression.

Thus, to solve $3n - 2n + 7 = 10$, substitute n for $3n - 2n$ to obtain $n + 7 = 10$. We now solve $n + 7 = 10$ to obtain 3, the root or solution of $3n - 2n + 7 = 10$.

Properties of an Equality

Three important properties of an equality are the following:

1. **Reflexive Property**: $a = a$.
   According to the reflexive property, an expression equals itself. Thus, $n + 7 = n + 7$
2. **Symmetric Property**: If $a = b$, then $b = a$.
   According to the symmetric property, we can interchange the sides of an equation. Thus, if $10 = n + 7$, then $n + 7 = 10$.
3. **Transitive Property**: If $a = b$ and $b = c$, then $a = c$.
   According to the transitive property, two expressions equal to a third expression are equal to each other. Thus, if $n + 7 = 10$ and $3 + 7 = 10$, then $n + 7 = 3 + 7$.

Checking or Verifying an Equation

Checking or *verifying* an equation is the process of testing to see if a given number is a root of an equation.

Thus, the equation $3n - 2n + 7 = 10$ may be checked to see if 3 or 5 is a root. (Read $\overset{?}{=}$ as "should equal." Read $\neq$ as "does not equal.")

<table>
<tr><td align="center">Check for $n = 3$</td><td align="center">Check for $n = 5$</td></tr>
<tr><td align="center">$3n - 2n + 7 = 10$</td><td align="center">$3n - 2n + 7 = 10$</td></tr>
<tr><td align="center">$3(3) - 2(3) + 7 \overset{?}{=} 10$</td><td align="center">$3(5) - 2(5) + 7 \overset{?}{=} 10$</td></tr>
<tr><td align="center">$9 - 6 + 7 \overset{?}{=} 10$</td><td align="center">$15 - 10 + 7 \overset{?}{=} 10$</td></tr>
<tr><td align="center">$10 = 10 \ \checkmark$</td><td align="center">$12 \neq 10$</td></tr>
</table>

The check shows that 3 is a root of $3n - 2n + 7 = 10$ but 5 is not.

1.1   CHECKING AN EQUATION TO DETERMINE A ROOT

By checking, determine which is a root of the equation:

(*a*) Check $2n + 3n = 25$ for $n = 5$ and $n = 6$ (*b*) Check $8x - 14 = 6x$ for $x = 6$ and $x = 7$

Solutions

<table>
<tr><td align="center">Check: $n = 5$</td><td align="center">$n = 6$</td><td align="center">Check: $x = 6$</td><td align="center">$x = 7$</td></tr>
<tr><td align="center">$2n + 3n = 25$</td><td align="center">$2n + 3n = 25$</td><td align="center">$8x - 14 = 6x$</td><td align="center">$8x - 14 = 6x$</td></tr>
<tr><td align="center">$2(5) + 3(5) \overset{?}{=} 25$</td><td align="center">$2(6) + 3(6) \overset{?}{=} 25$</td><td align="center">$8(6) - 14 \overset{?}{=} 6(6)$</td><td align="center">$8(7) - 14 \overset{?}{=} 6(7)$</td></tr>
<tr><td align="center">$10 + 15 \overset{?}{=} 25$</td><td align="center">$12 + 18 \overset{?}{=} 25$</td><td align="center">$48 - 14 \overset{?}{=} 36$</td><td align="center">$56 - 14 \overset{?}{=} 42$</td></tr>
<tr><td align="center">$25 = 25$</td><td align="center">$30 \neq 25$</td><td align="center">$34 \neq 36$</td><td align="center">$42 = 42$</td></tr>
</table>

Ans. 5 is a root of $2n + 3n = 25$ *Ans.* 7 is a root of $8x - 14 = 6x$

1.2   CHECKING AN IDENTITY TO SHOW THAT ANY GIVEN VALUE IS A ROOT

By checking the identity $4(x + 2) = 4x + 8$, show that it is satisfied by   (*a*) $x = 10$,   (*b*) $x = 6$, (*c*) $x = 4\frac{1}{2}$,   (*d*) $x = 3.2$

Solutions

(*a*) $4(x + 2) = 4x + 8$ (*b*) $4(x + 2) = 4x + 8$ (*c*) $4(x + 2) = 4x + 8$ (*d*) $4(x + 2) = 4x + 8$

$4(10 + 2) \overset{?}{=} 4(10) + 8$ $4(6 + 2) \overset{?}{=} 4(6) + 8$ $4(4\frac{1}{2} + 2) \overset{?}{=} 4(4\frac{1}{2}) + 8$ $4(3.2 + 2) \overset{?}{=} 4(3.2) + 8$

$4(12) \overset{?}{=} 40 + 8$ $4(8) \overset{?}{=} 24 + 8$ $4(6\frac{1}{2}) \overset{?}{=} 18 + 8$ $4(5.2) \overset{?}{=} 12.8 + 8$

$48 = 48$ $32 = 32$ $26 = 26$ $20.8 = 20.8$

2. TRANSLATING VERBAL PROBLEMS INTO EQUATIONS

In algebra, a simple verbal problem having one unknown is solved when the unknown is found. In the process, it is necessary to "translate" a verbal sentence into an equation. The first step in problems of this type is to let the unknown be represented by a variable.

Thus, if n represents the unknown in "Twice what number equals 12?", we obtain "$2n = 12$."

2.1   TRANSLATING INTO EQUATIONS

Translate into an equation, letting n represent the unknown number.   (*a*) 4 less than what number equals 8?   (*b*) One-half of what number equals 10?   (*c*) Ten times what number equals 20?   (*d*) What number increased by 12 equals 17?   (*e*) Twice what number added to 8 is 16?   (*f*) 15 less than three times what number is 27?   (*g*) The sum of what number and twice the same number is 18?   (*h*) What number and 4 more equals five times the number?   (*i*) Twice the sum of a certain number and five is 24. What is the number?

Illustrative Solution   (*e*) Use $2n$ for "twice what number."   *Ans.* $8 + 2n = 16$ or $2n + 8 = 16$.

Ans. (*a*) $n - 4 = 8$ (*c*) $10n = 20$ (*f*) $3n - 15 = 27$ (*h*) $n + 4 = 5n$

(*b*) $\dfrac{n}{2} = 10$ (*d*) $n + 12 = 17$ (*g*) $n + 2n = 18$ (*i*) $2(n + 5) = 24$

2.2   MATCHING SENTENCES AND EQUATIONS

Match a sentence in Column 1 with an equation in Column 2.

Column 1

1. The product of 8 and a number is 40.
2. A number increased by 8 is 40.
3. 8 less than a number equals 40.
4. Eight times a number, less 8, is 40.
5. Eight times the sum of a number and 8 is 40.
6. One-eighth of a number is 40.

Column 2

(*a*) $n - 8 = 40$
(*b*) $8(n + 8) = 40$
(*c*) $8n = 40$
(*d*) $n/8 = 40$
(*e*) $8n - 8 = 40$
(*f*) $n + 8 = 40$

Ans. 1. and (*c*)   2. and (*f*)   3. and (*a*)   4. and (*e*)   5. and (*b*)   6. and (*d*)

2.3   REPRESENTING UNKNOWNS

Represent the unknown by a variable and obtain an equation for each problem.   (*a*) A man worked for 5 hours and earned \$87.50. What was his hourly wage?   (*b*) How old is Henry now, if ten years ago, he was 23 years old?   (*c*) After gaining 12 lb, Mary weighed 120 lb. What was her previous weight?   (*d*) A baseball team won four times as many games as it lost. How many games did it lose, if it played a total of 100 games?

Solutions

(*a*) Let $w = $ his hourly wage in dollars.
Then, $5w = 87.50$

(*b*) Let $H = $ Henry's age now.
Then, $H - 10 = 23$

(*c*) Let $M = $ Mary's previous weight in lb.
Then, $M + 12 = 120$

(*d*) Let $n = $ no. of games lost, and
$4n = $ no. of games won.
Then, $n + 4n = 100$

3.   SOLVING SIMPLE EQUATIONS USING INVERSE OPERATIONS

In mathematics, you may think of inverse operations as operations that undo each other. To understand the functions of inverse operations, think how often you do something and then undo it. You earn money, then you spend it; you get up, then you lie down; you open a door, then you close it; you go out, then you go in. Of course, not everything that you do can be undone. Consider putting an egg together after you have fried it.

Rule 1: Addition and subtraction are inverse operations.

Rule 2: Multiplication and division are inverse operations.

Figure 8-1(*a*) illustrates how addition undoes subtraction, and how subtraction undoes addition. Figure 8-1(*b*) illustrates how multiplication undoes division, and how division undoes multiplication.

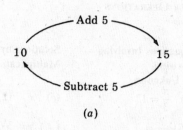

Fig. 8-1

(1) Addition and subtraction are shown to be inverse operations since adding 5 to 10, then subtracting 5 from the result leads back to 10; or, subtracting 5 from 15, then adding 5 to the result leads back to 15.

Suppose that in addition to having $10 you earn $5, making a total of $15. If now you spend $5, you will have the original amount of $10.

(2) Multiplication and division are shown to be inverse operations since multiplying 10 by 5, then dividing the result by 5 leads back to 10; or, dividing 50 by 5, then multiplying the result by 5 leads back to 50.

Suppose that the number of persons in a group of 10 is multiplied by 5, making a new group of 50 persons. If now the number is divided by 5, we will find that the number of persons is the original number 10.

An understanding of inverse operations helps us understand the solution of equations. To solve an equation, think of it as asking a question, as in each of the following examples.

EQUATION	QUESTION ASKED BY EQUATION	FINDING ROOT OF EQUATION
1. $n + 4 = 12$	What number plus 4 equals 12?	$n = 12 - 4 = 8$
2. $n - 4 = 12$	What number minus 4 equals 12?	$n = 12 + 4 = 16$
3. $4n = 12$	What number multiplied by 4 equals 12?	$n = 12 \div 4 = 3$
4. $n/4 = 12$	What number divided by 4 equals 12?	$n = 12 \cdot 4 = 48$

Note the inverse operations involved in the above cases.

1. The equation $n + 4 = 12$, involving *addition*, is solved by *subtracting* 4 from 12.
2. The equation $n - 4 = 12$, involving *subtraction*, is solved by *adding* 4 to 12.
3. The equation $4n = 12$, involving *multiplication*, is solved by *dividing* 4 into 12.
4. The equation $n/4 = 12$, involving *division*, is solved by *multiplying* 4 by 12.

3.1   RULE 1: ADDITION AND SUBTRACTION ARE INVERSE OPERATIONS

Solve the equations below.

Equations Involving Addition to Unknown	Solutions by Subtraction	Equations Involving Subtraction from Unknown	Solutions by Addition
(a) $x + 3 = 8$	(a) $x = 8 - 3$ or 5	(e) $x - 10 = 2$	(e) $x = 2 + 10$ or 12
(b) $5 + y = 13$	(b) $y = 13 - 5$ or 8	(f) $w - 20 = 12$	(f) $w = 12 + 20$ or 32
(c) $15 = a + 10$	(c) $a = 15 - 10$ or 5	(g) $18 = a - 13$	(g) $a = 18 + 13$ or 31
(d) $28 = 20 + b$	(d) $b = 28 - 20$ or 8	(h) $21 = b - 2$	(h) $b = 21 + 2$ or 23

3.2   RULE 2: MULTIPLICATION AND DIVISION ARE INVERSE OPERATIONS

Solve the equations below.

Equations Involving Multiplication of Unknown	Solutions by Division	Equations Involving Division of Unknown	Solutions by Multiplication
(a) $3x = 12$	(a) $x = \dfrac{12}{3}$ or 4	(e) $\dfrac{x}{3} = 12$	(e) $x = 12 \cdot 3$ or 36
(b) $12y = 3$	(b) $y = \dfrac{3}{12}$ or $\dfrac{1}{4}$	(f) $\dfrac{y}{12} = 3$	(f) $y = 3 \cdot 12$ or 36
(c) $35 = 7a$	(c) $a = \dfrac{35}{7}$ or 5	(g) $4 = \dfrac{a}{7}$	(g) $a = 4 \cdot 7$ or 28
(d) $7 = 35b$	(d) $b = \dfrac{7}{35}$ or $\dfrac{1}{5}$	(h) $7 = \dfrac{b}{4}$	(h) $b = 7 \cdot 4$ or 28

3.3   SOLVING BY USING INVERSE OPERATIONS

Solve each equation, showing the inverse operation used to solve.

(a) $x + 5 = 20$ (e) $10 + y = 30$ (i) $14 = a - 7$ (m) $8b = 2$

(b) $x - 5 = 20$ (f) $10y = 30$ (j) $14 = 7a$ (n) $\dfrac{b}{8} = 2$

(c) $5x - 20$ (g) $\dfrac{y}{10} = 30$ (k) $14 = \dfrac{a}{7}$ (o) $24 = 6 + c$

(d) $\dfrac{x}{5} = 20$ (h) $14 = a + 7$ (l) $b - 8 = 2$

Solutions

(a) $x = 20 - 5$ or 15   (f) $y = \dfrac{30}{10}$ or 3   (k) $a = 14(7)$ or 98

(b) $x = 20 + 5$ or 25   (g) $y = 30(10)$ or 300   (l) $b = 2 + 8$ or 10

(c) $x = \dfrac{20}{5}$ or 4   (h) $a = 14 - 7$ or 7   (m) $b = \dfrac{2}{8}$ or $\dfrac{1}{4}$

(d) $x = 20(5)$ or 100   (i) $a = 14 + 7$ or 21   (n) $b = 2(8)$ or 16

(e) $y = 30 - 10$ or 20   (j) $a = \dfrac{14}{7}$ or 2   (o) $c = 24 - 6$ or 18

4.   RULES FOR SOLVING EQUATIONS

Equivalent equations are equations having the same solution set; that is, the same root or roots. In the previous section, equations were solved by changing, or transforming, them into equivalent equations using inverse operations.

Thus, $n + 4 = 12$ and $n = 12 - 4$ are equivalent equations.

Rules of Equality for Solving Equations

1. **Addition Rule**: To change an equation into an equivalent equation, the same number may be added to both sides.
2. **Subtraction Rule**: To change an equation into an equivalent equation, the same number may be subtracted from both sides.
3. **Multiplication Rule**: To change an equation into an equivalent equation, both sides may be multiplied by the same number.
4. **Division Rule**: To change an equation into an equivalent equation, both sides may be divided by any number except zero.

These four rules of equality may be summed up in one rule:

The Rule of Equality for Fundamental Operations

To change an equation into an equivalent equation, perform the same fundamental operation on both sides using the same number, excepting division by zero.

To understand these rules of equality, think of an equality as a scale in balance (Fig. 8-2). If only one side of a balanced scale is changed, the scale becomes unbalanced. To balance the scale, exactly the same change must be made on the other side. Similarly, if only one side of an equation is changed, the result need not be an equivalent equation. To obtain an equivalent equation, the same fundamental operation should be performed on both sides using the same number.

BALANCED SCALES

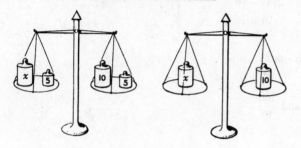

{ If 5 is subtracted from both sides of a balanced scale, the scale is still in balance.

EQUALITIES

If $x + 5 = 15$

then $x + 5 - 5 = 15 - 5$

and $x = 10$

{ If 5 is subtracted from both sides of an equation, the result is an equivalent equation.

Fig. 8-2

4.1   USING RULES OF EQUALITY

State the equality rule and the change needed to solve the equation:

(a) $x + 15 = 21$
$\underline{-15 = -15}$
$x = 6$

(b) $40 = r - 8$
$\underline{+8 = +8}$
$48 = r$

(c) $25 = 5m$
$\dfrac{25}{5} = \dfrac{5m}{5}$
$5 = m$

(d) $\dfrac{n}{8} = 3$
$8 \cdot \dfrac{n}{8} = 8 \cdot 3$
$n = 24$

(e) $24x = 8$
$\dfrac{24x}{24} = \dfrac{8}{24}$
$x = \dfrac{1}{3}$

Solutions

(a) Using the subtraction rule of equality, 15 is subtracted from each side.
(b) Using the addition rule of equality, 8 is added to each side.
(c) Using the division rule of equality, each side is divided by 5.
(d) Using the multiplication rule of equality, each side is multiplied by 8.
(e) Using the division rule of equality, each side is divided by 24.

5. USING DIVISION TO SOLVE AN EQUATION

To Solve an Equation Using the Division Rule of Equality (Section 4)

Solve:	(a) $2n = 16$	(b) $16n = 2$
PROCEDURE		SOLUTIONS

1. Divide both sides of the equation by the coefficient or multiplier of the variable:

   1. D_2 $\dfrac{2n}{2} = \dfrac{16}{2}$

      Ans. $n = 8$

   D_{16} $\dfrac{16n}{16} = \dfrac{2}{16}$

      Ans. $n = \dfrac{1}{8}$

2. Check the original equation:

   2. $\quad 2n = 16$

   $2(8) \overset{?}{=} 16$

   $16 = 16 \checkmark$

   $16n = 2$

   $16\left(\dfrac{1}{8}\right) \overset{?}{=} 2$

   $2 = 2 \checkmark$

Note 1: **D** is a convenient symbol for "dividing both sides."

   D_2 means "divide both sides by 2."

Note 2: A common factor may be eliminated in $\dfrac{\overset{1}{2}n}{\underset{}{2}}$ and $\dfrac{\overset{1}{16}n}{\underset{}{16}}$.

5.1 SOLVING EQUATIONS WITH NATURAL NUMBER COEFFICIENTS

Solve: (a) $7x = 35$,   (b) $35y = 7$,   (c) $33 = 11z$,   (d) $11 = 33w$.

Solutions

(a)	$7x = 35$	(b)	$35y = 7$	(c)	$33 = 11z$	(d)	$11 = 33w$
D_7	$\dfrac{7x}{7} = \dfrac{35}{7}$	D_{35}	$\dfrac{35y}{35} = \dfrac{7}{35}$	D_{11}	$\dfrac{33}{11} = \dfrac{11z}{11}$	D_{11}	$\dfrac{11}{33} = \dfrac{33w}{33}$
Ans.	$x = 5$	*Ans.*	$y = \dfrac{1}{5}$	*Ans.*	$3 = z$	*Ans.*	$\dfrac{1}{3} = w$

Check:	**Check:**	**Check:**	**Check:**
$7x = 35$	$35y = 7$	$33 = 11z$	$11 = 33w$
$7(5) \overset{?}{=} 35$	$35\left(\dfrac{1}{5}\right) \overset{?}{=} 7$	$33 \overset{?}{=} 11(3)$	$11 \overset{?}{=} 33\left(\dfrac{1}{3}\right)$
$35 = 35$	$7 = 7$	$33 = 33$	$11 = 11$

5.2 DIVISION IN EQUATIONS WITH DECIMAL COEFFICIENTS

Find each solution set: (a) $.3a = 9$,   (b) $1.2b = 48$,   (c) $15 = .05c$.

Solutions

(a)	$.3a = 9$	(b)	$1.2b = 48$	(c)	$15 = .05c$
$D_{.3}$	$\dfrac{.3a}{.3} = \dfrac{9}{.3}$	$D_{1.2}$	$\dfrac{1.2b}{1.2} = \dfrac{48}{1.2}$	$D_{.05}$	$\dfrac{15}{.05} = \dfrac{.05c}{.05}$
	$a = 30$		$b = 40$		$300 = c$
Ans. $\{30\}$		*Ans.* $\{40\}$		*Ans.* $\{300\}$	

Check: **Check:**   **Check:**

$$.3a = 9$$ $$1.2b = 48$$ $$15 = .05c$$
$$.3(30) \overset{?}{=} 9$$ $$1.2(40) \overset{?}{=} 48$$ $$15 \overset{?}{=} .05(300)$$
$$9 = 9$$ $$48 = 48$$ $$15 = 15$$

5.3   SOLVING EQUATIONS WITH PER CENTS AS COEFFICIENTS

Solve: (a) $22\% s = 88$,   (b) $75\% t = 18$,   (c) $72 = 2\% n$.

Solutions

(a) $22\% s = 88$ (b) $75\% t = 18$ (c) $72 = 2\% n$
Since $22\% = .22$,   Since $75\% = .75$,   Since $2\% = .02$,

$\mathbf{D}_{.22} \quad \dfrac{.22s}{.22} = \dfrac{88}{.22}$ $\mathbf{D}_{.75} \quad \dfrac{.75t}{.75} = \dfrac{18}{.75}$ $\mathbf{D}_{.02} \quad \dfrac{72}{.02} = \dfrac{.02n}{.02}$

Ans. $s = 400$ *Ans.* $t = 24$ *Ans.* $3600 = n$

Check: **Check:**   **Check:**

$$22\% = 88$$ $$75\% t = 18$$ $$72 = 2\% \, n$$
$$(.22)(400) \overset{?}{=} 88$$ $$(.75)(24) \overset{?}{=} 18$$ $$72 \overset{?}{=} (.02)(3600)$$
$$88 = 88$$ $$18 = 18$$ $$72 = 72$$

5.4   SOLVING EQUATIONS WITH LIKE TERMS ON ONE SIDE

Solve: (a) $60 = 7x - x$,   (b) $3x + 5x = 48$,   (c) $7x - 2x = 55$.

Solutions

(a) $60 = 7x = x$ (b) $3x + 5x = 48$ (c) $7x - 2x = 55$
$60 = 6x$ $8x = 48$ $5x = 55$

$\mathbf{D}_6 \quad \dfrac{60}{6} = \dfrac{6x}{6}$ $\mathbf{D}_8 \quad \dfrac{8x}{8} = \dfrac{48}{8}$ $\mathbf{D}_5 \quad \dfrac{5x}{5} = \dfrac{55}{5}$
Ans. $10 = x$ *Ans.* $x = 6$ *Ans.* $x = 11$

Check: **Check:**   **Check:**

$$60 = 7x - x$$ $$3x + 5x = 48$$ $$7x - 2x = 55$$
$$60 \overset{?}{=} 70 - 10$$ $$18 + 30 \overset{?}{=} 48$$ $$77 - 22 \overset{?}{=} 55$$
$$60 = 60$$ $$48 = 48$$ $$55 = 55$$

5.5   DIVISION RULE IN PROBLEM SOLVING

John worked 7 hr and earned \$85.40. What was his hourly wage?

Solution Let $h =$ his hourly wage in \$.   **Check** (the problem):
      Then, $7h = 85.40$ In 7 hr, John should earn \$85.40
                                                        Hence,
$\mathbf{D}_7 \qquad \dfrac{7h}{7} = \dfrac{85.40}{7}$ $7(\$12.20) \overset{?}{=} \$85.40$
                                                            $\$85.40 = \$85.40$
      $h = 12.20$

Ans.   John's hourly wage was \$12.20

5.6   DIVISION RULE IN PROBLEM SOLVING

Mr. Black's commission rate was 3%. If he earned $66 in commission, how much did he sell?

Solution   Let s = Mr. Black's sales in $.   **Check** (the problem):

Then, $3\% s = .03\,s = 66$ At 3%, Mr. Black's commission

$\mathbf{D}_{.03}$ $\dfrac{.03s}{.03} = \dfrac{66}{.03}$ should be $66. Hence,

                   3% of $\$2200 \stackrel{?}{=} \$66$

         $s = 2200$ $(.03)(\$2200) = \$66$

                       $\$66 = \$66$

Ans.   Mr. Black's sales were $2200.

6.   USING MULTIPLICATION TO SOLVE AN EQUATION

To Solve an Equation Using the Multiplication Rule of Equality (Section 4)

Solve: $(a)\ \dfrac{w}{3} = 5$ $(b)\ 10 = \dfrac{x}{7}$

PROCEDURE   **SOLUTIONS**

1. Multiply both sides of the equation by the divisor of the variable: 1. $\mathbf{M}_3$ $3 \cdot \dfrac{w}{3} = 3 \cdot 5$ $\mathbf{M}_7$ $7 \cdot 10 = 7 \cdot \dfrac{x}{7}$

                                  Ans. $w = 15$ *Ans.* $70 = x$

2. Check the original equation: 2. $\dfrac{w}{3} = 5$ $10 = \dfrac{x}{7}$

                                      $\dfrac{15}{3} \stackrel{?}{=} 5$ $10 \stackrel{?}{=} \dfrac{70}{7}$

                                        $5 = 5$ $10 = 10$

Note 1:   **M** is a convenient symbol for "multiplying both sides." $\mathbf{M}_3$ means "multiply both sides by 3."

Note 2:   **A** common factor may be eliminated in $3 \cdot \dfrac{w}{\overset{1}{3}}$ and $7 \cdot \dfrac{x}{\overset{1}{7}}$.

Solving an Equation Whose Variable Has a Fractional Coefficient

To divide by a fraction, invert the fraction and multiply (Chapter 4, Section 3).

To Solve an Equation Whose Variable Has a Fractional Coefficient

Solve: $(a)\ \dfrac{2}{3}x = 8$ $(b)\ \dfrac{5}{3}y = 25$

PROCEDURE

1. Multiply both sides of the equation by the fractional coefficient inverted:

2. Check the original equation:

SOLUTIONS

1. $M_{3/2}$ $\dfrac{3}{2} \cdot \dfrac{2}{3} x = \dfrac{3}{2} \cdot 8$ $M_{3/5}$ $\dfrac{3}{5} \cdot \dfrac{5}{3} y = \dfrac{3}{5} \cdot 25$

Ans. $x = 12$ *Ans.* $y = 15$

2. $\dfrac{2}{3} x = 8$ $\dfrac{5}{3} y = 25$

$\dfrac{2}{3} \cdot 12 \overset{?}{=} 8$ $\dfrac{5}{3} \cdot 15 \overset{?}{=} 25$

$8 = 8$ $25 = 25$

6.1 SOLVING EQUATIONS WITH NATURAL NUMBER DIVISORS

Solve:

(a) $\dfrac{x}{8} = 4$ (b) $\dfrac{1}{3} y = 12$ (c) $20 = \dfrac{z}{10}$ (d) $.2 = \dfrac{w}{40}$

Solutions

(a) $\dfrac{x}{8} = 4$ (b) $\dfrac{1}{3} y = 12$ (c) $20 = \dfrac{z}{10}$ (d) $.2 = \dfrac{w}{40}$

M_8 $8 \cdot \dfrac{x}{8} = 8 \cdot 4$ M_3 $3 \cdot \dfrac{1}{3} y = 3 \cdot 12$ M_{10} $10 \cdot 20 = 10 \cdot \dfrac{z}{10}$ M_{40} $40(.2) = 40 \cdot \dfrac{w}{40}$

Ans. $x = 32$ *Ans.* $y = 36$ *Ans.* $200 = z$ *Ans.* $8 = w$

Check: **Check:**   **Check:**   **Check:**

$\dfrac{x}{8} = 4$ $\dfrac{1}{3} y = 12$ $20 = \dfrac{z}{10}$ $.2 = \dfrac{w}{40}$

$\dfrac{32}{8} \overset{?}{=} 4$ $\dfrac{1}{3}(36) \overset{?}{=} 12$ $20 \overset{?}{=} \dfrac{200}{10}$ $.2 \overset{?}{=} \dfrac{8}{40}$

$4 = 4$ $12 = 12$ $20 = 20$ $.2 = .2$

6.2 SOLVING EQUATIONS WITH DECIMAL DIVISORS

Find each solution set:

(a) $\dfrac{a}{.5} = 4$ (b) $\dfrac{b}{.08} = 400$ (c) $1.5 = \dfrac{c}{1.2}$

Solutions

(a) $\dfrac{a}{.5} = 4$ (b) $\dfrac{b}{.08} = 400$ (c) $1.5 = \dfrac{c}{1.2}$

$M_{.5}$ $.5\left(\dfrac{a}{.5}\right) = .5(4)$ $M_{.08}$ $.08\left(\dfrac{b}{.08}\right) = .08(400)$ $M_{1.2}$ $1.2(1.5) = 1.2\left(\dfrac{c}{1.2}\right)$

$a = 2$ $b = 32$ $1.8 = c$

Ans. {2}   *Ans.* {32}   *Ans.* {1.8}

6.3   SOLVING EQUATIONS WITH FRACTIONAL COEFFICIENTS

Solve:

(a) $\dfrac{2}{5}x = 10$ (b) $1\dfrac{1}{3}w = 30$ (c) $c - \dfrac{1}{4}c = 24$

Solutions

(a) $\dfrac{2}{5}x = 10$ (b) $1\dfrac{1}{3}w = 30$ (c) $c - \dfrac{1}{4}c = 24$

                             $\dfrac{4}{3}w = 30$ $\dfrac{3}{4}c = 24$

$M_{5/2}$ $\dfrac{5}{2} \cdot \dfrac{2}{5}x = \dfrac{5}{2} \cdot 10$ $M_{3/4}$ $\dfrac{3}{4} \cdot \dfrac{4}{3}w = \dfrac{3}{4} \cdot 30$ $M_{4/3}$ $\dfrac{4}{3} \cdot \dfrac{3}{4}c = \dfrac{4}{3} \cdot 24$

Ans. $x = 25$ *Ans.* $w = 22\dfrac{1}{2}$ *Ans.* $c = 32$

6.4   SOLVING EQUATIONS WITH PER CENTS AS COEFFICIENTS

Solve: (a) $66\frac{2}{3}\% s = 22$,   (b) $87\frac{1}{2}\% t = 35$,   (c) $120\% w = 72$.

Solutions Replace a per cent by a fraction if the per cent equals an easy fraction.

(a) $66\frac{2}{3}\% s = 22$ (b) $87\frac{1}{2}\% t = 35$ (c) $120\% w = 72$

                $\dfrac{2}{3}s = 22$ $\dfrac{7}{8}t = 35$ $\dfrac{6}{5}w = 72$

$M_{3/2}$ $\dfrac{3}{2} \cdot \dfrac{2}{3}s = \dfrac{3}{2}(22)$ $M_{8/7}$ $\dfrac{8}{7} \cdot \dfrac{7}{8}t = \dfrac{8}{7}(35)$ $M_{5/6}$ $\dfrac{5}{6} \cdot \dfrac{6}{5}w = \dfrac{5}{6}(72)$

Ans. $s = 33$ *Ans.* $t = 40$ *Ans.* $w = 60$

6.5   MULTIPLICATION RULE IN PROBLEM SOLVING

Mr. White receives 5% on a stock investment. If his interest at the end of one year was \$140, how large was his investment?

Solution Let s = the invested sum \$.

    Then, $5\% s = \dfrac{s}{20} = 140$

    M_{20} $20 \cdot \dfrac{s}{20} = 20(140)$

                    $s = 2800$

Ans. The investment was \$2800.

Check (the problem):

5% of the investment should be \$140. Hence,

    $5\%(\$2800) \overset{?}{=} \$140$

    $\$140 = \$140$

7. USING SUBTRACTION TO SOLVE AN EQUATION

To Solve an Equation Using the Subtraction Rule of Equality (Section 4)

Solve: (a) $w + 12 = 19$ (b) $28 = 11 + x$

PROCEDURE		SOLUTIONS

1. Subtract from both sides the number added to the variable:

2. Check the original equation:

1.
$$w + 12 = 19$$
$$\mathbf{S}_{12} \quad \underline{-12 = -12}$$
$$Ans. \quad w \quad = \quad 7$$

2.
$$w + 12 = 19$$
$$7 + 12 \overset{?}{=} 19$$
$$19 = 19$$

$$28 = 11 + x$$
$$\mathbf{S}_{11} \quad \underline{-11 = -11}$$
$$Ans. \quad 17 = \quad x$$

$$28 = 11 + x$$
$$28 \overset{?}{=} 11 + 17$$
$$28 = 28$$

Note: **S** is a convenient symbol for "subtracting from both sides."
  S$_{11}$ means "subtract 11 from both sides."

7.1 **SUBTRACTION RULE IN EQUATIONS CONTAINING NATURAL NUMBERS**

Find each solution set: (*a*) $r + 8 = 13$,   (*b*) $15 + t = 60$,   (*c*) $110 = s + 20$.

Solutions

(*a*)
$$r + 8 = 13$$
$$\mathbf{S}_8 \quad \underline{-8 = -8}$$
$$r \quad = \quad 5$$
Ans. {5}

(*b*)
$$15 + t = 60$$
$$\mathbf{S}_{15} \quad \underline{-15 \quad = -15}$$
$$t = 45$$
Ans.   {45}

(*c*)
$$100 = s + 20$$
$$\mathbf{S}_{20} \quad \underline{-20 = \quad -20}$$
$$90 = s$$
Ans. {90}

Check:
$$r + 8 = 13$$
$$5 + 8 \overset{?}{=} 13$$
$$13 = 13$$

Check:
$$15 + t = 60$$
$$15 + 45 \overset{?}{=} 60$$
$$60 = 60$$

Check:
$$110 = s + 20$$
$$110 \overset{?}{=} 90 + 20$$
$$110 = 110$$

7.2 **SUBTRACTION RULE IN EQUATIONS CONTAINING FRACTIONS OR DECIMALS**

Solve: (*a*) $b + \frac{1}{3} = 3\frac{2}{3}$,   (*b*) $2\frac{3}{4} + c = 8\frac{1}{2}$,   (*c*) $20.8 = d + 6.9$.

Solutions

(*a*)
$$b + \frac{1}{3} = 3\frac{2}{3}$$
$$\mathbf{S}_{1/3} \quad \underline{-\frac{1}{3} = -\frac{1}{3}}$$
Ans. $b = 3\frac{1}{3}$

(*b*)
$$2\frac{3}{4} + c = 8\frac{1}{2}$$
$$\mathbf{S}_{2\frac{3}{4}} \quad \underline{-2\frac{3}{4} \quad = -2\frac{3}{4}}$$
Ans. $c = 5\frac{3}{4}$

(*c*)
$$20.8 = d + 6.9$$
$$\mathbf{S}_{6.9} \quad \underline{-6.9 = \quad -6.9}$$
Ans. $13.9 = d$

Check:
$$b + \frac{1}{3} = 3\frac{2}{3}$$
$$3\frac{1}{3} + \frac{1}{3} \overset{?}{=} 3\frac{2}{3}$$
$$3\frac{2}{3} = 3\frac{2}{3}$$

Check:
$$2\frac{3}{4} + c = 8\frac{1}{2}$$
$$2\frac{3}{4} + 5\frac{3}{4} \overset{?}{=} 8\frac{1}{2}$$
$$8\frac{1}{2} = 8\frac{1}{2}$$

Check:
$$20.8 = d + 6.9$$
$$20.8 \overset{?}{=} 13.9 + 6.9$$
$$20.8 = 20.8$$

7.3   SUBTRACTION RULE IN PROBLEM SOLVING

After an increase of 22¢, the price of eggs rose to $1.22. What was the original price?

Solution   Let p = original price in ¢.

Then, $p + 22 = 122$

S_{22} $\underline{\quad -22 = -\ 22\quad}$

$p \quad\ = 100$

Check (the problem):

After increasing 22¢, the new
price should be $1.22. Hence,
$$\$1.00 + 22¢ \overset{?}{=} \$1.22$$
$$\$1.22 = \$1.22$$

Ans.   The original price was $1.00

7.4   SUBTRACTION RULE IN PROBLEM SOLVING

Harold's height is 5 ft 3 in. If he is 9 in. taller than John, how tall is John?

Solution   Let J = John's height in ft.

Then, $J + \frac{3}{4} = 5\frac{1}{4}$ (9 in. $= \frac{3}{4}$ ft)

$S_{3/4}$ $\underline{\quad -\frac{3}{4} = -\frac{3}{4}\quad}$

$J \quad = 4\frac{1}{2}$

Check (the problem):

9 in. more than John's height
should equal 5 ft 3 in.
$$4\frac{1}{2}\ \text{ft} + \frac{3}{4}\ \text{ft} \overset{?}{=} 5\frac{1}{4}\ \text{ft}$$
$$5\frac{1}{4}\ \text{ft} = 5\frac{1}{4}\ \text{ft}$$

Ans.   John is $4\frac{1}{2}$ ft or 4 ft 6 in. tall.

8.   USING ADDITION TO SOLVE AN EQUATION

To Solve an Equation Using the Addition Rule of Equality (Section 4)

Solve: (*a*) $n - 19 = 21$ (*b*) $17 = m - 8$

PROCEDURE

1. Add to both sides the number
   subtracted from the variable:

2. Check the original equation:

SOLUTIONS

1. $n - 19 = \ \ 21$

A_{19} $\underline{\ \ +19 = +19\ \ }$

Ans. $n \quad\ \ \ = \ \ 40$

2. $n - 19 = 21$

$40 - 19 \overset{?}{=} 21$

$21 = 21$

$17 = m - 8$

A_s $\underline{\ + 8 = \ \ +8\ }$

Ans. $25 = m$

$17 = m - 8$

$17 \overset{?}{=} 25 - 8$

$17 = 17$

Note: **A** is a convenient symbol for "adding to both sides."
   A_{19} means "add 19 to both sides."

8.1   ADDITION RULE IN EQUATIONS CONTAINING NATURAL NUMBERS

Find each solution set: (*a*) $w - 10 = 19$,   (*b*) $x - 19 = 10$,   (*c*) $7 = y - 82$,   (*d*) $82 = z - 7$.

Solutions

(*a*) $w - 10 = \ \ 19$

A_{10} $\underline{\ + 10 = +10\ }$

$w \quad\ = \ \ 29$

Ans.   {29}

(*b*) $x - 19 = \ \ 10$

A_{19} $\underline{\ + 19 = +19\ }$

$x \quad\ = \ \ 29$

Ans.   {29}

(*c*) $7 = y - 82$

A_{82} $\underline{\ +82 = \ \ +82\ }$

$89 = y$

Ans.   {89}

(*d*) $82 = z - 7$

A_7 $\underline{\ +7 = \ \ +7\ }$

$89 = z$

Ans.   {89}

Check:	**Check:**	**Check:**	**Check:**
$w - 10 = 19$	$x - 19 = 10$	$7 = y - 82$	$82 = z - 7$
$29 - 10 \overset{?}{=} 19$	$29 - 19 \overset{?}{=} 10$	$7 \overset{?}{=} 89 - 82$	$82 \overset{?}{=} 89 - 7$
$19 = 19$	$10 = 10$	$7 = 7$	$82 = 82$

8.2   ADDITION RULE IN EQUATIONS CONTAINING FRACTIONS OR DECIMALS

Solve: (a) $h - \frac{3}{8} = 5\frac{1}{4}$,   (b) $j - 20\frac{7}{12} = 1\frac{1}{12}$,   (c) $12.5 = m - 2.9$.

Solutions

(a) $\qquad h - \frac{3}{8} = 5\frac{1}{4}$ (b) $\qquad j - 20\frac{7}{12} = 1\frac{1}{12}$ (c) $\qquad 12.5 = m - 2.9$

$\mathbf{A}_{3/8} \quad \underline{+\frac{3}{8} = \frac{3}{8}} \qquad \mathbf{A}_{20\frac{7}{12}} \quad \underline{+20\frac{7}{12} = 20\frac{7}{12}} \qquad \mathbf{A}_{2.9} \quad \underline{+2.9 = \quad +2.9}$

Ans. $\quad h \quad = 5\frac{5}{8}$ *Ans.* $\quad j \quad = 21\frac{2}{3}$ *Ans.* $\quad 15.4 = m$

8.3   ADDITION RULE IN PROBLEM SOLVING

A drop of 8° brought the temperature to 64°.   What was the original temperature?

Solution   Let $t =$ original temperature in °.   **Check** (the problem):

$$\begin{aligned} \text{Then,} \qquad t - 8 &= 64 \\ \mathbf{A}_8 \qquad \underline{+8} &= \underline{+8} \\ t &= 72 \end{aligned}$$

The original temperature, dropped 8°, should become 64°. Hence,

$$72° - 8° \overset{?}{=} 64°$$
$$64° = 64°$$

Ans.   The original temperature was 72°.

8.4   ADDITION RULE IN PROBLEM SOLVING

After giving 15 marbles to Sam, Joe has 43 left. How many did Joe have originally?

Solutions   Let $m =$ original no. of marbles.   **Check** (the probem):

$$\begin{aligned} \text{Then,} \qquad m - 15 &= 43 \\ \mathbf{A}_{15} \qquad \underline{+15} &= \underline{+15} \\ m &= 58 \end{aligned}$$

The original number of marbles, less 15, should be 43. Hence,

$$58 \text{ marbles} - 15 \text{ marbles} \overset{?}{=} 43 \text{ marbles}$$
$$43 \text{ marbles} = 43 \text{ marbles}$$

Ans.   Joe had 58 marbles at first.

9. USING TWO OR MORE OPERATIONS TO SOLVE AN EQUATION

In equations where two operations are performed upon the variable, two inverse operations may be needed to solve the equation.

Thus, in $2x + 7 = 19$, the two operations upon the variable are *multiplication and addition*. To solve, use *division and subtraction*, performing subtraction first.

Also, in $(x/3) - 5 = 2$, the two operations upon the variable are *division and subtraction*. To solve, use *multiplication and addition*, performing addition first.

To Solve Equations Using Two Inverse Operations

Solve: (a) $2x + 7 = 19$ (b) $\dfrac{x}{3} - 5 = 2$

PROCEDURE   **SOLUTIONS**

1. Perform addition to undo subtraction, or subtraction to undo addition:

$$\begin{array}{ll} \textbf{1.} & 2x + 7 = 19 \\ \mathbf{S_7} & \underline{-7 = -7} \\ & 2x = 12 \end{array}$$

$$\mathbf{A_5} \quad \begin{array}{l} \dfrac{x}{3} - 5 = 2 \\ \underline{\phantom{\dfrac{x}{3}}+5 = +5} \\ \dfrac{x}{3} = 7 \end{array}$$

2. Perform multiplication to undo division, or division to undo multiplication:

$$\textbf{2. } \mathbf{D_2} \quad \dfrac{2x}{2} = \dfrac{12}{2}$$

$$\mathbf{M_3} \quad 3 \cdot \dfrac{x}{3} = 3 \cdot 7$$

$$\textit{Ans.} \quad x = 6$$

$$\textit{Ans.} \quad x = 21$$

3. Check the original equation:

$$\begin{array}{ll} \textbf{3.} & 2x + 7 = 19 \\ & 2(6) + 7 \overset{?}{=} 19 \\ & 19 = 19 \end{array}$$

$$\begin{array}{l} \dfrac{x}{3} - 5 = 2 \\ \dfrac{21}{3} - 5 \overset{?}{=} 2 \\ 2 = 2 \end{array}$$

9.1   USING TWO INVERSE OPERATIONS TO SOLVE AN EQUATION

Solve:

(a) $2x + 7 = 11$ (b) $3x - 5 = 7$ (c) $\dfrac{x}{3} + 5 = 7$ (d) $\dfrac{x}{5} - 3 = 7$

Solutions

$$(a) \quad \begin{array}{l} 2x + 7 = 11 \\ \mathbf{S_7} \quad \underline{-7 = -7} \\ 2x = 4 \end{array}$$
$$\mathbf{D_2} \quad \dfrac{2x}{2} = \dfrac{4}{2}$$
$$\textit{Ans.} \quad x = 2$$

$$(b) \quad \begin{array}{l} 3x - 5 = 7 \\ \mathbf{A_5} \quad \underline{+5 = +5} \\ 3x = +12 \end{array}$$
$$\mathbf{D_3} \quad \dfrac{3x}{3} = \dfrac{12}{3}$$
$$\textit{Ans.} \quad x = 4$$

$$(c) \quad \begin{array}{l} \dfrac{x}{3} + 5 = 7 \\ \mathbf{S_5} \quad \underline{\phantom{\dfrac{x}{3}}-5 = -5} \\ \dfrac{x}{3} = 2 \end{array}$$
$$\mathbf{M_3} \quad 3 \cdot \dfrac{x}{3} = 3 \cdot 2$$
$$\textit{Ans.} \quad x = 6$$

$$(d) \quad \begin{array}{l} \dfrac{x}{5} - 3 = 7 \\ \mathbf{A_3} \quad \underline{\phantom{\dfrac{x}{5}}+3 = 3} \\ \dfrac{x}{5} = 10 \end{array}$$
$$\mathbf{M_5} \quad 5 \cdot \dfrac{x}{5} = 5 \cdot 10$$
$$\textit{Ans.} \quad x = 50$$

(a) **Check:**

$$\begin{array}{l} 2x + 7 = 11 \\ 2(2) + 7 \overset{?}{=} 11 \\ 4 + 7 \overset{?}{=} 11 \\ 11 = 11 \end{array}$$

(b) **Check:**

$$\begin{array}{l} 3x - 5 = 7 \\ 3(4) - 5 \overset{?}{=} 7 \\ 12 - 5 \overset{?}{=} 7 \\ 7 = 7 \end{array}$$

(c) **Check:**

$$\begin{array}{l} \dfrac{x}{3} + 5 = 7 \\ \dfrac{6}{3} + 5 \overset{?}{=} 7 \\ 2 + 5 \overset{?}{=} 7 \\ 7 = 7 \end{array}$$

(d) **Check:**

$$\begin{array}{l} \dfrac{x}{5} - 3 = 7 \\ \dfrac{50}{5} - 3 \overset{?}{=} 7 \\ 10 - 3 \overset{?}{=} 7 \\ 7 = 7 \end{array}$$

9.2   SOLVING EQUATIONS WITH LIKE TERMS ON THE SAME SIDE

Solve: (a) $8n + 4n - 3 = 9$,   (b) $13n + 4 + n = 39$,   (c) $10 = 7 + n - (n/2)$.

Solutions   Combine like terms first.

$$(a) \quad \begin{array}{l} 8n + 4n - 3 = 9 \\ 12n - 3 = 9 \\ \mathbf{A_3} \quad \underline{+3 = 3} \\ 12n = 12 \end{array}$$
$$\mathbf{D_{12}} \quad \dfrac{12n}{12} = \dfrac{12}{12}$$
$$\textit{Ans.} \quad n = 1$$

$$(b) \quad \begin{array}{l} 13n + 4 + n = 39 \\ 14n + 4 = 39 \\ \mathbf{S_4} \quad \underline{-4 = -4} \\ 14n = 35 \end{array}$$
$$\mathbf{D_{14}} \quad \dfrac{14n}{14} = \dfrac{35}{14}$$
$$\textit{Ans.} \quad n = 2\tfrac{1}{2}$$

$$(c) \quad \begin{array}{l} 10 = 7 + n - \dfrac{n}{2} \\ 10 = 7 + \dfrac{n}{2} \\ \mathbf{S_7} \quad \underline{-7 = -7} \\ 3 = \dfrac{n}{2} \end{array}$$
$$\mathbf{M_2} \quad 2 \cdot 3 = 2 \cdot \dfrac{n}{2}$$
$$\textit{Ans.} \quad 6 = n$$

9.3 SOLVING EQUATIONS WITH LIKE TERMS ON OPPOSITE SIDES

Find each solution set: (a) $5n = 40 - 3n$,   (b) $4u + 5 = 5u - 30$,   (c) $3r + 10 = 2r + 20$.

Solutions First, add or subtract to collect like terms on the same side.

(a)
$$5n = 40 - 3n$$
A_{3n} $\quad \dfrac{+3n = \qquad + 3n}{8n = 40}$

D_8 $\quad \dfrac{8n}{8} = \dfrac{40}{8}$

$$n = 5$$

Ans. $\{5\}$

(b)
$$4u + 5 = 5u - 30$$
A_{30} $\quad \dfrac{+ 30 = \qquad + 30}{4u + 35 = \quad 5u}$

S_{4u} $\quad \dfrac{-4u \qquad\quad = -4u}{35 = \quad u}$

Ans. $\{35\}$

(c)
$$3r + 10 = 2r + 20$$
S_{10} $\quad \dfrac{- 10 = \qquad - 10}{3r \quad = 2r + 10}$

S_{2r} $\quad \dfrac{-2r \qquad = -2r}{r \quad = \quad 10}$

Ans. $\{10\}$

9.4 SOLVING EQUATIONS IN WHICH THE VARIABLE IS A DIVISOR

Solve:

(a) $\dfrac{8}{x} = 2$ (b) $12 = \dfrac{3}{y}$ (c) $\dfrac{7}{x} = \dfrac{1}{5}$ (d) $\dfrac{1}{7} = \dfrac{3}{x}$

Solutions First multiply both sides by the variable.

(a)
$$\frac{8}{x} = 2$$
M_x $\quad \left(\dfrac{8}{x}\right)x = 2x$

$$8 = 2x$$
D_2 $\quad \dfrac{8}{2} = \dfrac{2x}{2}$

Ans. $4 = x$

(b)
$$12 = \frac{3}{y}$$
M_y $\quad 12y = \left(\dfrac{3}{y}\right)y$

$$12y = 3$$
D_{12} $\quad \dfrac{12y}{12} = \dfrac{3}{12}$

Ans. $y = 1/4$

(c)
$$\frac{7}{x} = \frac{1}{5}$$
M_x $\quad \dfrac{7}{x} \cdot x = \dfrac{1}{5}x$

$$7 = \frac{x}{5}$$
M_5 $\quad 5(7) = 5\left(\dfrac{x}{5}\right)$

Ans. $35 = x$

(d)
$$\frac{1}{7} = \frac{3}{x}$$
M_x $\quad \dfrac{1}{7}x = \dfrac{3}{x} \cdot x$

$$\frac{x}{7} = 3$$
M_7 $\quad 7 \cdot \dfrac{x}{7} = 7 \cdot 3$

Ans. $x = 21$

9.5 SOLVING EQUATIONS WHOSE VARIABLE HAS A FRACTIONAL COEFFICIENT

(a) Solve: $\dfrac{3}{8}x = 9$

Solutions

Using One Operation

(a)
$$\frac{3}{8}x = 9$$
$M_{8/3}$ $\quad \dfrac{8}{3} \cdot \dfrac{3}{8}x = \dfrac{8}{3} \cdot 9$

Ans. $x = 24$

Using Two Operations

(a)
$$\frac{3}{8}x = 9$$
M_8 $\quad 8 \cdot \dfrac{3}{8}x = 8 \cdot 9$

$$3x = 72$$
D_3 $\quad \dfrac{3x}{3} = \dfrac{72}{3}$

Ans. $x = 24$

(b) Solve: $25 = \dfrac{5}{4}x$

Solutions

Using One Operation

(b)
$$25 = \frac{5}{4}x$$
$M_{4/5}$ $\quad \dfrac{4}{5} \cdot 25 = \dfrac{4}{5} \cdot \dfrac{5}{4}x$

Ans. $20 = x$

Using Two Operations

(b)
$$25 = \frac{5}{4}x$$
M_4 $\quad 4(25) = 4 \cdot \dfrac{5}{4}x$

$$100 = 5x$$
D_5 $\quad \dfrac{100}{5} = \dfrac{5x}{5}$

Ans. $20 = x$

9.6   SOLVING MORE DIFFICULT EQUATIONS

Solve:

(a) $\dfrac{3}{4}y - 5 = 7$ (b) $8 + \dfrac{2}{7}b = 20$ (c) $48 - \dfrac{5}{3}w = 23$

Solutions

(a)
$$\frac{3}{4}y - 5 = 7$$

A_5 $\underline{+5 = +5}$

$$\frac{3}{4}y \;\;\;\; = 12$$

$M_{4/3}$ $\dfrac{4}{3} \cdot \dfrac{3}{4}y = \dfrac{4}{3} \cdot 12$

Ans. $y = 16$

(b)
$$8 + \frac{2}{7}b = 20$$

S_8 $\underline{-8 \qquad = -8}$

$$\frac{2}{7}b = 12$$

$M_{7/2}$ $\dfrac{7}{2} \cdot \dfrac{2}{7}b = \dfrac{7}{2} \cdot 12$

Ans. $b = 42$

(c)
$$48 - \frac{5}{3}w = 23$$

$A_{\frac{5}{3}w}$ $\underline{+\dfrac{5}{3}w = +\dfrac{5}{3}w}$

$$48 \;\;\; = \frac{5}{3}w + 23$$

S_{23} $\underline{-23 \qquad = \qquad -23}$

$$25 \;\;\; = \frac{5}{3}w$$

$M_{3/5}$ $\dfrac{3}{5} \cdot 25 = \dfrac{3}{5} \cdot \dfrac{5}{3}w$

Ans. $15 = w$

9.7   USING TWO OPERATIONS IN PROBLEM SOLVING

How many boys are there in a class of 36 pupils if the number of girls is   (a) 6 more?   (b) three times as many?

Solutions

(a)   Let b = number of boys
          Then $b + 6$ = number of girls
                $b + (b + 6) = 36$
S_6 $2b + 6 = 36$
D_2 $2b = 30$
                 $b = 15$

Ans.   There are 15 boys.

(b)   Let b = number of boys
          Then $3b$ = number of girls
                $b + 3b = 36$
D_4 $4b = 36$
                 $b = 9$

Ans.   There are 9 boys.

9.8   USING TWO OPERATIONS IN PROBLEM SOLVING

Paul has $26.00 in his bank. By adding equal deposits each week for 20 weeks, he hopes to have $78.00. How much should each weekly deposit be?

Solution   Let d = no. of dollars in each deposit
          Then, $20d + 26 = 78$
S_{260} $\underline{-26 = -26}$
D_{20} $\underline{20d \quad = 52}$
          $d \quad = 2.60$

Check (the problem):
   20 deposits and $26 should equal
   the total of $78. Hence,
   $20(\$2.60) + \$26 \overset{?}{=} \$78$
   $\$52 + \$26 \overset{?}{=} \$78$
   $\$78 = \$78$

Ans.   He must deposit $2.60 a week.

9.9   USING TWO OPERATIONS IN PROBLEM SOLVING

Mr. Richards sold his boat for $9000. His loss amounted to two-fifths of his cost. What did the boat cost him?

Solution   Let c = cost in $

Then $9000 = c - \dfrac{2}{5}c$

$9000 = \dfrac{3}{5}c$

$\mathbf{M}_{5/3}$ $\dfrac{5}{3} \cdot 9000 = \dfrac{5}{3} \cdot \dfrac{3}{5}c$

$15{,}000 = c$

Ans.   The cost was $15,000.

Check (the problem):
If $15,000 is the cost, the loss is $\frac{2}{5} \cdot$ $15,000 or $6,000. The selling price of $9000 should be the cost minus the loss. Hence,
$9000 \overset{?}{=} \$15{,}000 - \$6000$
$9000 = \$9000$

10.   DERIVING FORMULAS

To Derive a Formula for Related Quantities

Derive a formula relating the *distance (D)* traveled in a *time (T)* at a *rate of speed (R)*.

PROCEDURE

1. Obtain sets of values for these quantities, using convenient numbers:

2. Tabulate those sets of values:

SOLUTION

1. **Sets of Values**
   At 50 mph for 2 hr, 100 mi will be traveled.
   At 25 mph for 10 hr, 250 mi will be traveled.
   At 40 mph for 3 hr, 120 mi will be traveled.

2. **Table of Values.** (Place units above quantities.)

(mph) Rate (R)	(hr) Time (T)	(mi) Distance (D)
50	2	$50 \cdot 2 = 100$
25	10	$25 \cdot 10 = 250$
40	3	$40 \cdot 3 = 120$

3. State the rule that follows:

4. State the formula that expresses the rule:

3. **Rule:**   The product of the rate and time equals the distance.

4. **Formula:** = $RT = D$

Note: If D is in mi and T in hr, then R must be in mi per hr (mph). In general, *rate* must be in distance units per time unit.

Obtaining Formulas from a More General Formula

A formula such as $RT = D$ relates *three* quantities: time, rate and distance. Each of these quantities may vary in value; that is, they may have many values. However, in a problem, situation or discussion, one of these quantities may have a fixed value. When such is the case, this constant value may be used to obtain a formula relating the other *two* quantities.

Thus, $D = RT$ leads to $D = 30T$ if the rate of speed is fixed at 30 mph, 30 km/hr, etc.

10.1   DERIVING A COIN FORMULA

Derive a formula for the number of nickels (n) equivalent (equal in value) to q quarters.

Solution

1. Sets of Values   **2. Table of Values**

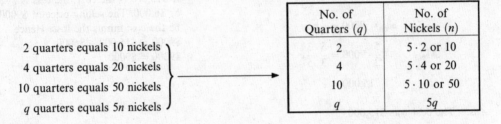

2 quarters equals 10 nickels
4 quarters equals 20 nickels
10 quarters equals 50 nickels
q quarters equals $5n$ nickels

No. of Quarters (q)	No. of Nickels (n)
2	$5 \cdot 2$ or 10
4	$5 \cdot 4$ or 20
10	$5 \cdot 10$ or 50
q	$5q$

3. Rule: The number of nickels equivalent to a number of quarters is five times that number.

4. Formula: $n = 5q$ *Ans.*

10.2   DERIVING A COIN FORMULA

Derive a formula for the value in cents (c) of d dimes and n nickels.

Solution

1. Sets of Values   **2. Table of Values**

3 dimes and 4 nickels equals 50¢
4 dimes and 2 nickels equals 50¢
5 dimes and 3 nickels equals 65¢
d dimes and n nickels equals $(10d + 5n)$¢

No. of Dimes (d)	No. of Nickels (n)	(cents) Value of Dimes & Nickels (c)
3	4	$10 \cdot 3 + 5 \cdot 4$ or 50
4	2	$10 \cdot 4 + 5 \cdot 2$ or 50
5	3	$10 \cdot 5 + 5 \cdot 3$ or 65
d	n	$10d + 5n$

3. Rule: The value in cents of dimes and nickels is ten times the number of dimes plus five times the number of nickels.

4. Formula: $c = 10d + 5n$ *Ans.*

10.3   DERIVING COIN FORMULAS

Derive a formula for   (*a*) the number of pennies (p) equivalent to q quarters,   (*b*) the number of nickels (n) equivalent to d dimes,   (*c*) the number of quarters (q) equivalent to D dollars,   (*d*) the number of pennies (p) equivalent to n nickels and q quarters,   (*e*) the number of nickels (n) equivalent to q quarters and d dimes.

Illustrative Solution   (*b*) Since 1 nickel is 5¢ and 1 dime is 10¢, if nickels are traded for dimes, the number of nickels must be twice the number of dimes.   *Ans.* $n = 2d$.

Ans.   (*a*) $p = 25q$ (*c*) $q = 4D$ (*d*) $p = 5n + 25q$ (*e*) $n = 5q + 2d$

10.4 ˙ DERIVING TIME FORMULAS

Derive a formula for   (a) the number of seconds (s) in m minutes,   (b) the number of hours (h) in d days,   (c) the number of weeks (w) in d days,   (d) the number of days (d) in w weeks and 5 days,   (e) the number of minutes (m) in h hours and 30 sec.

Illustrative Solution   (c) Since 1 week = 7 days, the number of weeks, w, equals 1/7 of the number of days, d.   *Ans.* $w = (1/7)d$ or $w = d/7$

Ans.   (a) $s = 60m$ (b) $h = 24d$ (d) $d = 7w + 5$ (e) $m = 60h + .5$ or $m = 60h + \frac{1}{2}$

10.5   DERIVING LENGTH FORMULAS

Derive a formula for   (a) the number of in. (i) in f ft,   (b) the number of ft (f) in y yd,   (c) the number of yd (y) in i in.,   (d) the number of ft (f) in m mi and 50 yd,   (e) the number of cm (c) in m meters.

Illustrative Solution   (b) Since 1 yd = 3 ft, the number of ft in y yd must be three times y.   *Ans.* $f = 3y$

Ans.   (a) $i = 12f$ (c) $y = i/36$ (d) $f = 5280m + 150$ (e) $c = \dfrac{m}{100}$

10.6   OBTAINING FORMULAS FROM $D = RT$

From $D = RT$, derive a formula relating   (a) the distance in mi and time in hr when the rate is 35 mph,   (b) the distance in ft and the time in sec when sound travels at 1100 ft per sec,   (c) the distance in mi and the time in sec if light travels at 186,000 mi per sec,   (d) the distance in mi and the rate in mph when the time of travel is 1 hr and 30 min,   (e) the rate in km per hr and the time in hr when the distance traveled is 125 km.

Illustrative Solutions

(d) If the distance is in miles and the rate in miles per hour (mph), the time of travel T, is in hours. Since $T = 1\frac{1}{2}$ hours, substitute $1\frac{1}{2}$ for T in $D = RT$.   *Ans.* $D = 1\frac{1}{2}R$
(e) Since the distance traveled, D, is 125, substitute 125 for D in $D = RT$.   *Ans.* $125 = RT$

Ans.   (a) $D = 35T$ (b) $D = 1100T$ (c) $D = 186,000T$

11.   TRANSFORMING FORMULAS

The *subject of a formula* is the variable that is expressed in terms of the other variables. Thus, in $p = 4s$, p is the subject of the formula.

Transforming a formula is the process of changing the subject of the formula. Thus, $p = 4s$ becomes $p/4 = s$ when both sides are divided by 4. In the transforming of the formula, the subject has changed from p to s.

In *Solving a formula* for a variable, the formula is transformed in order that the variable be made the subject. Thus, to solve $D = 5T$ for T, transform it into $T = D/5$.

Use of Inverse Operations to Transform Formulas

1. Use *division* to undo *multiplication*.
   Thus, $c = 25q$ becomes $c/25 = q$ by division.

2. Use *multiplication* to undo *division*.
   Thus, $w = d/7$ becomes $7w = d$ by multiplication.

3. Use *subtraction* to undo *addition*.
   Thus, $S = P + C$ becomes $S - P = C$ by subtraction.

4. Use *addition* to undo *subtraction*.
   Thus, $S = C - L$ becomes $S + L = C$ by addition.

Formulas may be transformed by transposing terms. In transposing a term, change its sign. Thus, $a + b = 180$ becomes $a = 180 - b$ when $+b$ is transposed. Actually, $+b$ has been subtracted from both sides to undo addition.

11.1 TRANSFORMATIONS REQUIRING DIVISION

Solve:

(a) $D = RT$ for R, (b) $D = RT$ for T, (c) $V = LWH$ for L, (d) $c = 10d$ for d, (e) $C = 2\pi r$ for r.

Solutions

(a)	$D = RT$	(b)	$D = RT$	(c)	$V = LWH$	(d)	$c = 10d$	(e)	$C = 2\pi r$
$\mathbf{D}_T$	$\dfrac{D}{T} = \dfrac{RT}{T}$	$\mathbf{D}_R$	$\dfrac{D}{R} = \dfrac{RT}{R}$	$\mathbf{D}_{WH}$	$\dfrac{V}{WH} = \dfrac{LWH}{WH}$	$\mathbf{D}_{10}$	$\dfrac{c}{10} = \dfrac{10d}{10}$	$\mathbf{D}_{2\pi}$	$\dfrac{C}{2\pi} = \dfrac{2\pi r}{2\pi}$
Ans.	$\dfrac{D}{T} = R$	Ans.	$\dfrac{D}{R} = T$	Ans.	$\dfrac{V}{WH} = L$	Ans.	$\dfrac{c}{10} = d$	Ans.	$\dfrac{C}{2\pi} = r$

11.2 TRANSFORMATIONS REQUIRING MULTIPLICATION

Solve:

(a) $\dfrac{i}{12} = f$ for i (b) $f = \dfrac{n}{d}$ for n (c) $\dfrac{V}{LW} = H$ for V (d) $\dfrac{b}{2} = \dfrac{A}{h}$ for A

Solutions

(a)	$\dfrac{i}{12} = f$	(b)	$f = \dfrac{n}{d}$	(c)	$\dfrac{V}{LW} = H$	(d)	$\dfrac{b}{2} = \dfrac{A}{h}$
$\mathbf{M}_{i2}$	$12\left(\dfrac{i}{12}\right) = 12f$	$\mathbf{M}_d$	$df = d\left(\dfrac{n}{d}\right)$	$\mathbf{M}_{LW}$	$LW\left(\dfrac{V}{LW}\right) = LWH$	$\mathbf{M}_h$	$\left(\dfrac{b}{2}\right)h = \left(\dfrac{A}{h}\right)h$
Ans.	$i = 12f$	Ans.	$df = n$	Ans.	$V = LWH$	Ans.	$\dfrac{bh}{2} = A$

11.3 TRANSFORMATIONS REQUIRING ADDITION OR SUBTRACTION (TRANSPOSING)

Solve:

(a) $a + b = 90$ for a (b) $a = b - 180$ for b (c) $a + c = b + 100d$ for b (d) $a - b - 25 = c$ for a

Solutions

(a) $a + b = 90$	(b) $a = b - 180$	(c) $a + c = b + 100d$	(d) $a - b - 25 = c$
Transpose b:	**Transpose** -180:	**Transpose** $100d$:	**Transpose** $-b - 25$:
Ans. $a = 90 - b$	Ans. $a + 180 = b$	Ans. $a + c - 100d = b$	Ans. $a = b + c + 25$

11.4 TRANSFORMATIONS REQUIRING TWO OPERATIONS

Solve:

(a) $P = 2a + b$ for a (b) $c = 10d + 25q$ for q (c) $F = \dfrac{9}{5}C + 32$ for C (d) $V = \dfrac{1}{3}Bh$ for B

Solutions

(a) $P = 2a + b$	(b) $c = 10d + 25q$	(c) $F = \dfrac{9}{5}C + 32$	(d) $V = \dfrac{1}{3}Bh.$
Transpose b:	**Transpose** $10d$:	**Transpose** 32:	$\mathbf{M}_3$ $3V = 3\left(\dfrac{1}{3}Bh\right)$
$\mathbf{D}_2$ $P - b = 2a$	$\mathbf{D}_{25}$ $c - 10d = 25q$		
		$\mathbf{M}_{5/9}$ $F - 32 = \dfrac{9}{5}C$ $\mathbf{D}_h$	$3V = Bh$
Ans. $\dfrac{P - b}{2} = a$	Ans. $\dfrac{c - 10d}{25} = q$	Ans. $\dfrac{5}{9}(F - 32) = C$ Ans.	$\dfrac{3V}{h} = B$

11.5 MORE DIFFICULT TRANSFORMATIONS

Solve:

(a) $A = \dfrac{h}{2}(b + b')$ for h (b) $S = \dfrac{n}{2}(a + l)$ for a (c) $l = a + (n - 1)d$ for n

Solutions

(a) $A = \dfrac{h}{2}(b + b')$ (b) $S = \dfrac{n}{2}(a + l)$ (c) $l = a + (n - 1)d$

$\mathbf{M_2}$ $2A = 2\left(\dfrac{h}{2}\right)(b + b')$ $\mathbf{M_2}$ $2S = n(a + l)$ **Transpose** a: $\mathbf{D_d}$ $l - a = (n - 1)d$

$\mathbf{D_{(b+b')}}$ $\dfrac{2A}{b + b'} = \dfrac{h(b + b')}{b + b'}$ $\mathbf{D_n}$ $\dfrac{2S}{n} = a + l$ $\dfrac{l - a}{d} = n - 1$

                                        Transpose $+l$:   **Transpose** -1:

Ans. $\dfrac{2A}{b + b'} = h$ Ans. $\dfrac{2S}{n} - l = a$ Ans. $\dfrac{l - a}{d} + 1 = n$

12. FINDING THE VALUE OF A VARIABLE IN A FORMULA

When the Variable Is the Subject of the Formula

To evaluate the subject of a formula, replace the other variables by their given values and evaluate the numerical expression that results.

Thus, in $A = bh$ if $b = 10$ and $h = 5$, then $A = 10(5)$ or 50.

When the Variable Is Not the Subject of the Formula

When a variable is to be found and it is not the subject of the formula, two methods may be used:

Method 1:   *Substitute first*, then solve.
Method 2:   *Transform the formula first* to make the variable to be found the subject of the formula, then substitute and evaluate.

Thus, in $p = 3s$ if $p = 27$, the value of s may be found

(1) by substituting first: $27 = 3s$, $s = 9$.
(2) by transforming first: transform $p = 3s$ into $s = \dfrac{p}{3}$. Then $s = \dfrac{27}{3}$ or 9.

12.1 TO FIND THE VALUE OF A VARIABLE THAT IS THE SUBJECT OF A FORMULA

(a) Find V if $V = lwh$ and $l = 10$, $w = 2$, $h = 3.2$.   (c) Find A if $A = p + prt$ and $p = 800$, $r = .04$, $t = 3$.

(b) Find S if $S = \dfrac{n}{2}(a + l)$ and $n = 8$, $a = 5$, $l - 12$.   (d) Find S if $S = \dfrac{1}{2}gt^2$ and $g = 32$, $t = 5$.

Solutions

(a) $V = lwh$ (b) $S = \dfrac{n}{2}(a + l)$ (c) $A = p + prt$ (d) $S = \dfrac{1}{2}gt^2$

         $V = 10(2)(3.2)$ $A = 800 + 800(.04)3$

Ans. $V = 64$ $S = \dfrac{8}{2}(5 + 12)$ Ans. $A = 896$ $S = \dfrac{1}{2} \cdot 32 \cdot 5^2$

                    Ans. $S = 68$ Ans. $S = 400$

12.2   To Find the Value of a Variable That is Not the Subject of a Formula

(a) Find h if $A = bh$, $b = 13$ and $A = 156$.

(b) Find a if $p = 2a + b$, $b = 20$ and $p = 74$.

(c) Find h if $V = \frac{1}{3}Bh$, $B = 240$ and $V = 960$.

Solutions

(a) **(1) By substitution first:**

$$A = bh$$
$$156 = 13h$$
Ans. $12 = h$

(b) **(1) By substitution first:**

$$p = 2a + b$$
$$74 = 2a + 20$$
Ans. $27 = a$

(c) **(1) By substitution first:**

$$V = \frac{1}{3}Bh$$
$$960 = \frac{1}{3}(240)h$$
Ans. $12 = h$

(2) By transformation first:

$$A = nh$$

Transform: $\dfrac{A}{b} = h$

Substitute: $\dfrac{156}{13} = h$

Ans. $12 = h$

(2) By transformation first:

$$p = 2a + b$$

Transform: $\dfrac{p - b}{2} = a$

Substitute: $\dfrac{74 - 20}{2} = a$

Ans. $27 = a$

(2) By transformation first:

$$V = \frac{1}{3}Bh$$

Transform: $\dfrac{3V}{B} = h$

Substitute: $\dfrac{3(960)}{240} = h$

Ans. $12 = h$

12.3   More Difficult Evaluations Using Transformations

(a) Find h if $V = \pi r^2 h$, $\pi = 3.14$, $V = 9420$ and $r = 10$.

(b) Find t if $A = p + prt$, $A = 864$, $p = 800$ and $r = 2$.

Solutions

(a) $$V = \pi r^2 h$$

Transform: $\dfrac{V}{\pi r^2} = h$

Substitute: $\dfrac{9420}{(3.14)(100)} = h$

$$\dfrac{9420}{314} = h$$

Ans. $30 = h$

(b) $$A = p + prt$$

Transform: $\dfrac{A - p}{pr} = t$

Substitute: $\dfrac{864 - 800}{800(2)} = t$

$$\dfrac{64}{1600} = t$$

Ans. $.04 = t$

12.4   Finding an Unknown in a Problem

A train takes 3 hours and 15 minutes to go a distance of 247 miles. Find its average speed.

Solution   Here, $D = RT$, $T = 3\frac{1}{4}$ hr and $D = 247$. To find R:

(1) By substitution first:

$$247 = \frac{13}{4}R$$

$\mathbf{M}_{4/13}$ $\dfrac{4}{13} \cdot \overset{19}{247} = R$

$$76 = R$$

(2) By transformation first:

$$D = RT$$

Transform: $\dfrac{D}{T} = R$

$$247 \div \frac{13}{4} = R$$

$$247 \cdot \frac{4}{13} = R$$

Ans.   Average speed is 76 mph.

Supplementary Problems

8.1. By checking, determine which value is a root of the equation: (a) $3x + 4x = 42$ for **(1.1)**
$x = 4$, 6 and 8; (b) $3n + 14 = 47$ for $n = 9$, 10 and 11; (c) $6y - 48 = 2y$ for $y = 8$, 10 and 12.

Ans. (a) $x = 6$ (b) $n = 11$ (c) $y = 12$

8.2. By checking, show that x may have any of the following values in the identity **(1.2)**
$2(x - 3) = 2x - 6$: (a) $x = 10$, (b) $x = 6$, (c) $x = 4\frac{1}{2}$, (d) $x = 3.1$.

8.3. Translate into an equation, letting n represent the number: (a) What number **(2.1)**
diminished by 8 equals 13? (b) Two-thirds of what number equals 10? (c) Three times the sum
of a number and six is 33. What is the number? (d) What number increased by 20 equals three
times the same number? (e) What number increased by 5 equals twice the same number,
decreased by 4?

Ans. (a) $n - 8 = 13$ (b) $\frac{2}{3}n = 10$ (c) $3(n + 6) = 33$ (d) $n + 20 = 3n$ (e) $n + 5 = 2n - 4$

8.4. Match the statements in Column 1 with the equations in Column 2. **(2.2)**

Column 1	Column 2
1. The sum of 8 and twice a number is 18.	(a) $\frac{1}{8}n + 2 = 18$
2. Twice a number, less 8, is 18.	(b) $8(n - 2) = 18$
3. Twice the sum of a number and 8 is 18.	(c) $\frac{1}{2}(8 - n) = 18$
4. Eight times the difference of a number and 2 is 18.	(d) $2(n + 8) = 18$
5. One-half the difference of 8 and a number is 18.	(e) $2n + 8 = 18$
6. 2 more than one-eighth of a number is 18.	(f) $\frac{1}{2}n - 8 = 18$
7. 8 less than half a number is 18.	(g) $2n - 8 = 18$

Ans. 1. and (e)   2. and (g)   3. and (d)   4. and (b)   5. and (c)   6. and (a)   7. and (f)

8.5. Letting n represent the number of games lost, obtain an equation for each problem: **(2.3)**
(a) A team won three times as many games as it lost. It played a total of 52 games. (b) A team
won 20 games more than it lost. It played a total of 84 games. (c) A team won 15 games less
than twice the number lost. It played a total of 78 games.

Ans. (a) $n + 3n = 52$ (b) $n + n + 20 = 84$ (c) $n + 2n - 15 = 78$

8.6. Solve each equation: **(3.1)**

(a) $a + 5 = 9$ (e) $x + 11 = 21 + 8$ (i) $45 = m - 13$
(b) $7 + b = 15$ (f) $27 + 13 = 18 + y$ (j) $22 = n - 50$
(c) $20 = c + 12$ (g) $h - 6 = 14$ (k) $x - 42 = 80 - 75$
(d) $75 = 55 + d$ (h) $k - 14 = 6$ (l) $100 - 31 = y - 84$

Ans. (a) $a = 4$ (c) $c = 8$ (e) $x = 18$ (g) $h = 20$ (i) $m = 58$ (k) $x = 47$
(b) $b = 8$ (d) $d = 20$ (f) $y = 22$ (h) $k = 20$ (j) $n = 72$ (l) $y = 153$

8.7. Solve each equation: (3.2)

(a) $4p = 48$ (d) $4n = 2$ (g) $\dfrac{t}{5} = 6$ (i) $\dfrac{x}{15} = 4$ (k) $\dfrac{a}{10} = \dfrac{2}{5}$

(b) $10r = 160$ (e) $12w = 4$

(c) $25s = 35$ (f) $24x = 21$ (h) $\dfrac{u}{65} = 1$ (j) $\dfrac{y}{12} = \dfrac{3}{2}$ (l) $\dfrac{1}{3}b = \dfrac{5}{6}$

Ans.   (a) $p = 12$ (d) $n = 1/2$ (g) $t = 30$ (j) $y = 18$
       (b) $r = 16$ (e) $w = 1/3$ (h) $u = 65$ (k) $a = 4$
       (c) $s = 7/5$ or $1\frac{2}{5}$ (f) $x = 7/8$ (i) $x = 60$ (l) $b = 5/2$ or $2\frac{1}{2}$

8.8. Solve each equation: (3.3)

(a) $n + 8 = 24$ (d) $\dfrac{n}{8} = 24$ (g) $15 = 3y$ (j) $16 = \dfrac{y}{20}$ (m) $x + \dfrac{1}{3} = 9$ (p) $\dfrac{x}{9} = \dfrac{1}{3}$

(b) $n - 8 = 24$ (e) $3 + y = 15$ (h) $15 = \dfrac{y}{3}$ (k) $\dfrac{y}{20} = 16$ (n) $x - \dfrac{1}{3} = 9$

(c) $8n = 24$ (f) $15 = y - 3$ (i) $16 = y - 20$ (l) $16 + y = 20$ (o) $\dfrac{1}{3}x = 9$

Ans.   (a) $n = 16$ (d) $n = 192$ (g) $y = 5$ (j) $y = 320$ (m) $x = 8\frac{2}{3}$ (p) $x = 3$
       (b) $n = 32$ (e) $y = 12$ (h) $y = 45$ (k) $y = 320$ (n) $x = 9\frac{1}{3}$
       (c) $n = 3$ (f) $y = 18$ (i) $y = 36$ (l) $y = 4$ (o) $x = 27$

8.9. Solve each equation: (3.3)

(a) $x + 11 = 14$ (f) $h - 3 = 7\frac{1}{2}$ (k) $11r = 55$ (p) $6\frac{1}{2} = \dfrac{l}{2}$

(b) $11 + y = 24$ (g) $35 = m - 20\frac{1}{3}$ (l) $44s = 44$ (q) $1.7 = \dfrac{n}{3}$

(c) $22 = 13 + a$ (h) $17\frac{3}{4} = n - 2\frac{1}{4}$ (m) $10t = 5$ (r) $100 = \dfrac{h}{.7}$

(d) $45 = b + 33$ (i) $x + 1.2 = 5.7$ (n) $8x = 3$ (s) $24 = \dfrac{t}{.5}$

(e) $z - 9 = 3$ (j) $10.8 = y - 3.2$ (o) $3y = 0$ (t) $.009 = \dfrac{x}{1000}$

Ans.   (a) $x = 3$ (d) $b = 12$ (g) $m = 55\frac{1}{3}$ (j) $y = 14$ (m) $t = 1/2$ (p) $l = 13$ (s) $t = 12$
       (b) $y = 13$ (e) $z = 12$ (h) $n = 20$ (k) $r = 5$ (n) $x = 3/8$ (q) $n = 5.1$ (t) $x = 9$
       (c) $a = 9$ (f) $h = 10\frac{1}{2}$ (i) $x = 4.5$ (l) $s = 1$ (o) $y = 0$ (r) $h = 70$

8.10. State the equality rule used in each: (4.1)

(a) $6r = 30$ (b) $30 = r - 6$ (c) $30 = \dfrac{r}{6}$

$\dfrac{6r}{6} = \dfrac{30}{6}$ $\dfrac{+6 = +6}{36 = r}$ $6(30) = 6\dfrac{r}{6}$

$r = 5$ $180 = r$

(d)
$$30 = 6 + r$$
$$\underline{-6 = -6}$$
$$24 = r$$

(e)
$$100x = 5$$
$$\frac{100x}{100} = \frac{5}{100}$$
$$x = \frac{1}{20}$$

(f)
$$100 = \frac{y}{5}$$
$$5(100) = 5\left(\frac{y}{5}\right)$$
$$500 = y$$

Ans. (*a*) division (*b*) addition (*c*) multiplication (*d*) subtraction (*e*) division (*f*) multiplication

8.11. Solve each equation: **(5.1)**

(*a*) $12x = 60$ (*c*) $24 = 2z$ (*e*) $6r = 9$ (*g*) $10 = 4t$
(*b*) $60y = 12$ (*d*)$2 = 24w$ (*f*) $9s = 6$ (*h*) $4 = 10u$

Ans. (*a*) $x = 5$ (*c*) $12 = z$ (*e*) $r = 3/2$ (*g*) $5/2 = t$
        (*b*) $y = 1/5$ (*d*) $1/12 = w$ (*f*) $s = 2/3$ (*h*) $2/5 = u$

8.12. Find each solution set: **(5.2)**

(*a*) $.7a = 21$ (*c*) $24 = .06c$ (*e*) $.1h = 100$ (*g*) $25.2 = .12k$
(*b*) $1.1b = 55$ (*d*)$18 = .009d$ (*f*) $.6j = .96$ (*h*) $7.5 = .015m$

Ans. (*a*) {30}   (*c*) {400}   (*e*) {1000}   (*g*) {210}
        (*b*) {50}   (*d*) {2000}   (*f*) {1.6}   (*h*) {500}

8.13. Solve each equation: **(5.3)**

(*a*) $10\%s = 7$ (*c*) $18 = 3\%n$ (*e*) $5\%m = 13$ (*g*) $.23 = 1\%y$
(*b*) $25\%t = 3$ (*d*)$14 = 70\%w$ (*f*) $17\%x = 6.8$ (*h*) $3.69 = 90\%z$

Ans. (*a*) $s = 70$ (*c*) $600 = n$ (*e*) $m = 260$ (*g*) $23 = y$
        (*b*) $t = 12$ (*d*) $20 = w$ (*f*) $x = 40$ (*h*) $4.1 = z$

8.14. Solve each equation: **(5.4)**

(*a*) $14 = 3x - x$ (*c*) $8z - 3z = 45$ (*e*) $24 = 4\frac{1}{2}x - \frac{1}{2}x$ (*g*) $7\frac{1}{2}z - 7z = 28$
(*b*) $7y + 3y = 50$ (*d*)$132 = 10w + 3w - w$ (*f*) $4y + 15y = 57$ (*h*) $15w - 3w - 2w = 85$

Ans. (*a*) $7 = x$ (*c*) $z = 9$ (*e*) $6 = x$ (*g*) $z = 56$
        (*b*) $y = 5$ (*d*) $11 = w$ (*f*) $y = 3$ (*h*) $w = 8\frac{1}{2}$

8.15. Harry earned \$28.89. What was his hourly wage if he worked (*a*) 3 hr, (*b*) 9 hr,   **(5.5)**
(*c*) $\frac{1}{2}$ hr?

Ans. (*a*) \$9.63   (*b*) \$3.21   (*c*) \$57.78

8.16. Mr. Brown's commission rate is 5%. How much did he sell if his commissions were   **(5.6)**
(*a*) \$85,   (*b*) \$750,   (*c*) \$6.20?

Ans. (*a*) \$1700   (*b*) \$15,000   (*c*) \$124

8.17. Solve each equation: **(6.1)**

(*a*) $\dfrac{x}{3} = 2$ (*b*) $\dfrac{1}{7}y = 12$ (*c*) $16 = \dfrac{z}{5}$ (*d*) $3 = \dfrac{1}{50}w$ (*e*) $\dfrac{a}{2} = 3$ (*f*) $\dfrac{1}{30}b = 20$ (*g*) $.6 = \dfrac{c}{10}$

Ans. (*a*) $x = 6$ (*b*) $y = 84$ (*c*) $80 = z$ (*d*) $150 = w$ (*e*) $a = 6$ (*f*) $b = 600$ (*g*) $6 = c$

8.18. Find each solution set: (6.2)

(a) $\dfrac{a}{.7} = 10$ (b) $\dfrac{b}{.02} = 600$ (c) $30 = \dfrac{c}{2.4}$ (d) $11 = \dfrac{d}{.05}$ (e) $\dfrac{m}{.4} = 220$ (f) $\dfrac{n}{.01} = 3$

Ans. (a) $\{7\}$ (b) $\{12\}$ (c) $\{72\}$ (d) $\{.55\}$ (e) $\{88\}$ (f) $\{.03\}$

8.19. Solve each equation: (6.3)

(a) $\dfrac{3}{4}x = 21$ (b) $\dfrac{4}{3}y = 32$ (c) $\dfrac{3x}{2} = 9$ (d) $45 = \dfrac{5}{9}y$ (e) $2\frac{1}{5}z = 55$ (f) $2c + \dfrac{1}{2}c = 10$

Ans. (a) $x = 28$ (b) $y = 24$ (c) $x = 6$ (d) $y = 81$ (e) $z = 25$ (f) $c = 4$

8.20. Solve each equation: (6.4)

(a) $37\frac{1}{2}\%s = 15$ (b) $60\%t = 60$ (c) $16\frac{2}{3}\%n = 14$ (d) $150\%r = 15$ (e) $83\frac{1}{3}\%w = 35$

Hint: $37\frac{1}{2}\% = 3/8$ $60\% = 3/5$ $16\frac{2}{3}\% = 1/6$ $150\% = 1\frac{1}{2}$ or $3/2$ $83\frac{1}{3}\% = 5/6$

Ans. (a) $s = 40$ (b) $t = 100$ (c) $n = 84$ (d) $r = 10$ (e) $w = 42$

8.21. On a trip, John covered a distance of 35 km. What was the total distance of the trip if   (6.5)
the distance traveled was   (a) 5/6 of the total distance,   (b) 70% of the total distance?

Ans. (a) 42 km.   (b) 50 km.

8.22. Mr. Reynolds receives 7% per year on a stock investment. How large is his investment   (6.5)
if, at the end of one year, his interest is   (a) \$28,   (b) \$350,   (c) \$42.70?

Ans. (a) \$400   (b) \$5000   (c) \$610

8.23. Find each solution set: (7.1)

(a) $r + 25 = 70$ (c) $18 = s + 3$ (e) $x + 130 = 754$ (g) $259 = s + 237$
(b) $31 + t = 140$ (d) $842 = 720 + u$ (f) $116 + y = 807$ (h) $901 = 857 + w$

Ans. (a) $\{45\}$ (c) $\{15\}$ (e) $\{624\}$ (g) $\{22\}$
        (b) $\{109\}$ (d) $\{122\}$ (f) $\{691\}$ (h) $\{44\}$

8.24. Solve each equation: (7.2)

(a) $b + 2/3 = 7\frac{2}{3}$ (c) $35.4 = d + 23.2$ (e) $f + 5/8 = 3\frac{1}{2}$ (g) $7.28 = m + .79$
(b) $1\frac{1}{2} + c = 8\frac{3}{4}$ (d) $87.4 = 80.6 + e$ (f) $8\frac{1}{6} + g = 10\frac{5}{6}$ (h) $15.87 = 6.41 + n$

Ans. (a) $b = 7$ (c) $12.2 = d$ (e) $f = 2\frac{7}{8}$ (g) $6.49 = m$
        (b) $c = 7\frac{1}{4}$ (d) $6.8 = e$ (f) $g = 2\frac{2}{3}$ (h) $9.46 = n$

8.25. The price of eggs rose 29¢. What was the original price if the new price is   (a) \$1.29   (7.3)
(b) \$1.41?

Ans. (a) \$1.00   (b) \$1.12

8.26. Will is 8 in. taller than George. How tall is George if Will's height is   (a) 5 ft 2 in.,   (7.4)
(b) 4 ft 3 in.?

Ans. (a) 4 ft 6 in.   (b) 3 ft 7 in.

8.27. Find each solution set: (8.1)

(a) $w - 8 = 22$ (c) $40 = y - 3$ (e) $m - 140 = 25$ (g) $158 = p - 317$
(b) $x - 22 = 8$ (d) $3 = z - 40$ (f) $n - 200 = 41$ (h) $256 = r - 781$

Ans. (a) {30}   (c) {43}   (e) {165}   (g) {475}
      (b) {30}   (d) {43}   (f) {241}   (h) {1037}

8.28. Solve each equation: (8.2)

(a) $h - \frac{7}{8} = 8\frac{3}{4}$ (c) $28.4 = m - 13.9$ (e) $p - 1\frac{5}{12} = 1\frac{7}{12}$ (g) $.03 = s - 2.07$

(b) $j - 34\frac{1}{2} = 65$ (d) $.37 = n - 8.96$ (f) $r - 14\frac{2}{3} = 5\frac{1}{3}$ (h) $5.84 = t - 3.06$

Ans. (a) $h = 9\frac{5}{8}$ (c) $42.3 = m$ (e) $p = 3$ (g) $2.10 = s$

      (b) $j = 99\frac{1}{2}$ (d) $9.33 = n$ (f) $r = 20$ (h) $8.90 = t$

8.29. What was the original temperature if a drop of $12°$ brought the temperature to   (8.3)

(a) $75°$,   (b) $14\frac{1}{2}°$,   (c) $6\frac{1}{4}°$?

Ans. (a) $87°$ (b) $26\frac{1}{2}°$ (c) $18\frac{1}{4}°$

8.30. How many marbles did Sam have originally if after giving 35 marbles to Jim, he found   (8.4)
that the number of marbles he had left was   (a) 5,   (b) 12,   (c) 15,   (d) 35,   (e) 75?

Ans. (a) 40   (b) 47   (c) 50   (d) 70   (e) 110

8.31. Solve each equation: (9.1)

(a) $2x + 5 = 9$ (e) $2x - 5 = 9$ (i) $\frac{x}{4} + 3 = 7$ (m) $\frac{x}{4} - 3 = 7$

(b) $4x + 11 = 21$ (f) $4x - 11 = 21$ (j) $\frac{x}{5} + 2 = 10$ (n) $\frac{x}{5} - 2 = 10$

(c) $20 = 3x + 8$ (g) $60 = 10x - 20$ (k) $17 = \frac{x}{2} + 15$ (o) $3 = \frac{x}{12} - 7\frac{1}{4}$

(d) $13 = 6 + 7x$ (h) $11 = 6x - 16$ (l) $25 = \frac{x}{10} + 2$ (p) $5\frac{1}{2} = \frac{x}{8} - 4$

Ans. (a) $x = 2$ (e) $x = 7$ (i) $x = 16$ (m) $x = 40$
      (b) $x = 2\frac{1}{2}$ (f) $x = 8$ (j) $x = 40$ (n) $x = 60$
      (c) $x = 4$ (g) $x = 8$ (k) $x = 4$ (o) $x = 123$
      (d) $x = 1$ (h) $x = 4\frac{1}{2}$ (l) $x = 230$ (p) $x = 76$

8.32. Solve each equation: (9.2)

(a) $10n + 5n - 6 = 9$ (d) $35 = 6p + 8 + 3p$ (g) $40 = 25t + 22 - 13t$
(b) $7m + 10 - 2m = 45$ (e) $19n - 10 + n = 80$ (h) $145 = 10 + 7.6s - 3.1s$
(c) $25 = 19 + 20n - 18n$ (f) $3\frac{1}{2}r + r + 2 = 20$

Ans. (a) $n = 1$ (c) $n = 3$ (e) $n = 4\frac{1}{2}$ (g) $t = 3/2$ or $1\frac{1}{2}$
      (b) $m = 7$ (d) $p = 3$ (f) $r = 4$ (h) $s = 30$

8.33. Find each solution set: (9.3)

(a) $5r = 2r + 27$ (c) $10r - 11 = 8r$ (e) $13b = 15 + 3b$ (g) $9u = 16u - 105$

(b) $2r = 90 - 7r$ (d) $18 - 5a = a$ (f) $100 + 3\frac{1}{2}t = 23\frac{1}{2}t$ (h) $5x + 3 - 2x = x + 8$

Ans. (a) {9}   (b) {10}   (c) {$5\frac{1}{2}$}   (d) {3}   (e) {$1\frac{1}{2}$}   (f) {5}   (g) {15}   (h) {$2\frac{1}{2}$}

8.34. Solve each equation: (9.4)

(a) $\dfrac{40}{x} = 5$ (c) $14 = \dfrac{28}{y}$ (e) $\dfrac{32}{n} = 8$ (g) $4 = \dfrac{15}{w}$

(b) $\dfrac{5}{x} = 40$ (d) $28 = \dfrac{14}{y}$ (f) $\dfrac{3}{n} = 2$ (h) $15 = \dfrac{90}{w}$

Ans. (a) $x = 8$ (c) $y = 2$ (e) $n = 4$ (g) $w = 3\frac{3}{4}$
     (b) $x = 1/8$ (d) $y = 1/2$ (f) $n = 1\frac{1}{2}$ (h) $w = 6$

8.35. Find each solution set: (9.5, 9.6)

(a) $\dfrac{7}{8}x = 21$ (d) $1\frac{1}{2}w = 15$ (g) $\dfrac{4}{5}n + 6 = 22$ (j) $10 = \dfrac{2}{9}r + 8$

(a) $\dfrac{3}{4}x = 39$ (e) $2\frac{1}{3}b = 35$ (h) $10 + \dfrac{6}{5}m = 52$ (k) $6 = 16 - \dfrac{5t}{3}$

(c) $\dfrac{5}{4}y = 15$ (f) $2c + 2\frac{1}{2}c = 54$ (i) $30 - \dfrac{3}{2}p = 24$ (l) $3s + \dfrac{s}{3} - 7 = 5$

Ans. (a) {24}   (d) {10}   (g) {20}   (j) {9}
     (b) {52}   (e) {15}   (h) {35}   (k) {6}
     (c) {12}   (f) {12}   (i) {4}   (l) {3.6}

8.36. Find each solution set: (9.1 to 9.6)

(a) $20 = 3x - 10$ (f) $\dfrac{15}{x} = \dfrac{5}{4}$ (k) $\dfrac{x}{2} + 27 = 30$ (p) $12b - 5 = .28 + b$

(b) $20 = \dfrac{x}{3} - 10$ (g) $5c = 2c + 4.5$ (l) $8x + 3 = 43$ (q) $6d - .8 = 2d$

(c) $15 = \dfrac{3}{4}y$ (h) $.30 - g = .13$ (m) $21 = \dfrac{7}{5}w$ (r) $40 - .5h = 5$

(d) $17 = 24 - z$ (i) $\frac{3}{4}n + 11\frac{1}{2} = 20\frac{1}{2}$ (n) $60 = 66 - 12w$ (s) $40 - \dfrac{3}{5}m = 37$

(e) $3 = \dfrac{39}{x}$ (j) $6w + 5w - 8 = 8w$ (o) $10b - 3b = 49$ (t) $12t - 2t + 10 = 9t + 12$

Ans. (a) {10}   (d) {7}   (g) {1.5}   (j) {$2\frac{2}{3}$}   (m) {15}   (p) {.48}   (s) {5}
     (b) {90}   (e) {13}   (h) {.17}   (k) {6}   (n) {$\frac{1}{2}$}   (q) {.2}   (t) {2}
     (c) {20}   (f) {12}   (i) {12}   (l) {5}   (o) {7}   (r) {70}

8.37. How many girls are there in a class of 30 pupils if   (a) the number of boys is 10, (9.7)
(b) the number of boys is four times as many?

Ans. (a) 20 girls   (b) 6 girls

8.38. Charles has \$37 in his bank and hopes to increase this to \$100 by making equal deposits (9.8)
each week. How much should he deposit if he deposits money for   (a) 14 wk,   (b) 5 wk?

Ans. (a) \$4.50   (b) \$12.60

8.39. Mr. Barr sold his house for $120,000. How much did the house cost him if his loss **(9.9)** was   (*a*) 1/3 of the cost,   (*b*) 20% of the cost?

Ans.   (*a*) $180,000   (*b*) $150,000

8.40. Derive a formula for   (*a*) the no. of pennies (*p*) equivalent to *d* dimes,   (*b*) the no.   **(10.1)** of dimes (*d*) equivalent to *p* pennies,   (*c*) the no. of nickels (*n*) equivalent to *D* dollars,   (*d*) the no. of half dollars (*h*) equivalent to *q* quarters,   (*e*) the no. of quarters (*q*) equivalent to *d* dimes.

Ans.   (*a*) $p = 10d$ (*b*) $d = \dfrac{p}{10}$ (*c*) $n = 20D$ (*d*) $h = \dfrac{1}{2}q$ (*e*) $q = \dfrac{2}{5}d$ or $\dfrac{2d}{5}$

8.41. Derive a formula for   (*a*) the value in cents (*c*) of *d* dimes and *q* quarters,   **(10.2, 10.3)** (*b*) the value in cents (*c*) of *n* nickels and *D* dollars,   (*c*) the no. of nickels (*n*) equivalent to *d* dimes and *h* half dollars,   (*d*) the no. of dimes (*d*) equivalent to *n* nickels and *p* pennies,   (*e*) the no. of quarters (*q*) equivalent to *D* dollars and *n* nickels.

Ans.   (*a*) $c = 10d + 25q$ (*c*) $n = 2d + 10h$ (*e*) $q = 4D + \dfrac{n}{5}$

     (*b*) $c = 5n + 100D$ (*d*) $d = \dfrac{n}{2} + \dfrac{p}{10}$

8.42. Derive a formula for   (*a*) the no. of sec (*s*) in *h* hr;   (*b*) the no. of hr (*h*) in *m*   **(10.4)** min;   (*c*) the no. of hr (*h*) in *w* wk;   (*d*) the no. of da (*d*) in *M* mo of 30 days;   (*e*) the no. of da (*d*) in *M* mo of 30 days, *w* wk and 5 da;   (*f*) the no. of min (*m*) in *h* hr and 30 sec;   (*g*) the no. of da (*d*) in *y* yr of 365 da and 3 wk.

Ans.   (*a*) $s = 3600h$ (*c*) $h = 168w$ (*e*) $d = 30M + 7w + 5$ (*g*) $d = 365y + 21$

     (*b*) $h = \dfrac{m}{60}$ (*d*) $d = 30M$ (*f*) $m = 60h + .5$ or $m = 60h + \tfrac{1}{2}$

8.43. Derive a formula for   (*a*) the no. of in (*i*) in *y* yd,   (*b*) the no. of yd (*y*) in *f*   **(10.5)** ft,   (*c*) the no. of meters (*m*) in *c* centimeters.

Ans.   (*a*) $i = 36y$ (*b*) $y = f/3$ (*c*) $m = c/100$

8.44. From $D = RT$, obtain a formula relating   (*a*) distance in mi and rate in mph for a   **(10.6)** time of 5 hr,   (*b*) distance in mi and rate in mph for a time of 30 min,   (*c*) time in hr and rate in mph for a distance of 25 mi,   (*d*) time in sec and rate in ft per sec for a distance of 100 yd,   (*e*) distance in ft and time in min for a rate of 20 ft per min,   (*f*) distance in ft and time in min for a rate of 20 ft per sec.

Ans.   (*a*) $D = 5R$ (*b*) $D = R/2$ (*c*) $RT = 25$ (*d*) $RT = 300$ (*e*) $D = 20T$ (*f*) $D = 1200T$

8.45. Solve: **(11.1)**

(*a*) $d = 2r$ for *r*   (*e*) $c = \pi d$ for *d*   (*i*) $V = LWH$ for *H*
(*b*) $p = 5s$ for *s*   (*f*) $c = \pi d$ for π (*j*) $V = 2\pi r^2 h$ for *h*
(*c*) $D = 30T$ for *T*   (*g*) $NP = C$ for *N*   (*k*) $9C = 5(F - 32)$ for *C*
(*d*) $25W = A$ for *W*   (*h*) $I = PR$ for *R*   (*l*) $2A = h(b + b')$ for *h*

Ans.   (*a*) $\dfrac{d}{2} = r$ (*c*) $\dfrac{D}{30} = T$ (*e*) $\dfrac{c}{\pi} = d$ (*g*) $N = \dfrac{C}{P}$ (*i*) $\dfrac{V}{LW} = H$ (*k*) $C = \dfrac{5(F - 32)}{9}$

     (*b*) $\dfrac{p}{5} = s$ (*d*) $W = \dfrac{A}{25}$ (*f*) $\pi = \dfrac{c}{d}$ (*h*) $\dfrac{I}{P} = R$ (*j*) $\dfrac{V}{2\pi r^2} = h$ (*l*) $\dfrac{2A}{b + b'} = h$

8.46. Solve: **(11.2)**

(a) $\frac{p}{10} = s$ for p (e) $\pi = \frac{c}{2r}$ for c (i) $\frac{V}{3LW} = H$ for V

(b) $R = \frac{D}{15}$ for D (f) $\frac{M}{D} = F$ for M (j) $\frac{T}{14RS} = \frac{1}{2}$ for T

(c) $W = \frac{A}{8}$ for A (g) $P = \frac{A}{2F}$ for A (k) $\frac{L}{KA} = V^2$ for L

(d) $w = \frac{d}{7}$ for d (h) $\frac{T}{Q} = 5R$ for T (l) $\frac{V}{\pi r^2} = \frac{h}{3}$ for V

Ans. (a) $p = 10s$ (c) $8W = A$ (e) $2\pi r = c$ (g) $2PF = A$ (i) $V = 3LWH$ (k) $L = KAV^2$

        (b) $15R = D$ (d) $7w = d$ (f) $M = FD$ (h) $T = 5RQ$ (j) $T = 7RS$ (l) $V = \frac{\pi r^2 h}{3}$

8.47. Solve: **(11.3)**

(a) $a + b = 60$ for a (d) $3m = 4n + p$ for p (g) $5a + b = c - d$ for b

(b) $3c + g = 85$ for g (e) $10r = s - 5t$ for s (h) $5a - 4c = 3e + f$ for f

(c) $h - 10r = l$ for h (f) $4g + h - 12 = j$ for h (i) $\frac{b}{2} - 10 + c = 100p$ for c

Ans. (a) $a = 60 - b$ (d) $3m - 4n = p$ (g) $b = c - d - 5a$

        (b) $g = 85 - 3c$ (e) $10r + 5t = s$ (h) $5a - 4c - 3e = f$

        (c) $h = l + 10r$ (f) $h = j + 12 - 4g$ (i) $c = 100p + 10 - \frac{b}{2}$

8.48. Solve: **(11.4)**

(a) $4P - 3R = 40$ for P (d) $A = \frac{1}{2}bh$ for b (g) $\frac{R}{2} - 4S = T$ for R

(b) $36 - 5i = 12f$ for i (e) $V = \frac{1}{3}\pi r^2 h$ for h (h) $8h - \frac{k}{5} = 12$ for k

(c) $\frac{P}{2} + R = S$ for P (f) $A = \frac{1}{2}h(b + b')$ for h (i) $20p - \frac{2}{3}q = 8t$ for q

Ans. (a) $P = \frac{3R + 40}{4}$ (d) $\frac{2A}{h} = b$ (g) $R = 8S + 2T$

      (b) $\frac{36 - 12f}{5} = i$ (e) $\frac{3V}{\pi r^2} = h$ (h) $5(8h - 12) = k$ or $40h - 60 = k$

      (c) $P = 2S - 2R$ (f) $\frac{2A}{b + b'} = h$ (i) $\frac{3}{2}(20p - 8t) = q$ or $30p - 12t = q$

8.49. Solve: (a) $l = a + (n - 1)d$ for d (b) $s = \frac{n}{2}(a + l)$ for l (c) $F = \frac{9}{5}C + 32$ for C **(11.5)**

Ans. (a) $\frac{l - a}{n - 1} = d$ (b) $l - \frac{2s}{n} - a$ or $\frac{2s - an}{n}$ (c) $\frac{5}{9}(F - 32) = C$

8.50. Find: **(12.1)**

(a) l if $l = prt$ and $p = 3000$, $r = .05$, $t = 2$;

(b) t if $t = \frac{l}{pr}$ and $l = 40$, $p = 2000$, $r = .01$;

(c) F if $F = \frac{9}{5}C + 32$ and $C = 55$;

(d) F if $F = \frac{9}{5}C + 32$ and $C = -40$;

(e) C if $C = \frac{5}{9}(F - 32)$ and $F = 212$;

(f) S if $S = \frac{1}{2}gt^2$ and $g = 32$, $t = 8$;

(g) g if $g = \dfrac{2S}{t^2}$ and $S = 800$, $t = 10$;

(h) S if $S = \dfrac{a - lr}{1 - r}$ and $a = 5$, $l = 40$, $r = -1$.

Ans.   (a) . 300   (c) 131   (e)   100   (g) 16
       (b) 2   (d) −40   (f) 1024   (h) $22\frac{1}{2}$

8.51.   Find: **(12.2, 12.3)**
    (a) R if $D = RT$ and $D = 30$, $T = 4$;
    (b) b if $A = bh$ and $A = 22$, $h = 2.2$;
    (c) a if $P = 2a + b$ and $P = 12$, $b = 3$;
    (d) w if $P = 2l + 2w$ and $P = 68$, $l = 21$;
    (e) c if $p = a + 2b + c$ and $p = 33$, $a = 11$, $b = 3\frac{1}{2}$;
    (f) h if $2A = h(b + b')$ and $A = 70$, $b = 3.3$, $b' = 6.7$;
    (g) B if $V = \frac{1}{3}Bh$ and $V = 480$, $h = 12$;
    (h) a if $l = a + (n - 1)d$ and $l = 140$, $n = 8$, $d = 3$.

Ans.   (a) $7\frac{1}{2}$ (c) $4\frac{1}{2}$ (e)   15   (g) 120
       (b) 10   (d) 13   (f) 14   (h) 119

8.52.   (a) A train takes 5 hours and 20 minutes to go a distance of 304 kilometers. Find its   **(12.4)**
average speed.   (b) A rectangle has a perimeter of 2 m and a length of .4 m. Find its width.

Ans.   (a) 57 kmph   (b) 0.6 m

Chapter 9

Ratios, Proportions, and Rates

1. RATIOS OF TWO QUANTITIES

The question "How does a dollar compare with a quarter?" can be answered in two ways:

    (1) The dollar is 75¢ more than a quarter.
    (2) The dollar is 4 times as much as a quarter.

The first of the two answers is obtained by subtracting 25¢ from 100¢:

$$100¢ - 25¢ = 75¢$$

Hence, the dollar is 75¢ more than a quarter.

The second of the two answers is obtained by dividing 100¢ by 25¢:

$$\frac{100¢}{25¢} = \frac{100}{25} \text{ (Eliminate the common unit, cents; then reduce to lowest terms.)} = \frac{4}{1}$$

Hence, the dollar is 4 times as much as a quarter. The result 4, obtained by division, is the *ratio* of a dollar to a quarter.

The *ratio of two quantities expressed in the same unit* is the first quantity divided by the second. Thus, the ratio of a dime to a half-dollar is 1/5, found by dividing 10¢ by 50¢:

$$10¢ \div 50¢ = \frac{10}{50} = \frac{1}{5}$$

Expressing a Ratio

METHODS	STATEMENTS EXPRESSING A RATIO
1. Using "to":	1. The ratio of a dime to a half-dollar is 1 to 5.
2. Using a colon:	2. The ratio of a dime to a half-dollar is 1:5.
3. As a common fraction:	3. A dime is 1/5 or one-fifth of a half-dollar.
4. As a decimal:	4. A dime is .2 or two-tenths of a half-dollar.
5. As a per cent:	5. A dime is 20% or twenty per cent of a half-dollar.

Ratio Rules

Rule 1: To obtain a ratio of two quantities having the same unit of measure, *eliminate the common unit of measure* before simplifying.

Thus, the ratio of \$3 to \$10 is $\frac{\$3}{\$10} = \frac{3 \times \$1}{10 \times \$1} = \frac{3}{10}$ or $3:10$

Rule 2: To obtain a ratio of two quantities having different units of measure, *change to the same unit of measure*; then eliminate the common unit of measure before simplifying.

Thus, the ratio of 1 ft to 7 in. is $\frac{12 \text{ in.}}{7 \text{ in.}} = \frac{12}{7}$ or $12:7$

Rule 3: To express a ratio in simplest terms, *eliminate any common factor of its terms.*

Thus, the ratio of $30x$ to $40x$ is $\frac{30x}{40x} = \frac{3 \cdot 10x}{4 \cdot 10x} = \frac{3}{4}$ or $3:4$

Rule 4: To express a ratio in simplest terms, *eliminate any fraction or decimal* in its terms in order to obtain terms that are whole numbers.

Thus, the ratio of 1.5 to $\frac{1}{4}$ is $\frac{1.5}{.25} = \frac{1.50}{.25} = 6$ or $6:1$

To Find the Ratio of Two Quantities Having Different Units of Measure

Express each ratio in simplest terms: (a) $1\frac{1}{2}$ years to 4 months   (b) 4.5 seconds to 2 minutes

PROCEDURE   **SOLUTIONS**

1. Express each quantity in the same unit: $\dfrac{1\frac{1}{2}\,\text{yr}}{4\,\text{mo}} = \dfrac{18\,\text{mo}}{4\,\text{mo}}$ $\dfrac{4.5\,\text{sec}}{2\,\text{min}} = \dfrac{4.5\,\text{sec}}{120\,\text{sec}}$
   (Rule 2)

2. Eliminate the common unit of measure: $= \dfrac{18}{4}$ $= \dfrac{4.5}{120}$
   (Rule 1)

3. Express in simplest terms: $= \dfrac{9 \cdot 2}{2 \cdot 2}$ $= \dfrac{45}{1200}$
   (Rules 3 and 4)

   $= \dfrac{9}{2}$ or $9:2$ *Ans.* $= \dfrac{3 \cdot 15}{80 \cdot 15}$

   $= \dfrac{3}{80}$ or $3:80$ *Ans.*

1.1   RATIOS OF TWO QUANTITIES HAVING THE SAME UNIT

Express each ratio in simplest terms:

(a) 9 lb to 12 lb   (d) 5 mo to 10 mo   (g) $2.50 to $1.50
(b) 9 sec to 15 sec   (e) 50 yr to 100 yr   (h) 12.5 ft to 3.75 ft
(c) 120 min to 40 min   (f) 200 da to 180 da   (i) $1\frac{1}{2}$ cm to 12 cm

Solutions   Apply Rules 1, 3, and 4. Each answer is in fractional form.

(a) $\dfrac{9}{12} = \dfrac{3}{4}$ (c) $\dfrac{120}{40} = \dfrac{3}{1}$ (e) $\dfrac{50}{100} = \dfrac{1}{2}$ (g) $\dfrac{2.50}{1.50} = \dfrac{5}{3}$ (i) $1\frac{1}{2} \div 12 = \dfrac{3}{2} \times \dfrac{1}{12} = \dfrac{1}{8}$

(b) $\dfrac{9}{15} = \dfrac{3}{5}$ (d) $\dfrac{5}{10} = \dfrac{1}{2}$ (f) $\dfrac{200}{180} = \dfrac{10}{9}$ (h) $\dfrac{12.5}{3.75} = \dfrac{10}{3}$

1.2   RATIOS OF TWO QUANTITIES HAVING DIFFERENT UNITS

Express each ratio in simplest terms:

(a) 5 min to 20 sec   (c) 20 mph to 2 mi per min   (e) 40 cm to 2 m
(b) 45¢ to $1.65   (d) 8 in. to $1\frac{1}{2}$ ft   (f) 4 oz to 4 lb

Solutions   Apply Rules 2, 3, and 4. Each answer is in fractional form.

(a) 5 min = 300 sec   (c) 2 mi per min = 120 mph   (e) 2 m = 200 cm

$\dfrac{300}{20} = \dfrac{15}{1}$ $\dfrac{20}{120} = \dfrac{1}{6}$ $\dfrac{40}{200} = \dfrac{1}{5}$

(b) $1.65 = 165¢   (d) $1\frac{1}{2}$ ft = 18 in.   (f) 4 lb = 64 oz

$\dfrac{45}{165} = \dfrac{3}{11}$ $\dfrac{8}{18} = \dfrac{4}{9}$ $\dfrac{4}{64} = \dfrac{1}{16}$

1.3   REDUCING RATIOS TO SIMPLEST TERMS

Express each ratio in simplest terms:

(a) .3 to 6   (c) .003 to .6   (e) 100% to 250%   (g) 6 to 1/4   (i) 30% to .7
(b) 30 to .06   (d) 80% to 60%   (f) $6\frac{1}{4}$% to .25%   (h) $2\frac{1}{2}$ to $8\frac{3}{4}$

Solutions   Apply Rules 3 and 4.

(a) $\dfrac{.3}{6} = \dfrac{.3}{6.0} = \dfrac{3}{60} = \dfrac{1}{20}$ (b) $\dfrac{30}{.06} = \dfrac{30.00}{.06} = \dfrac{3000}{6} = \dfrac{500}{1}$ or 500   (c) $\dfrac{.003}{.6} = \dfrac{.003}{.600} = \dfrac{3}{600} = \dfrac{1}{200}$

$(d)\ \dfrac{80\%}{60\%} = \dfrac{80}{60} = \dfrac{4}{3}$ $(f)\ \dfrac{6\frac{1}{4}\%}{.25\%} = \dfrac{6\frac{1}{4}}{.25} = \dfrac{25}{4} \div \dfrac{1}{4} = \dfrac{25}{1}$ or 25 $(h)\ \dfrac{5}{2} \div \dfrac{35}{4} = \dfrac{5}{2} \cdot \dfrac{4}{35} = \dfrac{2}{7}$

$(e)\ \dfrac{100\%}{250\%} = \dfrac{100}{250} = \dfrac{2}{5}$ $(g)\ 6 \div \dfrac{1}{4} = 6 \times \dfrac{4}{1} = \dfrac{24}{1}$ or 24 $(i)\ \dfrac{.30}{.7} = \dfrac{.30}{.70} = \dfrac{30}{70} = \dfrac{3}{7}$

1.4 **SIMPLIFYING ALGEBRAIC RATIOS**

Express each ratio in simplest terms:

(a) $5x$ to $6x$	(d) πr to πR	(g) y^2 to y
(b) $3x$ to $6y$	(e) $120x$ to $100x$	(h) x to x^2
(c) $15ab$ to $20ac$	(f) $25\%x$ to $35\%x$	(i) z^3 to z^2

Solutions

(a) Eliminate x. *Ans.* $\dfrac{5}{6}$ (d) Eliminate π. *Ans.* $\dfrac{r}{R}$ (g) Eliminate y. *Ans.* $\dfrac{y}{1}$ or y

(b) Eliminate 3. *Ans.* $\dfrac{x}{2y}$ (e) Eliminate $20x$. *Ans.* $\dfrac{6}{5}$ (h) Eliminate x. *Ans.* $\dfrac{1}{x}$

(c) Eliminate $5a$. *Ans.* $\dfrac{3b}{4c}$ (f) Eliminate $5\%x$. *Ans.* $\dfrac{5}{7}$ (i) Eliminate z^2. *Ans.* $\dfrac{z}{1}$ or z

2. EXPRESSING RATIOS AS FRACTIONS, DECIMALS, AND PER CENTS

A ratio can be expressed as a common fraction, as a decimal, and as a per cent. Thus, the ratio of $4:5$ can be expressed as the fraction 4/5, the decimal .8, and the per cent 80%.

To Express a Ratio as a Fraction, a Decimal, or a Per Cent

In each, express the ratio of games won to games played as (1) as a fraction in simplest terms, (2) a decimal to nearest thousandth, and (3) a per cent.

(a) Team A won 20 games while losing 5.

(b) Team B won 10 games while losing 20.

PROCEDURE

SOLUTIONS

1. Express the ratio as a fraction in simplest terms:

1. $\dfrac{\text{Games won}}{\text{Games played}} = \dfrac{20}{20 + 5}$
   $= \dfrac{20}{25} = \dfrac{4}{5}$

   $\dfrac{\text{Games won}}{\text{Games played}} = \dfrac{10}{10 + 20}$
   $= \dfrac{10}{30} = \dfrac{1}{3}$

2. Express the fraction as a decimal:

2. $5)\overline{4.000}$ quotient $.800$

   $3)\overline{1.000}$ quotient $.333\frac{1}{3} = .333$
   to nearest thousandth.

3. Express the decimal as a per cent:

3. $.80 = 80\%$

   $.33\frac{1}{3} = 33\frac{1}{3}\%$

Ans. $\dfrac{4}{5}$, .800, 80%

Ans. $\dfrac{1}{3}$, .333, $33\frac{1}{3}\%$

2.1 **EXPRESSING RATIOS AS FRACTIONS, DECIMALS, AND PER CENTS**

Express each ratio as a fraction in simplest terms, as a decimal, and as a per cent:

(a) $71:100$	(c) $500:800$	(e) $7:200$	(g) $17:500$	(i) $123:10,000$
(b) $3:1.2$	(d) $70:20$	(f) $9:50$	(h) $123:1,000$	(j) $895:500$

Solutions   See Table 9-1.

Table 9-1

Ratio	Fraction	Decimal	Per Cent	Ratio	Fraction	Decimal	Per Cent
(a)	$\dfrac{71}{100}$	.71	71%	(f)	$\dfrac{9}{50}$	1.18	18%
(b)	$\dfrac{3}{1.2} = \dfrac{5}{2}$	2.5	250%	(g)	$\dfrac{17}{500}$	.034	3.4%
(c)	$\dfrac{5}{8}$	.625	62.5%	(h)	$\dfrac{123}{1,000}$	.123	12.3%
(d)	$\dfrac{7}{2}$	3.5	350%	(i)	$\dfrac{123}{10,000}$	.0123	1.23%
(e)	$\dfrac{7}{200}$	.035	3.5%	(j)	$\dfrac{895}{500}$	1.79	179%

2.2   PROBLEM SOLVING

Find the ratio of the number of men to the number of students in a class and express it as a fraction, as a decimal, and as a per cent if, in the class, there are (a) 10 men and 10 women, (b) 20 men and 5 women, (c) 3 men and 17 women.

Solutions   See Table 9-2.

Table 9-2

	Ratio	Fraction	Decimal	Per Cent
(a)	10 : 20	1/2	.5	50%
(b)	20 : 25	4/5	.8	80%
(c)	3 : 20	3/20	.15	15%

3.   PROPORTIONS: EQUAL RATIOS

A *proportion* is an equality of two ratios. Thus,

$$2 : 5 = 4 : 10 \text{ or } \frac{2}{5} = \frac{4}{10}$$

is a proportion.

The fourth term of a proportion is the *fourth proportional* to the other three taken in order. Thus, in $2 : 3 = 4 : x$, x is the fourth proportional to 2, 3, and 4.

The *means* of a proportion are its middle terms; that is, its second and third terms. The *extremes* of a proportion are its outside terms; that is, its first and fourth terms. Thus, in $a : b = c : d$, the means are b and c, and the extremes are a and d.

If both sides of the proportion

$$\frac{a}{b} = \frac{c}{d}$$

are multiplied by bd (see Chapter 8, Section 4), the result is $ad = bc$. This result may be stated in two forms:

Rule (fraction form): In any proportion, the cross products are equal.

Rule (colon form): In any proportion, the product of the means equals the product of the extremes.

To Find an Unknown in a Proportion in Fraction Form or in Colon Form

Solve for x:
(a) $\dfrac{20}{35} = \dfrac{4}{x}$
(b) $20 : 30 = x : 7$

PROCEDURE

1. Apply the Proportion Rule:

2. Solve the resulting equation:

3. Check the results in the original proportion:

SOLUTIONS

1. Equating cross products:

$$20x = 35(4)$$

2. D_{20} $\dfrac{20x}{20} = \dfrac{140}{20}$

$$x = 7 \quad Ans.$$

3. $\dfrac{20}{35} = \dfrac{4}{x}$

$\dfrac{20}{35} \overset{?}{=} \dfrac{4}{7}$

$\dfrac{4}{7} = \dfrac{4}{7}$

Equating products of means and extremes:

$$35x = 20(7)$$

D_{35} $\dfrac{35x}{35} = \dfrac{140}{35}$

$$x = 4 \quad Ans.$$

$20 : 35 = x : 7$

$20 : 35 \overset{?}{=} 4 : 7$

$4 : 7 = 4 : 7$

3.1 SOLVING PROPORTIONS IN FRACTION FORM

Solve for x in each:

(a) $\dfrac{x}{8} = \dfrac{3}{4}$
(b) $\dfrac{6}{x} = \dfrac{2}{5}$
(c) $\dfrac{x}{x+6} = \dfrac{3}{5}$
(d) $\dfrac{x-4}{x+2} = \dfrac{2}{5}$
(e) $\dfrac{x}{a} = \dfrac{b}{c}$

Solutions Equate cross products in each.

(a) $4x = 24$ (b) $2x = 30$ (c) $5x = 3x + 18$ (d) $5x - 20 = 2x + 4$ (e) $cx = ab$
$\quad x = 6$ $\qquad x = 15$ $\qquad 2x = 18$ $\qquad\qquad 3x = 24$ $\qquad\qquad x = \dfrac{ab}{c}$
$\qquad\qquad\qquad\qquad\qquad\quad x = 9$ $\qquad\qquad\qquad x = 8$

3.2 SOLVING PROPORTIONS IN COLON FORM

Solve for x in each:

(a) $x : 5 = 12 : 30$
(b) $6 : (x+1) = 3 : 5$
(c) $(x+2) : 4 = x : 3$
(d) $3 : 2x = 4 : (x+5)$

Solutions Equate the product of the means and the product of the extremes in each:

(a) $30x = 60$ (b) $3x + 3 = 20$ (c) $4x = 3x + 6$ (d) $8x = 3x + 15$
$\quad x = 2$ $\qquad\quad 3x = 27$ $\qquad\quad x = 6$ $\qquad\quad 5x = 15$
$\qquad\qquad\quad x = 9$ $\qquad\qquad\qquad\qquad\qquad\quad x = 3$

3.3 FINDING FOURTH PROPORTIONALS

Find the fourth proportional to the three given numbers in each:

(a) $1, 2, 3$
(b) $2, 3, 1$
(c) $\frac{1}{2}, 3, 6$
(d) $3, \frac{1}{2}, 12$
(e) a, b, c
(f) $2a, 3a, 6b$

Solutions

(a) $1 : 2 = 3 : x$ (c) $\frac{1}{2} : 3 = 6 : x$ (e) $a : b = c : x$
$\quad x = 6$ $\qquad\quad \frac{1}{2}x = 18$ $\qquad\quad ax = bc$
$\qquad\qquad\quad x = 36$ $\qquad\qquad\quad x = \dfrac{bc}{a}$

(b) $2 : 3 = 1 : x$ (d) $3 : \frac{1}{2} = 12 : x$ (f) $2a : 3a = 6b : x$
$\quad 2x = 3$ $\qquad\quad 3x = 6$ $\qquad\quad 2ax = 18ab$
$\quad x = 1\frac{1}{2}$ $\qquad\quad x = 2$ $\qquad\quad x = 9b$

4. PROBLEMS INVOLVING UNKNOWNS IN A GIVEN RATIO

Unknowns in the ratio of $3:4$ may be represented by $3x$ and $4x$, because

$$\frac{3x}{4x} = \frac{3}{4}$$

To Solve Problems Involving Unknowns in a Given Ratio

(a) Find two numbers in the ratio of $3:4$ whose sum is 77.

(b) Find two numbers in the ratio of $8:3$ whose difference is 14 more than the smaller.

PROCEDURE

1. Represent the unknowns using x as a common factor:
2. Obtain an equation relating the unknowns:

3. Solve the equation to find x:

4. Find the unknowns:

5. Check the results in the original problem:

SOLUTIONS

1. Let $3x$ = smaller number
       $4x$ = greater number
2. Their sum is 77. Thus,
   $$3x + 4x = 77$$

3. $\mathbf{D_7}$ $7x = 77$
            $x = 11$

4. $3x = 33,\ 4x = 44$
   Ans.   Numbers are 33 and 44.

5. Ratio of numbers is $3:4$.
$$\frac{33}{44} \overset{?}{=} \frac{3}{4}$$
$$\frac{3}{4} = \frac{3}{4}$$

Sum of numbers is 77.

$$33 + 44 \overset{?}{=} 77$$
$$77 = 77$$

Let $8x$ = greater number
    $3x$ = smaller number
Their difference is 14 more than the smaller. Thus,
$$8x - 3x = 3x + 14$$
$\mathbf{S_{3x}}$ $5x = 3x + 14$
$\mathbf{D_2}$ $2x = 14$
         $x = 7$

$8x = 56,\ 3x = 21$
Ans.   Numbers are 56 and 21.

Ratio of numbers is $8:3$.
$$\frac{56}{21} = \frac{8}{3}$$
$$\frac{8}{3} = \frac{8}{3}$$

Difference of numbers is 14 more than the smaller.
$$56 - 21 \overset{?}{=} 21 + 14$$
$$35 = 35$$

4.1 SOLVING NUMBER PROBLEMS INVOLVING RATIOS

Two numbers in the ratio of $7:3$ are represented by $7x$ and $3x$. Express each relationship as an equation; then find x and the numbers: (a) the sum of the numbers is 90; (b) the difference of the numbers is 14; (c) the larger is 10 more than twice the smaller; (d) three times the smaller exceeds the larger by 24; (e) if both are increased 2, the new ratio is $2:1$.

Solutions   See Table 9-3.

Table 9-3

	Equations	x	Numbers
(a)	$10x = 90$	9	63, 27
(b)	$4x = 14$	$3\frac{1}{2}$	$24\frac{1}{2},\ 10\frac{1}{2}$
(c)	$7x = 6x + 10$	10	70, 30
(d)	$9x = 7x + 24$	12	84, 36
(e)	$\dfrac{7x + 2}{3x + 2} = \dfrac{2}{1}$	2	14, 6

4.2   PERIMETER PROBLEM INVOLVING RATIOS

The ratio of the length of a rectangle to its width is $5:3$. Find each dimension if   (*a*) the perimeter of the rectangle is 88;   (*b*) the semiperimeter of the rectangle is 42;   (*c*) the new perimeter of the rectangle is 140 if each dimension is increased by 5;   (*d*) increasing the width by 14 makes the rectangle a square.

Solutions   Represent the length and width by $5x$ and $3x$, respectively, and proceed as in Table 9-4.

Table 9-4

	Equation	x	Length	Width
(*a*)	$16x = 88$	$5\frac{1}{2}$	$27\frac{1}{2}$	$16\frac{1}{2}$
(*b*)	$8x = 42$	$5\frac{1}{4}$	$26\frac{1}{4}$	$15\frac{3}{4}$
(*c*)	$16x + 20 = 140$	$7\frac{1}{2}$	$37\frac{1}{2}$	$22\frac{1}{2}$
(*d*)	$5x = 3x + 14$	7	35	21

5.   RATIOS OF THREE OR MORE QUANTITIES: CONTINUED RATIOS

The ratios of three or more quantities may be expressed as a *continued ratio*. A continued ratio is an enlarged ratio statement combining two or more separate ratios.

The ratio $a:b:c$ combines the ratios $a:b$, $b:c$, and $a:c$. But these three ratios are not independent, because any two determine the third. For example, $a:b$ and $b:c$ determine $a:c$, since

$$\frac{a}{c} = \frac{a}{b} \cdot \frac{b}{c}$$

Express a continued ratio in simplest terms in the same way that ratios are simplified:

(1) Eliminate any common unit of measure.
   Thus, $3\,\text{lb}:5\,\text{lb}:8\,\text{lb} = 3:5:8$.

(2) Eliminate any common factor.
   Thus, $30:50:80 = 3:5:8$. Also, $3x:5x:8x = 3:5:8$.

(3) Eliminate any decimals, fractions, or per cents.
   Thus, $30\%:50\%:80\% = 3:5:8$. Also, $.03:.05:.08 = 3:5:8$.

5.1   EXPRESSING CONTINUED RATIOS IN SIMPLEST FORM

Express in simplest form:
(*a*) $2\,\text{oz}:7\,\text{oz}:15\,\text{oz}$ (*c*) $3\cancel{c}:9\cancel{c}:12\cancel{c}$ (*e*) $5\%:50\%:75\%$
(*b*) $2\,\text{oz}:7\,\text{oz}:2\,\text{lb}$ (*d*) $\$3:9\cancel{c}:12\cancel{c}$ (*f*) $.03x:.3x:.3x$

Solutions   In (*a*) to (*d*) eliminate the common unit of measurement.

(*a*) $2:7:15$ (*c*) $3:9:12 = 1:3:4$ (*e*) $5:50:75 = 1:10:15$
(*b*) $2:7:32$ (*d*) $300:9:12 = 100:3:4$ (*f*) $3:30:300 = 1:10:100$

5.2   PROBLEM SOLVING INVOLVING A CONTINUED RATIO

Three partners in a business share profits in the ratio of 2 : 3 : 5. Find the share of each partner if the profits are (1) $25, (2) $450, (3) $7,250.

Solutions   Let $2x$, $3x$, and $5x$ represent the shares of the partners in dollars. Then

$$2x + 3x + 5x = 10x$$

(*a*) $10x = 25$, $x = 2.50$
$2x = 5$, $3x = 7.50$, $5x = 12.50$
Ans. $5, $7.50, $12.50

(*c*) $10x = 7,250$, $x = 725$
$2x = 1,450$, $3x = 2,175$, $5x = 3,625$
Ans. $1,450, $2,175, $3,625

(*b*) $10x = 450$, $x = 45$
$2x = 90$, $3x = 135$, $5x = 225$
Ans. $90, $135, $225

6. RATES OF SPEED AND PRICES

Average Rate of Speed

To find the average rate of speed of a traveler, divide the distance traveled by the time spent traveling; that is,

$$\text{Average rate of speed} = \frac{\text{Distance traveled}}{\text{Travel time}}$$

or

$$r = \frac{d}{t}$$

Thus, if a person travels 120 miles in 2 hours, his average rate of speed is found to be 60 miles per hour, as follows:

$$r = \frac{d}{t}$$

$$= \frac{120 \text{ miles}}{2 \text{ hours}}$$

$$= \frac{60 \text{ miles}}{1 \text{ hour}} \text{ or } 60 \text{ miles per hour, abbreviated } 60 \text{ mph}$$

Similarly, a person who travels 120 feet in 3 seconds is traveling at an average rate of speed of 40 feet per second, abbreviated 40 fps.

Price or Unit Cost

The *price* or *unit cost* of an item is the cost of 1 item. Thus, to find the price or unit cost of an item, divide the cost by the number of items:

$$\text{Price} = \frac{\text{Cost}}{\text{Number}}$$

or

$$p = \frac{c}{n}$$

Thus, if 3 similar cans of fruit cost $1.50, the price of unit cost is found to be $.50 per can, as follows:

$$p = \frac{c}{n}$$

$$= \frac{\$1.50}{3}$$

$$= \frac{\$.50}{1} \text{ or } 50¢ \text{ per can}$$

Similarly, if 5 pounds of apples cost $1.25, the price of the apples is 25¢ per lb.

Solving Rate of Speed Problems Using Proportions

Each of the following problems involves two situations in which average rates of speed are equal. Note the use of a proportion to solve each problem.

To Solve a Rate of Speed Problem Using a Proportion

A distance of 1800 yards is covered in 5 minutes. At *the same average rate*, (*a*) how many yards are covered in 12 minutes? (*b*) in how many minutes will a distance of 1500 yards be covered?

PROCEDURE   **SOLUTIONS**

1. Let x be the measure of the unknown. Form a proportion equating both average rates:

   1. $\dfrac{x \text{ yd}}{12 \text{ min}} = \dfrac{1800 \text{ yd}}{5 \text{ min}}$ $\dfrac{1500 \text{ yd}}{x \text{ min}} = \dfrac{1800 \text{ yd}}{5 \text{ min}}$

2. Eliminate the common units of measure:

   2. $\dfrac{x}{12} = \dfrac{1800}{5}$ $\dfrac{1500}{x} = \dfrac{1800}{5}$

3. Solve the proportion:

   3. $\mathbf{M_{12}} \quad x = 12\left(\dfrac{1800}{5}\right)$ Equate cross products.
   $1800x = 5(1500)$

   $\qquad\qquad = 4{,}320$ $x = \dfrac{5(1500)}{1800} = 4\frac{1}{6}$

   Ans.   4,320 yd   *Ans.* $4\frac{1}{6}$ min

Important Notes: In both proportions, there is a common unit of measure in the numerators and another common unit of measure in the denominators. Such common units of measure may be eliminated.

Comparing Rates of Speed or Prices

As shown in the following problems, rates of speed or prices can be compared by using subtraction or division.

To Compare Rates of Speed or Prices

(*a*) Compare the average rate of speed of a person traveling 80 miles in 5 hours with that of another person traveling 96 miles in 4 hours.

(*b*) Compare the price of 4 cans of soup 84¢ with the price of 5 cans of soup costing 70¢.

PROCEDURE   **SOLUTIONS**

1. Find the average rates using $r = d/t$, or the prices using $p = c/n$:

   1. 1st rate $= \dfrac{80 \text{ mi}}{5 \text{ hr}} = 16$ mph   1st price $= \dfrac{84¢}{4} = 21¢$ per can

   2nd rate $= \dfrac{96 \text{ mi}}{4 \text{ hr}} = 24$ mph   2nd price $= \dfrac{70¢}{5} = 14¢$ per can

2. Compare the results found in step 1 using subtraction:

   2. Since 24 mph -16 mph $= 8$ mph, the 1st rate is 8 mph less than the 2nd rate.   Since 21¢ $- 14¢ = 7¢$, the 1st price is 7¢ more than the 2nd price.

3. Compare the results found in step 1 using division:

   3. Since $\dfrac{16 \text{ mph}}{24 \text{ mph}} = \dfrac{2}{3}$, the 1st rate is two-thirds of the 2nd.   Since $\dfrac{21¢}{14¢} = \dfrac{3}{2}$, the 1st price is three-halves of the 2nd.

6.1   SOLVING PROPORTIONS INVOLVING RATES AND PRICES

Find x in each:

(a) $\dfrac{x\,\text{ft}}{5\,\text{sec}} = \dfrac{450\,\text{ft}}{15\,\text{sec}}$ (c) $\dfrac{630\,\text{mi}}{15\,\text{hr}} = \dfrac{336\,\text{mi}}{x\,\text{hr}}$ (e) $\dfrac{\$x}{12\,\text{lb}} = \dfrac{\$2.16}{9\,\text{lb}}$

(b) $\dfrac{750\,\text{yd}}{3\,\text{min}} = \dfrac{x\,\text{yd}}{7\,\text{min}}$ (d) $\dfrac{x\cancel{c}}{9\,\text{oz}} = \dfrac{60\cancel{c}}{5\,\text{oz}}$ (f) $\dfrac{\$3.50}{x\,\text{kg}} = \dfrac{\$4.20}{6\,\text{kg}}$

Solutions

(a) $x = 5\left(\dfrac{450}{15}\right) = 150$ (c) $x = \dfrac{15(336)}{630} = 8$ (e) $x = 12\left(\dfrac{2.16}{9}\right) = 2.88$

(b) $x = 7\left(\dfrac{750}{3}\right) = 1{,}750$ (d) $x = 9\left(\dfrac{60}{5}\right) = 108$ (f) $x = \dfrac{6(3.50)}{4.20} = 5$

6.2   SOLVING RATE AND PRICE PROBLEMS USING PROPORTIONS

(a) A distance of 240 miles is covered in 10 hours. At the same average rate, how many miles are covered in 7 hours?

(b) The cost of 6 kilograms of sugar is \$4.50. At the same price, how many kilograms can be bought for \$7.50?

Solutions

(a) By equating rates, $\dfrac{x\,\text{mi}}{7\,\text{hr}} = \dfrac{240\,\text{mi}}{10\,\text{hr}}$ Hence, $x = 7\left(\dfrac{240}{10}\right) = 168$

(b) By equating prices, $\dfrac{\$7.50}{x\,\text{kg}} = \dfrac{\$4.50}{6\,\text{kg}}$ Hence, $x = 6\left(\dfrac{7.50}{4.50}\right) = 10$

6.3   COMPARING RATES OF SPEED OR PRICES

(a) Compare the average rate of speed needed to travel 125 miles in 5 hours with that needed to travel 80 miles in 4 hours.

(b) Compare the price in a purchase of 3 cans of fruit for \$.96 with the price in a purchase of 4 cans for \$1.44.

Solutions

(a) Since the first average rate is 25 mph and the second is 20 mph, then (1) the first average rate is 5 mph more than the second, and (2) five-fourths as much.

(b) Since the first price is 32¢ per can and the second price is 36¢ per can, then (1) the first price is 4¢ less than the second, and (2) eight-ninths as much.

Supplementary Problems

9.1.   Express each ratio in simplest terms: **(1.1)**

    (a) 10 oz to 15 oz   (d) 4 yr to 8 yr   (g) \$125 to \$50

    (b) 8 hr to 14 hr   (e) 9 da to 12 da   (h) 3.9 in. to 6.5 in.

    (c) 80 sec to 20 sec   (f) 45 min to 33 min   (i) $2\frac{1}{2}$ km to $7\frac{1}{2}$ km.

    Ans.   (a) 2/3   (b) 4/7   (c) 4   (d) 1/2   (e) 3/4   (f) 15/11   (g) 5/2   (h) 3/5   (i) 1/3

9.2. Express each ratio in simplest terms: **(1.2)**

    (a) 30 sec to 2 min   (d) 6 in. to $1\frac{1}{4}$ ft   (g) 80 cm to 2 m

    (b) \$1.75 to 35¢   (e) 6 oz to 3 lb   (h) 1 kg to 100 g

    (c) 15¢ to \$15   (f) 1 ton to 1600 lb   (i) 120 mph to 1 mi per min

Ans. (a) 1/4   (b) 5   (c) 1/100   (d) 2/5   (e) 1/8   (f) 5/4   (g) 2/5   (h) 10   (i) 2

9.3. Express each ratio in simplest terms: **(1.3)**

    (a) .2 to 1.6   (d) 300% to 100%   (g) 1/2 to 8

    (b) 2 to 1.6   (e) 75% to 45%   (h) $8\frac{1}{2}$ to $4\frac{1}{4}$

    (c) 20 to .16   (f) 3.5% to 10.5%   (i) 60% to .06

Ans. (a) 1/8   (b) 5/4   (c) 125   (d) 3   (e) 5/3   (f) 1/3   (g) 1/16   (h) 2   (i) 10

9.4. Express each ratio in simplest terms: **(1.4)**

    (a) $4x$ to $7x$ (d) $30\%a$ to $40\%b$ (g) y^2 to y^3

    (b) $6y$ to y (e) $.8c$ to $.5d$ (h) πr^2 to $2\pi r$

    (c) $15xy$ to $25xy$ (f) $3x^2$ to $9x$ (i) x^2y to xy^2

Ans. (a) $\dfrac{4}{7}$ (b) 6   (c) $\dfrac{3}{5}$ (d) $\dfrac{3a}{4b}$ (e) $\dfrac{8c}{5d}$ (f) $\dfrac{x}{3}$ (g) $\dfrac{1}{y}$ (h) $\dfrac{r}{2}$ (i) $\dfrac{x}{y}$

9.5. Express each ratio as a fraction in simplest terms, as a decimal, and as a per cent: **(2.1)**

    (a) 27 : 100   (c) 12 : 1   (e) 9 : 2   (g) 77 : 1,000   (i) 13 : 500

    (b) 3 : 12   (d) 8 : 5   (f) 7 : 1,000 (h) 11 : 50   (j) 111 : 200

Ans. See Table 9-5.

Table 9-5

Ratio	Fraction	Decimal	Per Cent	Ratio	Fraction	Decimal	Per Cent
(a)	27/100	.27	27%	(f)	7/1,000	.007	.7%
(b)	1/4	.25	25%	(g)	77/1,000	.077	7.7%
(c)	12/1	12.	1200%	(h)	11/50	.22	22%
(d)	8/5	1.6	160%	(i)	13/500	.026	2.6%
(e)	9/2	4.5	450%	(j)	111/200	.555	55.5%

9.6. In each, express the ratio of games won to games played as a fraction, a decimal to the **(2.2)** nearest thousandth, and a per cent: (a) team A won 15 and lost 15, (b) team B won 50 and lost 10, (c) team C won 28 and lost 21.

Ans. (a) 1/2, .500, 50%   (b) 5/6, .833, $83\frac{1}{3}$%   (c) 4/7, .571, $57\frac{1}{7}$%

9.7. Solve for x in each: **(3.1)**

    (a) $\dfrac{x}{9} = \dfrac{2}{3}$ (d) $\dfrac{8}{9} = \dfrac{32}{x}$ (g) $\dfrac{2x}{3} = \dfrac{4}{5}$ (j) $\dfrac{3x}{4x-14} = \dfrac{6}{7}$

    (b) $\dfrac{48}{x} = \dfrac{24}{25}$ (e) $\dfrac{x}{x+3} = \dfrac{4}{5}$ (h) $\dfrac{2}{2x+3} = \dfrac{1}{5}$ (k) $\dfrac{x}{a} = \dfrac{c}{b}$

    (c) $\dfrac{7}{20} = \dfrac{x}{100}$ (f) $\dfrac{x-4}{x+6} = \dfrac{4}{5}$ (i) $\dfrac{5}{3x-3} = \dfrac{1}{3}$ (l) $\dfrac{p}{x} = \dfrac{q}{r}$

Ans. (a) 6   (c) 35   (e) 12   (g) 1.2 or 6/5   (i) 6   (k) ac/b

          (b) 50   (d) 36   (f) 44   (h) $3\frac{1}{2}$ (j) 28   (l) pr/q

9.8. Solve for x in each: **(3.2)**

(a) $x : 6 = 7 : 3$ (c) $(x - 2) : 40 = 7 : 10$ (e) $x : (x + 4) = 3 : 4$ (g) $4 : (x - 5) = 6 : x$

(b) $9 : (x + 1) = 3 : 4$ (d) $5 : 8 = 15 : 3x$ (f) $x : 5 = (x + 3) : 10$ (h) $x : r = q : p$

Ans.   (a) 14   (b) 11   (c) 30   (d) 8   (e) 12   (f) 3   (g) 15   (h) qr/p

9.9. Find the fourth proportional to the three numbers in each given set: **(3.3)**

(a) 2, 3, 6   (c) 6, 3, 2   (e) $\frac{1}{2}$, 2, 8   (g) a, b, c (i) c, a, b

(b) 3, 6, 2   (d) $2, \frac{1}{2}, 8$ (f) $8, \frac{1}{2}, 2$ (h) b, a, c

Ans.   (a) 9   (b) 4   (c) 1   (d) 2   (e) 32   (f) 1/8   (g) bc/a (h) ac/b (i) ab/c

9.10. Two numbers in the ratio of 5 : 3 are represented by $5x$ and $3x$. Express each **(4.1)** relationship as an equation; then find x and the numbers: (a) the sum of the numbers is 56;   (b) the difference of the numbers is 30;   (c) the smaller is 20 less than the greater;   (d) twice the greater exceeds the smaller by 28;   (e) if each of the numbers is increased 5, the new ratio of the numbers is 3 : 2.

Ans.   See Table 9-6.

Table 9-6

	Equations	x	Numbers
(a)	$8x = 56$	7	35, 21
(b)	$2x = 30$	15	75, 45
(c)	$3x = 5x - 20$	10	50, 30
(d)	$10x - 3x = 28$	4	20, 12
(e)	$\dfrac{5x + 5}{3x + 5} = \dfrac{3}{2}$	5	25, 15

9.11. The ratio of the length of a rectangle to its width is 7 : 2. Find each dimension **(4.2)** if   (a) the perimeter of the rectangle is 180,   (b) the semiperimeter of the rectangle is 54,   (c) the perimeter of the new rectangle is 92 if each dimension is increased 5,   (d) the semiperimeter of the new rectangle is 52 if each dimension is decreased 10,   (e) the semiperimeter equals the perimeter of a square whose side is 5 more than the width of the rectangle.

Ans.   See Table 9-7.

Table 9-7

	Equations	x	Length, Width
(a)	$18x = 180$	10	70, 20
(b)	$9x = 54$	6	42, 12
(c)	$18x + 20 = 92$	4	28, 8
(d)	$9x - 20 = 52$	8	56, 16
(e)	$9x = 4(2x + 5)$	20	140, 40

9.12. Express in simplest form: **(5.1)**

    (*a*) 2 ft : 9 ft : 12 ft   (*d*) \$2 : 10¢ : 50¢   (*g*) 77 : 88 : 99   (*j*) $4\frac{1}{2} : 5\frac{1}{2} : 6\frac{1}{2}$

    (*b*) 2 ft : 3 yd : 4 yd   (*e*) 1 oz : 8 oz : 12 oz   (*h*) $7x : 8x : 9x$ (*k*) 20% : 25% : 35%

    (*c*) 2¢ : 10¢ : 50¢   (*f*) 1 lb : 8 oz : 12 oz   (*i*) .4 : .5 : .6   (*l*) $.05y : .5y : 5y$

    Ans. (*a*) 2 : 9 : 12   (*c*) 1 : 5 : 10   (*e*) 1 : 8 : 12   (*g*) 7 : 8 : 9   (*i*) 4 : 5 : 6   (*k*) 4 : 5 : 7

            (*b*) 2 : 9 : 12   (*d*) 20 : 1 : 5   (*f*) 4 : 2 : 3   (*h*) 7 : 8 : 9   (*j*) 9 : 11 : 13   (*l*) 1 : 10 : 100

9.13. Three partners in a business shared a sum of \$180 in profits. Find the share of each if   **(5.2)** profits are shared in the ratio of (*a*) 1 : 2 : 3, (*b*) 2 : 3 : 5, (*c*) 6 : 5 : 1.

    Ans. (*a*) \$30, \$60, \$90   (*b*) \$36, \$54, \$90   (*c*) \$90, \$75, \$15

9.14. Find x in each: **(6.1)**

    (*a*) $\dfrac{x \text{ mi}}{6 \text{ hr}} = \dfrac{50 \text{ mi}}{15 \text{ hr}}$ (*c*) $\dfrac{75 \text{ in.}}{2 \text{ min}} = \dfrac{225 \text{ in.}}{x \text{ min}}$ (*e*) $\dfrac{\$4.75}{x \text{ km}} = \dfrac{\$1.33}{7 \text{ km}}$

    (*b*) $\dfrac{80 \text{ ft}}{10 \text{ min}} = \dfrac{x \text{ ft}}{35 \text{ min}}$ (*d*) $\dfrac{\$x}{15 \text{ lb}} = \dfrac{\$50}{3 \text{ lb}}$ (*f*) $\dfrac{\$4.50}{10 \text{ ft}} = \dfrac{\$1.80}{x \text{ ft}}$

    Ans. (*a*) 20   (*b*) 280   (*c*) 6   (*d*) 250   (*e*) 25   (*f*) 4

9.15. (*a*) A distance of 320 miles is traveled in 20 hours. At the same average rate, how   **(6.2)** many miles are traveled in $8\frac{1}{2}$ hours?

      (*b*) The cost of 9 kilograms of cheese is \$13.50. At the same price, how much must be paid for 15 kilograms of the same cheese?

    Ans. (*a*) 136 mi   (*b*) \$22.50

9.16. (*a*) Compare the average rate of speed needed to travel 140 miles in 4 hours with that   **(6.3)** needed to travel 100 miles in 4 hours.

      (*b*) Compare the price in a purchase of 5 cans of a vegetable for \$1.35 with the price in a purchase of 3 cans of the same vegetable for \$1.62.

    Ans. (*a*) The first average rate is 10 mph greater than the second, and seven-fifths as much.

         (*b*) The first price is 27¢ less than the second, and one-half as much.

Chapter 10

Fundamentals of Geometry

1. FUNDAMENTAL TERMS: POINT, LINE, SURFACE, SOLID, LINE SEGMENT, RAY

The fundamental terms of geometry are *point*, *line*, *surface*, and *solid*. These terms begin the process of definition and underlie the definitions of all other geometric terms.

Point

A point has position only. It has no length, width, or thickness. A point has no size.

A point is represented by a dot. Keep in mind, however, that the dot represents a point but is not a point, just as a dot on a map may represent a locality but is not the locality. A dot, unlike a point, has size. Another representation of a point is the tip of a needle.

A point is named by a capital letter next to the dot; thus, $\cdot P$ indicates point P.

Line

A line has length but has no width or thickness.

A line may be straight, curved, or a combination of these. To understand how lines differ, think of a line as being generated by a moving point.

A *straight line*, such as ——→, is generated by a point moving in a fixed direction.

A *curved line*, such as ⌒, is generated by a point moving in a continuously changing direction.

A *broken line*, such as ⋀, is a combination of straight lines.

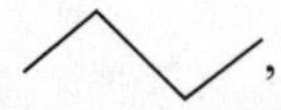

Rule: One and only one straight line can be drawn through any two points.

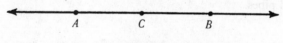

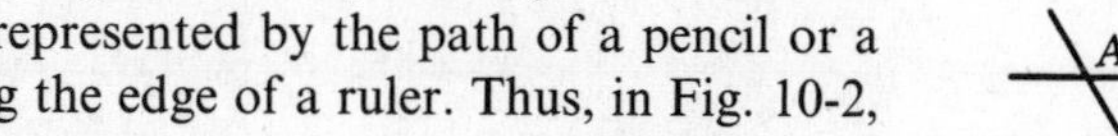

Fig. 10-1

Thus, in Fig. 10-1 one and only one straight line can be drawn through points A and B. The line drawn through A and B is said to be "determined by" A and B. The rule may be restated as "Two points determine a straight line."

A straight line may be represented by the edge of a ruler. A straight line may also be represented by the path of a pencil or a piece of chalk drawn along the edge of a ruler. Thus, in Fig. 10-2, the paths through points A, B, and C are representations of straight lines.

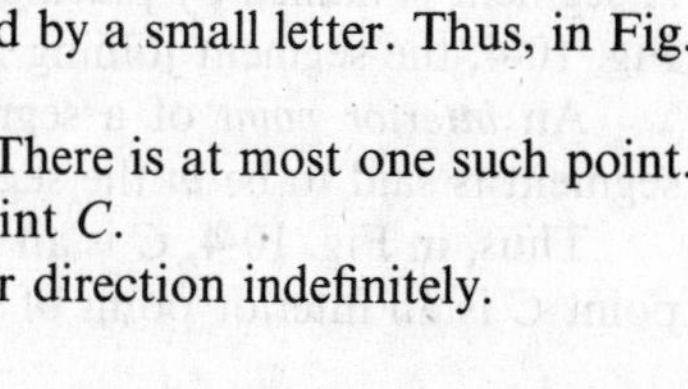

If the meaning is clear, "straight line" may be shortened to "line." Thus, "the line through points A and B" means "the straight line through points A and B," unless otherwise indicated.

Fig. 10-2

A line is named by placing a double-headed arrow over the letters of any two of its points in either order. A line may also be named by a small letter. Thus, in Fig. 10-2, the line through points A and B may be named $\overleftrightarrow{AB}$, $\overleftrightarrow{BA}$ or a.

The *intersection* of two lines is a point that they have in common. There is at most one such point. Thus, in Fig. 10-2, the intersection of $\overleftrightarrow{AC}$ and $\overleftrightarrow{BC}$ is their common point C.

A straight line is unlimited in extent and may be extended in either direction indefinitely.

229

Surface

A surface has length and width but no thickness. It may be represented by a blackboard, a side of a box or the outside of a ball; these are representations of a surface but are not surfaces.

A *plane surface* or a *plane* is a surface such that a straight line connecting any two of its points lies entirely in it. A plane is a flat surface and may be represented by the surface of a flat mirror or the top of a desk.

Plane geometry is the geometry that deals with figures that may be drawn on a plane surface. Such figures are called *planar* figures.

Solid

A *solid* is an enclosed portion of space bounded by plane or curved surfaces. Several solids are represented in Fig. 10-3.

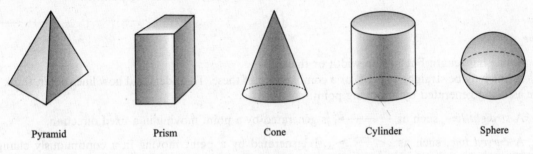

| Pyramid | Prism | Cone | Cylinder | Sphere |

Fig. 10-3

A solid has length, width, and thickness. Models useful for representing a solid are boxes, bricks, blocks, crystals, balls, and pipes.

1.1   ILLUSTRATING FUNDAMENTAL TERMS

Point, line, and *surface* are fundamental terms. Which of these terms is illustrated by each of the following?

(*a*) a light ray   (*c*) a projection screen   (*e*) a stretched thread
(*b*) the top of a desk   (*d*) a ruler's edge   (*f*) the tip of a pin

Ans.   (*a*) line   (*b*) surface   (*c*) surface   (*d*) line   (*e*) line   (*f*) point

Line Segments

The expression *straight line segment* may be shortened to *line segment* or *segment*. Unless otherwise indicated, *line segment* or *segment* means a straight line segment.

A segment joins two points and is part of a line. The joined points are the *endpoints* of the segment. A segment is named by placing a bar over the letters of the endpoints, written in either order. Thus, in Fig. 10-4, the segment joining A and B may be named either $\overline{AB}$ or $\overline{BA}$.

An *interior point* of a segment is any point of the segment between its endpoints. A point of a segment is said to be *in* the segment, *on* the segment, or to *belong to* the segment.

Thus, in Fig. 10-4, C is an interior point of segment $\overline{AB}$. The notation $\overline{ACB}$ is used to indicate that point C is an interior point of segment $\overline{AB}$.

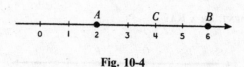

Fig. 10-4

Length of a Line Segment: Equal Segments

The *length* of a line segment is the distance between its endpoints. The distance is the measure of the line segment, the number of units in the line segment. The notation AB denotes the measure or length of $\overline{AB}$. Thus, in Fig. 10-4, $AB = 6 - 2 = 4$.

Congruent segments are segments having the same measure. Thus, in Fig. 10-4 since $AC = 2$ and $CB = 2$, then $\overline{AC}$ is congruent to $\overline{CB}$, written $\overline{AC} \cong \overline{CB}$.

If a line divides a segment into two equal segments, as in Fig. 10-5, then:

1. The point of division is the *midpoint* of the segment.
2. The line is said to *bisect* the given segment.

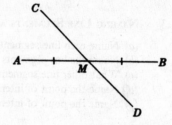

Thus if $AM = MB$, then M is the midpoint of $\overline{AB}$, and $\overleftrightarrow{CD}$ bisects $\overline{AB}$.

Equal line segments may be shown by crossing them with the same number of strokes. Note that $\overline{AM}$ and $\overline{MB}$ are crossed by a single stroke in Fig. 10-5.

Fig. 10-5

Rays

Imagine the tiniest of searchlights casting a streak of light that is endless. The streak of light suggests a geometric figure called a *ray*, as in Fig. 10-6.

Fig. 10-6

A ray has only one endpoint, its initial point. Two points are used to name a ray; first, its endpoint, and then any other point of the ray. A single-headed arrow is placed over the letters. Thus, in Fig. 10-6, $\overrightarrow{AB}$ is the ray on the left having endpoint A and passing through B. $\overrightarrow{BA}$ is the ray on the right having endpoint B and passing through A.

Summary of Line Notation

$\overline{AB}$ denotes the line segment between points A and B.
AB denotes the length or the measure of line segment $\overline{AB}$.
$\overleftrightarrow{AB}$ denotes the line determined by A and B.
$\overrightarrow{AB}$ denotes the ray having endpoint A and passing through B; $\overrightarrow{BA}$ denotes the ray having endpoint B and passing through A.
$\overline{ACB}$ denotes line segment $\overline{AB}$ with C between A and B.
$\cong$ denotes congruency.

1.2   NAMING A LINE, A SEGMENT, OR A RAY

Fig. 10-7

In Fig. 10-7, state the meaning of (a) PQ,   (b) $\overrightarrow{QR}$,   (c) $\overline{RS}$,   (d) $\overline{QRS}$,   (e) $\overleftrightarrow{QR}$,   (f) $\overrightarrow{SQ}$.

Ans.   (a)   measure or length of segment $\overline{PQ}$
      (b)   ray having endpoint Q and passing through R
      (c)   line segment having endpoints R and S
      (d)   line segment $\overline{QS}$ with interior point R between Q and S
      (e)   line passing through or determined by Q and R
      (f)   ray having endpoint S and passing through Q

1.3   NAMING LINE SEGMENTS AND POINTS

(a) Name each line segment shown in Fig. 10-8.
(b) Name the line segments that intersect at A.
(c) What other line segment can be drawn?
(d) Name the point of intersection of $\overline{CD}$ and $\overline{AD}$.
(e) Name the point of intersection of $\overline{BC}$, $\overline{AC}$ and $\overline{CD}$.

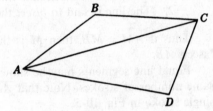

Fig. 10-8

Solutions

(a) $\overline{AB}$, $\overline{BC}$, $\overline{CD}$, $\overline{AC}$, and $\overline{AD}$. These segments may also be named by interchanging the letters; thus $\overline{BA}$, $\overline{CB}$, $\overline{DC}$, $\overline{CA}$, and $\overline{DA}$ are also correct.
(b) $\overline{AB}$, $\overline{AC}$, and $\overline{AD}$ (c) $\overline{BD}$ (d) D (e) C

1.4   FINDING MEASURES, MIDPOINTS, AND BISECTORS

(a) State the lengths of $\overline{AB}$, $\overline{AC}$, and $\overline{AF}$ in Fig. 10-9.
(b) Name two midpoints.
(c) Name two bisectors.

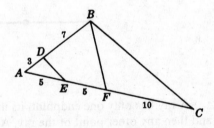

Solutions

(a) $AB = 3 + 7 = 10$, $AC = 5 + 5 + 10 = 20$, $AF = 5 + 5 = 10$.
(b) E is midpoint of $\overline{AF}$, and F is midpoint of $\overline{AC}$.
(c) $\overline{DE}$ is bisector of $\overline{AF}$ since $AE = EF = 5$. $\overline{BF}$ is bisector of $\overline{AC}$ since $AF = FC = 10$.

Fig. 10-9

2.   ANGLES

An *angle* is the figure formed by two rays having a common endpoint. Thus, in Fig. 10-10 angle a is formed by rays $\overrightarrow{AB}$ and $\overrightarrow{AC}$ having a common endpoint A.

Each of the rays of an angle is one of its *sides*. The common endpoint is the *vertex* of the angle.

Naming an Angle

An angle may be named in any of the following ways:

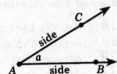

Fig. 10-10

   1.   By the vertex letter, if there is only one angle having this vertex, as $\angle B$ in Fig. 10-11(a).

2. By a small letter or a number placed between the sides of the angle and near the vertex, as $\angle a$ or $\angle 1$ in Fig. 10-11(b).

3. By three capital letters, with the vertex letter between two others on the sides of the angle. Referring to Fig. 10-11(c), $\angle E$ may be named $\angle DEG$ or $\angle GED$. $\angle G$ may be named $\angle EGH$ or $\angle HGE$.

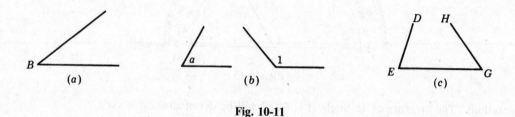

Fig. 10-11

The Degree, a Unit of Measure for Angles

The *degree*, symbolized °, is a unit of measure for angles. If a complete turn about a fixed point is made, the number of degrees in the complete turn or rotation is 360. In Fig. 10-12 the circle is divided into 360 equal parts. If lines were drawn from the center to each of the points of division, 360 equal angles would be formed. Each of these angles would be an angle of 1°.

The numbers shown in Fig. 10-12 indicate the number of degrees of rotation. The number 120 indicates that the number of degrees in a counterclockwise rotation to that point from 0 is 120.

The Measure of an Angle

The *measure of an angle* is the number of degrees in the angle. The measure of an angle depends on the extent to which one side of the angle must be rotated about the vertex until the rotated side meets the other side; it does not depend on either side.

Thus the protractor in Fig. 10-13 shows that the measure of $\angle A$ is 60. If $\overrightarrow{AC}$ were rotated about the vertex A until it met $\overrightarrow{AB}$, the amount of turn would be 60°.

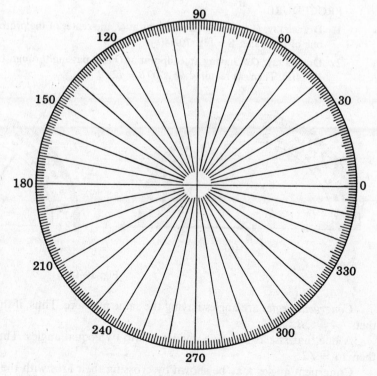

Fig. 10-12

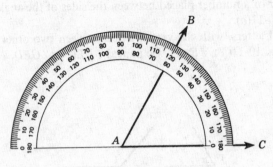

Fig. 10-13

Notation: "The measure of an angle A is $60°$," will be denoted $m \angle A = 60°$.

Using a Protractor to Measure an Angle

1. Place the vertex of the angle at the center of the protractor. Thus, in Fig. 10-13, vertex A of $\angle BAC$ is placed at the center of the protractor.
2. Pair one ray of the angle with 0, zero, on one of the protractor scales. Thus, ray $\overrightarrow{AC}$ is paired with 0 on the inner scale.
3. Pair the other ray of the angle with a number between 0 and 180 on the same scale used in step 2. Thus, ray AB is paired with 60 on the inner scale. Hence, $m \angle BAC = 60°$.

Using a Protractor to Draw an Angle

Draw an angle of measure $124°$.

PROCEDURE

1. Draw a ray, $\overrightarrow{OA}$, having an endpoint at O, the center of the protractor, and passing through the zero of one of the scales. See Fig. 10-14(a).
2. Draw a ray, $\overrightarrow{OB}$, having an endpoint at O and passing through 124 of the scale used in step 1. See Fig. 10-14(b). Then, as required, $m \angle AOB = 124°$.

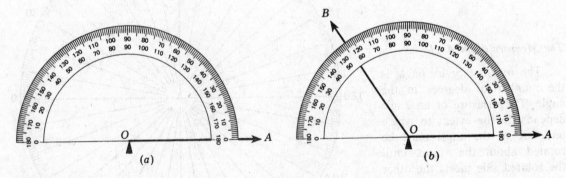

(a)   (b)

Fig. 10-14

Congruent angles are angles having the same measure. Thus, if the measure of angles A and B is $90°$, then $\angle A \cong \angle B$.

A line that *bisects* an angle divides it into two equal angles. Thus, in Fig. 10-15, if $\overline{AD}$ bisects $\angle A$, then $\angle 1 \cong \angle 2$.

Congruent angles may be shown by crossing their arcs with the same number of strokes. Here the arcs of $\angle 1$ and $\angle 2$ are crossed by a single stroke.

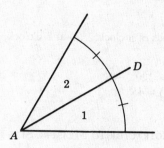

Fig. 10-15

2.1   Naming Angles

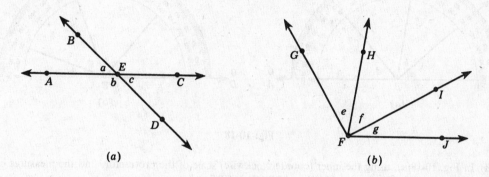

(a)   (b)

Fig. 10-16

In Fig. 10-16(a), using three capital letters, name   (a) ∠a,   (b) ∠b,   (c) ∠c,   (d) a straight angle (see Section 4) of which ∠b and ∠c are parts. In Fig. 10-16(b), name   (e) ∠e,   (f) ∠f,   (g) ∠g,   (h) an angle of which ∠e and ∠g are parts.

Ans.   (a) ∠BEA   (b) ∠AED   (c) ∠CED   (d) ∠AEC   (e) ∠GFH   (f) ∠HFI   (g) ∠IFJ   (h) ∠GEJ
(In each of the answers, the end letters may be interchanged.)

2.2   Finding Rotations or Turns

In a half hour, what turn or rotation is made   (a) by the minute hand,   (b) by the hour hand?   What rotation is needed to turn: (c) from north to southeast in a clockwise direction.

Solutions

(a) In one hour, a minute hand completes a full circle of 360°. Hence in a half hour it turns 180°.
(b) In one hour, an hour hand turns 1/12 of 360° or 30°. Hence in a half hour it turns 15°.
(c) Add turns of 90° from north to east and 45° from east to southeast: 90° + 45° = 135°.

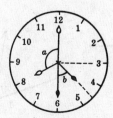

Fig. 10-17

2.3 FINDING MEASURES OF ANGLES

Find the angle formed by the hands of a clock: (*a*) at 8 o'clock, (*b*) at 4:30 o'clock. See Fig. 10-17.

Solutions

(*a*) At 8 o'clock, $m \angle a = \frac{1}{3}(360°) = 120°$.
(*b*) At 4:30 o'clock, $m \angle b = \frac{1}{2}(90°) = 45°$.

2.4 USING BOTH SCALES OF THE PROTRACTOR TO MEASURE SETS OF ANGLES

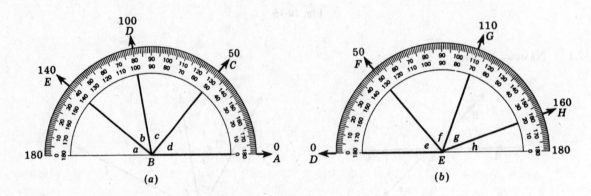

Fig. 10-18

(*a*) In Fig. 10-18(*a*), using the inner (counterclockwise) scale of the protractor, find the measures of:
   (1) $\angle a$, (2) $\angle b$, (3) $\angle c$, (4) $\angle d$, (5) $\angle ABD$, (6) $\angle CBE$.
(*b*) In Fig. 10-18(*b*), using the outer (clockwise) scale of the protractor, find the measures of: (1) $\angle e$, (2) $\angle f$,
   (3) $\angle g$, (4) $\angle h$, (5) $\angle DEH$, (6) $\angle FEH$.

Solutions Find the differences of the coordinates of the rays of the angle being measured.

(*a*) (1) $m \angle a = 180° - 140° = 40°$	(3) $m \angle c = 100° - 50° = 50°$	(5) $m \angle ABD = 100°$
(2) $m \angle b = 140° - 100° = 40°$	(4) $m \angle d = 50°$	(6) $m \angle CBE = 140° - 50° = 90°$
(*b*) (1) $m \angle e = 50°$	(3) $m \angle g = 160° - 110° = 50°$	(5) $m \angle DEH = 160°$
(2) $m \angle f = 110° - 50° = 60°$	(4) $m \angle h = 180° - 160° = 20°$	(6) $m \angle FEH = 160° - 50° = 110°$

2.5 USING A PROTRACTOR TO MEASURE ANGLES

Using a protractor with Fig. 10-19, find the measure of (*a*) $\angle a$, (*b*) $\angle b$, (*c*) $\angle c$, (*d*) $\angle d$, (*e*) $\angle e$,
(*f*) $\angle f$, (*g*) $\angle g$, (*h*) $\angle f + \angle g$ or $\angle BAC$.

Solutions Be sure to place the vertex of the angle at the center of the protractor. Then pair each ray of the angle with a direction number of the same protractor scale.

(*a*) 30   (*b*) 140   (*c*) 90   (*d*) 60   (*e*) 105   (*f*) 45   (*g*) 45   (*h*) 90

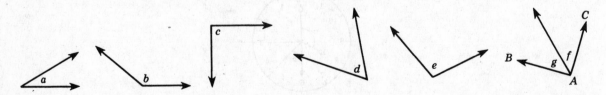

Fig. 10-19

2.6   USING THE PROTRACTOR TO DRAW ANGLES

(a) Draw the following angles using the inner (counterclockwise) scale of the protractor: (1) 25° (2) 50° (3) 95° (4) 160°

(b) Draw the following angles using the outer (clockwise) scale of the protractor: (1) 30° (2) 65° (3) 110° (4) 175°

Solutions   Refer to Fig. 10-20

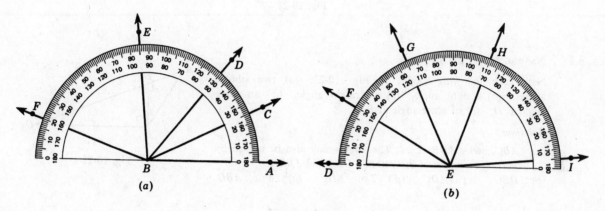

Fig. 10-20

(a) (1) $m\angle ABC = 25°$ (2) $m\angle ABD = 50°$ (3) $m\angle ABE = 95°$ (4) $m\angle ABF = 160°$

(b) (1) $m\angle DEF = 30°$ (2) $m\angle DEG = 65°$ (3) $m\angle DEH = 110°$ (4) $m\angle DEI = 175°$

3. KINDS OF ANGLES

1. **Acute angle**. An acute angle is an angle whose measure is more than 0 and less than 90, Fig. 10-21(a).

   Thus, 0° is less than $a°$ and $a°$ is less than 90°; this is denoted by $0 < a < 90$.

2. **Right angle**. A right angle is an angle whose measure is 90, Fig. 10-21(b).

   Thus, rt. $m\angle A = 90°$. The square corner denotes a right angle.

3. **Obtuse angle**. An obtuse angle is an angle whose measure is more than 90 and less than 180, Fig. 10-21(c).

   Thus, 90° is less than $b°$ and $b°$ is less than 180°; this is denoted by $90 < b < 180$.

4. **Straight angle**. A straight angle is an angle whose measure is 180, Fig. 10-21(d).

   Thus, st. $m\angle B = 180°$. Note that the sides of a straight angle lie in the same straight line. However, do not confuse a straight angle with a straight line!

5. **Reflex angle**. A reflex angle is an angle whose measure is more than 180 and less than 360, Fig. 10-21(e).

   Thus, 180° is less than $c°$ and $c°$ is less than 360°; this is symbolized by $180 < c < 360$.

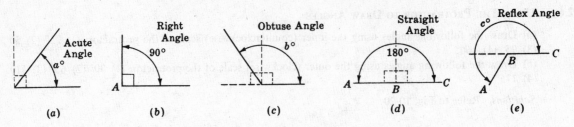

Fig. 10-21

3.1 NAMING KINDS OF ANGLES

Name the following angles in Fig. 10-22: (*a*) two obtuse angles, (*b*) a right angle, (*c*) a straight angle, (*d*) an acute angle at *D*, (*e*) an acute angle at *B*.

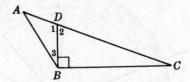

Solutions

(*a*) ∠*ABC*, and ∠*ADB* or ∠1. The angles may also be named by reversing the order of the letters: ∠*CBA* and ∠*BDA*.

(*b*) ∠*DBC*   (*c*) ∠*ADC*   (*d*) ∠2 or ∠*BDC*   (*e*) ∠3 or ∠*ABD*

Fig. 10-22

3.2 PARTS OF A RIGHT ANGLE OR A STRAIGHT ANGLE

Find the number of degrees in (*a*) 1/9 of a right angle, (*b*) 40% of a right angle, (*c*) 3/5 of a straight angle, (*d*) 75% of a straight angle, (*e*) 125% of a right angle.

Solutions

(*a*) $\frac{1}{9}(90°) = 10°$ (*b*) $\frac{2}{5}(90°) = 36°$ (*c*) $\frac{3}{5}(180°) = 108°$ (*d*) $\frac{3}{4}(180°) = 135°$ (*e*) $\frac{5}{4}(90°) = 112\frac{1}{2}°$

3.3 DETERMINING KINDS OF ANGLES

Classify ∠*a* through ∠*h* in Fig. 10-23.

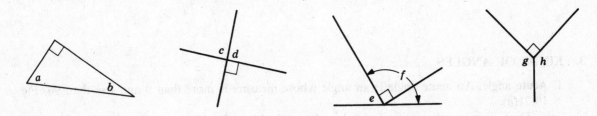

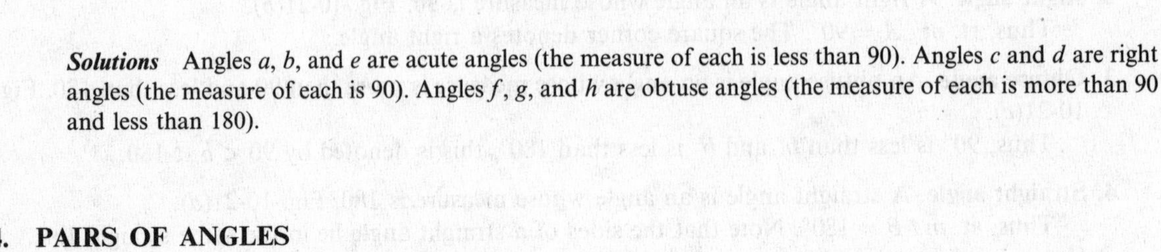

Fig. 10-23

Solutions   Angles *a*, *b*, and *e* are acute angles (the measure of each is less than 90). Angles *c* and *d* are right angles (the measure of each is 90). Angles *f*, *g*, and *h* are obtuse angles (the measure of each is more than 90 and less than 180).

4. PAIRS OF ANGLES

Adjacent Angles

Adjacent angles are two angles having the same vertex and a common side between them.

Rule: If an angle of *c*° consists of adjacent angles of *a*° and *b*°, as in Fig. 10-24(*a*), then *c* = *a* + *b*.

Complementary Angles

Complementary angles are two angles the sum of whose measures is 90. Thus, in Fig. 10-24(*b*), $a + b = 90$. Note that complementary angles may or may not be adjacent.

Supplementary Angles

Supplementary angles are two angles the sum of whose measures is 180. Thus, in Fig. 10-24(*c*), $a + b = 180$. Note that supplementary angles may or may not be adjacent.

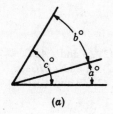   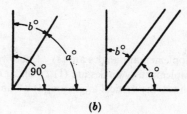   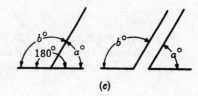

     (a)   (b)   (c)

Fig. 10-24

Vertical Angles

Vertical angles are two angles formed by two intersecting lines and having only one point in common. In Fig. 10-25, the intersecting lines $\overleftrightarrow{AB}$ and $\overleftrightarrow{CD}$ form two separate pairs of vertical angles.

    Rule: Vertical angles have equal measures.

       Thus, in Fig. 10-25, $a = b$ and $c = d$.

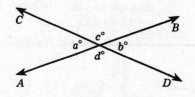

Fig. 10-25

4.1   SELECTING COMPLEMENTARY AND SUPPLEMENTARY ANGLES

$$A = \{10°, 30°, 55°, 70°, 100°, 120°, 155°\}$$
$$B = \{35°, 60°, 80°, 110°, 125°, 150°, 170°\}$$

(*a*) For each angle in set *A*, find its supplement in set *B*   (*b*) Which angles in sets *A* and *B* are complementary angles?

Solutions

(*a*) Angles are supplementary if their angle-sum is 180.

Ans.   10° and 170°   55° and 125°   100° and 80°
       30° and 150°   70° and 110°   120° and 60°

(*b*) Angles are complementary if their angle-sum is 90.

Ans.   10° and 80°   30° and 60°   35° and 55°

4.2   NAMING PAIRS OF ANGLES

(*a*) In Fig. 10-26(*a*), name two pairs of supplementary angles.
(*b*) In Fig. 10-26(*b*), name two pairs of complementary angles.
(*c*) In Fig. 10-26(*c*), name two pairs of vertical angles.

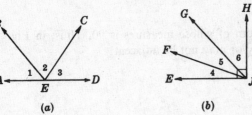 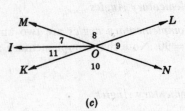

Fig. 10-26

Solutions

(a) Since their angle-sum is 180, the supplementary angles are (1) $\angle 1$ and $\angle BED$, (2) $\angle 3$ and $\angle AEC$.

(b) Since their angle-sum is 90, the complementary angles are (1) $\angle 4$ and $\angle FJH$, (2) $\angle 6$ and $\angle EJG$.

(c) Since $\overleftrightarrow{KL}$ and $\overleftrightarrow{MN}$ are intersecting lines, the vertical anges are (1) $\angle 8$ and $\angle 10$, (2) $\angle 9$ and $\angle MOK$.

4.3   USING ALGEBRA TO FIND PAIRS OF ANGLES

Find the angles in each. (a) The angles are supplementary and the larger is twice the smaller. (b) The angles are complementary and the larger is 20° more than the smaller. (c) The angles are adjacent and form an angle of 120°. The larger is 20° less than three times the smaller. (d) The angles are vertical and complementary.

Solutions   Refer to Fig. 10-27, in which x denotes the measure of the smaller angle.

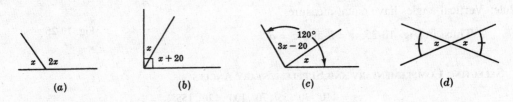

Fig. 10-27

(a) $x + 2x = 180$
        $3x = 180$
        $x = 60, 2x = 120$
    Ans. 60° and 120°

(b) $x + (x + 20) = 90$
        $2x + 20 = 90$
        $x = 35, x + 20 = 55$
    Ans. 35° and 55°

(c) $x + (3x - 20) = 120$
        $4x - 20 = 120$
        $x = 35, 3x - 20 = 85$
    Ans. 35° and 85°

(d) $x + x = 90$
        $2x = 90$
        $x = 45$
    Ans. 45° and 45°

4.4   ADDITION OR SUBTRACTION INVOLVING PAIRS OF ANGLES

Find the number of degrees in the measures of the following angles pictured in Fig. 10-28: (a) $\angle a$, (b) $\angle b$, (c) $\angle c$, (d) $\angle d$, (e) $\angle e$ (f) $\angle f$, (g) $\angle g$, (h) $\angle h$.

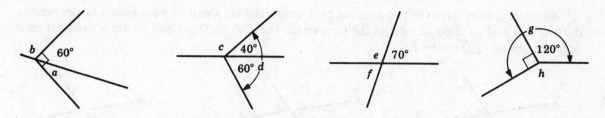

Fig. 10-28

Solutions   (*a*) Using complementary angles, $m \angle a = 90° - 60° = 30°$.   (*b*) Using supplementary angles, $m \angle b = 180° - 60° = 120°$.   (*c*) Using supplementary angles, $m \angle c = 180° - 40° = 140°$.   (*d*) Using adjacent angles, $m \angle d = 40° + 60° = 100°$.   (*e*) Using supplementary angles, $m \angle e = 180° - 70° = 110°$.   (*f*) Using vertical angles, $m \angle f = 70°$.   (*g*) Using adjacent angles, $m \angle g = 120° + 90° = 210°$.   (*h*) Since a complete rotation contains 360°, $m \angle h = 360° - 210° = 150°$.

5. PERPENDICULAR AND PARALLEL LINES; TRANSVERSALS

Perpendicular Lines

Perpendicular lines are lines that intersect at right angles. Also, we may have perpendicular rays and perpendicular segments. The symbol ⊥ means "is perpendicular to."

In Fig. 10-29, $\overleftrightarrow{AC}$ and $\overleftrightarrow{BD}$ are perpendicular lines. Hence, $\overleftrightarrow{AC} \perp \overleftrightarrow{BD}$. Also, since $\overline{PC}$ and $\overline{PB}$ are perpendicular segments, $\overline{PC} \perp \overline{PB}$. The square corner in a figure indicates a right angle.

Fig. 10-29   **Fig.10-30**

Transversals

A *transversal* of two or more lines is a line that intersects or cuts the lines. Thus, in Fig. 10-30, $\overleftrightarrow{EF}$ is a transversal of $\overleftrightarrow{AB}$ and $\overleftrightarrow{CD}$.

Angles Formed by Two Lines Cut by a Transversal

Interior angles formed by two lines cut by a transversal are the angles between the two lines, while exterior angles are those on the outside. In Fig. 10-31, of the eight angles formed by $\overleftrightarrow{AB}$ and $\overleftrightarrow{CD}$ cut by $\overleftrightarrow{EF}$: the interior angles are $\angle 1$, $\angle 2$, $\angle 3$ and $\angle 4$; the exterior angles are $\angle 5$, $\angle 6$, $\angle 7$ and $\angle 8$.

Corresponding angles of two lines cut by a transversal are a pair of angles on the same side of the transversal and on the same side of the lines. For Fig. 10-32, $\angle 1$ and $\angle 2$ are corresponding angles of $\overleftrightarrow{AB}$ and $\overleftrightarrow{CD}$ cut by transversal $\overleftrightarrow{EF}$. Note that the two angles are to the right of the transversal and below the lines.

Alternate interior angles of two lines cut by a transversal are a pair of nonadjacent angles between the two lines and on opposite sides of the transversal. For Fig. 10-33, $\angle 1$ and $\angle 2$ are alternate interior angles of $\overleftrightarrow{AB}$ and $\overleftrightarrow{CD}$ cut by $\overleftrightarrow{EF}$.

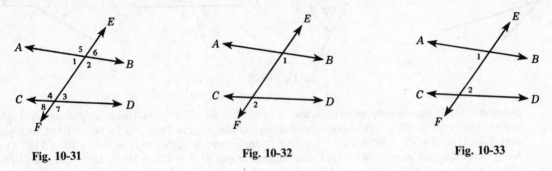

Fig. 10-31   Fig. 10-32   Fig. 10-33

Parallel Lines

Parallel lines are lines in a plane that will not intersect no matter how far they are extended. The symbol ‖ means "is parallel to."

Rule 1: If two parallel lines are cut by a transversal, corresponding angles have equal measures.

Thus, in Fig. 10-34, if $\overleftrightarrow{AB} \parallel \overleftrightarrow{CD}$, then $a = b$.

Rule 2: If corresponding angles of two lines cut by a transversal have equal measures, then the lines are parallel.

Thus, in Fig. 10-34, if $a = b$, then $\overleftrightarrow{AB} \parallel \overleftrightarrow{CD}$.

Rule 3: If two parallel lines are cut by a transversal, alternate interior angles have equal measures.

Thus, in Fig. 10-35, if $\overleftrightarrow{AB} \parallel \overleftrightarrow{CD}$, then $c = d$.

Rule 4: If alternate interior angles of two lines cut by a transversal have equal measures, then the lines are parallel.

Thus, in Fig. 10-35, if $c = d$, then $\overleftrightarrow{AB} \parallel \overleftrightarrow{CD}$.

Rule 5: If lines are perpendicular to the same line, then they are parallel to each other.

Thus, in Fig. 10-36, if $\overleftrightarrow{AB} \perp \overleftrightarrow{EF}$ and $\overleftrightarrow{CD} \perp \overleftrightarrow{EF}$, then $\overleftrightarrow{AB} \parallel \overleftrightarrow{CD}$.

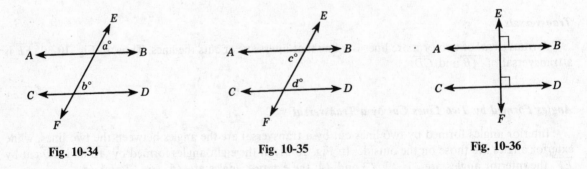

Fig. 10-34   Fig. 10-35   Fig. 10-36

To Draw a Line Parallel to a Given Line through a Point
Not on the Line

In Fig. 10-37, draw a line passing through P and parallel to $\overleftrightarrow{AB}$.

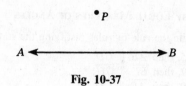

Fig. 10-37

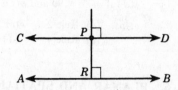

PROCEDURE
Use the protractor to draw two perpendiculars, as shown in Fig. 10-38.

1. Draw a transversal $\overleftrightarrow{PR}$ perpendicular to $\overleftrightarrow{AB}$.
2. Through P, draw a line $\overleftrightarrow{CD}$ perpendicular to $\overleftrightarrow{PR}$. By Rule 5, $\overleftrightarrow{CD}$ is the required line.

Fig. 10-38

5.1 APPLYING PARALLEL LINE RULES

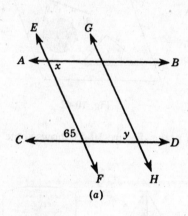

(a)

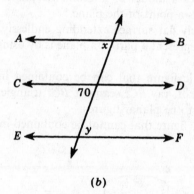

(b)

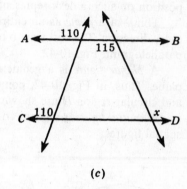

(c)

Fig. 10-39

(a) In Fig. 10-39(a), $\overleftrightarrow{AB} \parallel \overleftrightarrow{CD}$ and $\overleftrightarrow{EF} \parallel \overleftrightarrow{GH}$. Find $m\angle x$ and $m\angle y$. (b) In Fig. 10-39(b), $\overleftrightarrow{AB} \parallel \overleftrightarrow{CD} \parallel \overleftrightarrow{EF}$. Find x and y. (c) In Fig. 10-39(c), show that $\overleftrightarrow{AB} \parallel \overleftrightarrow{CD}$ and find $m\angle x$.

Solutions

(a) $m\angle x = 65$ by Rule 3; $m\angle y = 65$ by Rule 1.
(b) $m\angle x = 70$ by Rule 1; $m\angle y = 70$ by Rule 3.
(c) $\overleftrightarrow{AB} \parallel \overleftrightarrow{CD}$ by Rule 2; $m\angle x = 115$ by Rule 3.

5.2 FINDING MEASURES OF ANGLES FORMED BY PARALLEL LINES

In Fig. 10-40, $\overleftrightarrow{AB} \parallel \overleftrightarrow{CD}$ and $\overleftrightarrow{EF} \parallel \overleftrightarrow{GH}$.

(a) If $m\angle e = 100$, find (1) $m\angle a$, (2) $m\angle c$, (3) $m\angle d$.
(b) If $m\angle d = 70$, find (1) $m\angle f$, (2) $m\angle h$, (3) $m\angle g$.
(c) If $m\angle g = 120$, find (1) $m\angle p$, (2) $m\angle r$, (3) $m\angle s$.

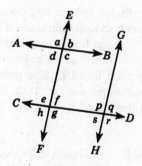

Fig. 10-40

Ans. (a) (1) $m\angle a = 100$ (by Rule 1, $a = e$)
            (2) $m\angle c = 100$ (by Rule 3, $c = e$)
            (3) $m\angle d = 80$ ($a + d = 180$)

      (b) (1) $m\angle f = 70$ (by Rule 3, $f = d$)
            (2) $m\angle h = 70$ (by Rule 1, $h = d$)
            (3) $m\angle g = 110$ ($f + g = 180$)

      (c) (1) $m\angle p = 120$ (by Rule 3, $g = p$)
            (2) $m\angle r = 120$ (by Rule 1, $r = g$)
            (3) $m\angle s = 60$ ($s + p = 180$)

5.3 OBTAINING PARALLEL LINES BY EQUAL MEASURES OF ANGLES

Refer to Fig. 10-41. In each state the rule or rules justifying the statement:

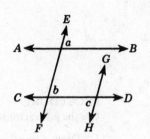

(a) If $m \angle a = 85$ and $m \angle b = 85$, then $\overleftrightarrow{AB} \parallel \overleftrightarrow{CD}$.
(b) If $m \angle b = 89$ and $m \angle c = 89$, then $\overleftrightarrow{EF} \parallel \overleftrightarrow{GH}$.
(c) If $m \angle a = 90$ and $m \angle b = 90$, then $\overleftrightarrow{AB} \parallel \overleftrightarrow{CD}$.
(d) If $m \angle b = 90$ and $m \angle c = 90$, then $\overleftrightarrow{EF} \parallel \overleftrightarrow{GH}$.

Ans. (a) Rule 2   (c) Rule 2 or 5
        (b) Rule 4   (d) Rule 4 or 5

Fig. 10-41

6. PLANAR AND SPATIAL FIGURES

Imagine a tiny bug on the flat top of a very large desk. To the bug, the top of the desk seems endless in every direction. The thought of a desk with an endless top provides a model of a geometric plane. A position on such a desk represents a point of the plane.

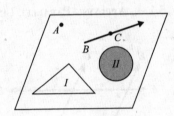

Think of a *plane* as an entirely flat surface extending endlessly in every direction. The best way to represent a part of a plane is by using a parallelogram, Fig. 10-42.

A *planar figure* is a geometric figure that can be contained in a plane. Thus, in Fig. 10-42, point A, ray $\overrightarrow{BC}$, segment $\overline{BC}$, triangle I, and circular region II are shown to be planar figures.

Fig. 10-42

A *spatial figure* is a geometric figure that cannot be contained in a plane. Figure 10-43 shows some spatial figures.

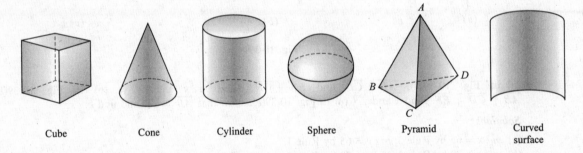

| Cube | Cone | Cylinder | Sphere | Pyramid | Curved surface |

Fig. 10-43

A *solid* is a closed spatial figure, bounded by plane or curved surfaces, or both. Thus, the cube, cone, cylinder, sphere, and pyramid shown in Fig. 10-43 are solids. However, the curved surface shown is an open spatial figure, not a solid.

Note that the pyramid in Fig. 10-43 is bounded by four plane surfaces. Each of these plane surfaces passes through three of the vertices of the pyramid. It can be shown that one and only one plane can be passed through three points which do not all lie on one straight line; that is, *three noncollinear points determine a plane*. Thus, there is a plane through points A, B, and C; a second through A, B, and D; a third through A, C, and D; and a fourth through B, C, and D.

6.1 SPATIAL AND PLANAR FIGURES

Is the figure planar or spatial? (a) a cylinder, (b) the curved surface of a cylinder, (c) each circular base of a cylinder, (d) a pyramid, (e) a face of a pyramid, (f) a vertex of a pyramid, (g) a cross section of a pyramid cut by a plane, (h) a sphere, (i) a region of a sphere, (j) a circle on the surface of a sphere, (k) two parallel lines.

Ans.   (*a*) spatial   (*b*) spatial (however, it can be "unrolled" into a planar region)
(*c*) planar   (*d*) spatial   (*e*) planar   (*f*) planar   (*g*) planar   (*h*) spatial   (*i*) spatial
(*j*) planar   (*k*) planar

7. POLYGONS

A *polygon* is a closed planar figure bounded by line segments that meet at their endpoints, do not cross, and enclose one and only one region.

Thus, in Fig. 10-44, only (*a*), (*b*), and (*c*) are polygons. Figure (*d*) is not a polygon since two regions are enclosed. Figure (*e*) is not a polygon since one of the sides is curved. Figure (*f*) is not a polygon since line segments cross.

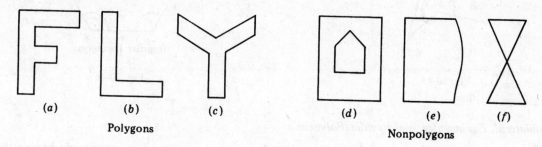

| (*a*) | (*b*) | (*c*) | (*d*) | (*e*) | (*f*) |
| Polygons | | | Nonpolygons | | |

Fig. 10-44

Names of Polygons According to the Number of Sides

Each of the bounding segments of a polygon is a *side* of the polygon. Each segment endpoint is a *vertex* of the polygon. A *vertex angle* of a polygon has sides that are sides of the polygon meeting at a vertex of the polygon. Note in Table 10-1 how names are given polygons according to the number of their sides.

Table 10-1

No. of Sides	Polygon	No. of Sides	Polygon
3	Triangle	8	Octagon
4	Quadrilateral	9	Nonagon
5	Pentagon	10	Decagon
6	Hexagon	12	Dodecagon
7	Septagon	n	n-gon

Properties of Polygons

1. The number of vertices of a polygon equals the number of its sides. Thus, a quadrilateral has 4 sides and also 4 vertices.
2. The number of vertex angles of a polygon equals the number of its sides. Thus, a pentagon has 5 sides and also 5 vertex angles.
3. Each side of a polygon is a side of two vertex angles.
4. A vertex angle of a polygon is *not* a straight angle.

A vertex angle may be an angle greater than 180°, as in Fig. 10-45. However, our discussion of polygons will not include those whose sides when extended penetrate the region enclosed by the polygon.

Important Note:

As in Fig. 10-46, congruent sides may be indicated by using the *same letter*, such as *s*; and congruent angles may be indicated by using the *same number of strokes*, such as a single stroke in this case.

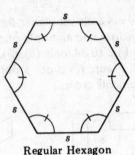

Regular Hexagon

Fig. 10-45

Fig. 10-46

Equilateral, Equiangular, and Regular Polygons

1. An *equilateral polygon* is a polygon having congruent sides. Thus, a triangle having three congruent sides is an equilateral triangle.
2. An *equiangular polygon* is a polygon having congruent angles. Thus, a rectangle whose angles are right angles is an equiangular polygon.
3. A *regular polygon* is both an equilateral and an equiangular polygon. Thus, in Fig. 10-46, the regular hexagon shown is an equilateral and also an equiangular polygon of six sides.

7.1   PROPERTIES OF POLYGONS

(*a*) How many vertex angles has a (1) hexagon, (2) octagon, (3) nonagon?
(*b*) How many vertices has a (1) pentagon, (2) decagon, (3) dodecagon?
(*c*) What is the value of *n* if the *n*-gon is a (1) triangle, (2) quadrilateral, (3) septagon?

Ans.   (*a*) (1) 6   (2) 8   (3) 9   (*c*) (1) 3   (2) 4   (3) 7
         (*b*) (1) 5   (2) 10   (3) 12

7.2   DETERMINING IF A FIGURE IS A POLYGON

Name each figure in Fig. 10-47 that is a polygon and for each figure that is not a polygon, state the reason why it is not a polygon.

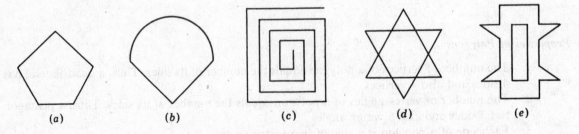

(*a*)   (*b*)   (*c*)   (*d*)   (*e*)

Fig. 10-47

Ans. (*a*) pentagon   (*c*) nonpolygon (open figure)   (*e*) nonpolygon (encloses
        (*b*) nonpolygon (curved side)   (*d*) nonpolygon (crossing segments)   more than 1 region)

7.3 EQUILATERAL, EQUIANGULAR, AND REGULAR POLYGONS

State whether each polygon in Fig. 10-48 is equilateral, equiangular, regular, or none of these. (Sides having the same letter are congruent sides and angles having the same number of strokes are congruent angles.)

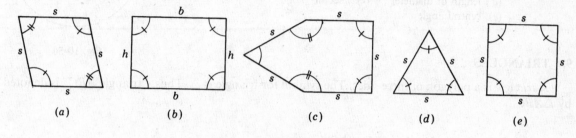

(*a*)   (*b*)   (*c*)   (*d*)   (*e*)

Fig. 10-48

Ans. (*a*) equilateral   (*b*) equiangular   (*c*) equilateral   (*d*) regular   (*e*) regular

8. CIRCLES

1. A *circle* is a closed curve in a plane every point of which is the same distance from a given point in the plane, its *center*. A circle is named by naming its center. The symbol ⊙ is used for a circle; thus, ⊙ *O* in Fig. 10-49.

2. A *radius* of a circle is any segment joining the center of the circle to a point on the circle, or the length of that segment. The context will determine which meaning of radius is to be used. Thus, in Fig. 10-49, segments $\overline{OA}$, $\overline{OB}$, and $\overline{OC}$, are radii of ⊙ *O*. In the rule "Radii of a circle are congruent," the meaning to be given "radii" is that of length of the segments. Hence $OA = OB = OC$, and therefore $\overline{OA} \cong \overline{OB} \cong \overline{OC}$.

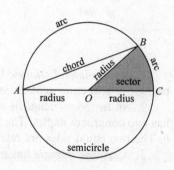

Fig. 10-49

3. The *circumference* of a circle may be thought of as the "perimeter" of the circle, or the "distance" around the circle.

4. A *chord* of a circle is a line segment joining any two points of the circle. Thus, $\overline{AB}$ is a chord of ⊙ *O* in Fig. 10-49.

5. A *diameter* of a circle is any chord containing the center of the circle, or the length of such a chord. Thus, the diameter of ⊙ *O* in Fig. 10-49 is chord $\overline{AC}$ or its length *AC*. In the rule "A diameter of a circle is twice its radius," the lengths of diameter and radius are meant.

6. An *arc* is part of a circle. In Fig. 10-49, "arc *AB*" is denoted by $\overset{\frown}{AB}$.

7. A *semicircle* is an arc that is one-half a circle. A diameter of a circle divides the circle into two semicircles. Thus, $\overset{\frown}{ABC}$ is the upper semicircle of ⊙ *O* in Fig. 10-49.

8. A *minor arc* is an arc that is less than a semicircle; a *major arc* is an arc that is greater than a semicircle. Thus, $\overset{\frown}{BC}$ is a minor arc and $\overset{\frown}{ACB}$ is a major arc of ⊙ *O* in Fig. 10-49.

9. A *central angle* of a circle is an angle whose vertex is at the center of the circle. Thus, ∠*BOC* and ∠*AOB* are central angles of ⊙ *O* in Fig. 10-49.

10. A *sector* of a circle is a part of the inner region of a circle bounded by an arc and two radii drawn to the endpoints of the arc. Thus, in Fig. 10-49, the shaded sector is bounded by radius $\overline{OB}$, radius $\overline{OC}$, and arc $\overset{\frown}{BC}$.

8.1   CIRCLE TERMINOLOGY

In Fig. 10-50, name: (a) $\overline{OQ}$,   (b) $\overline{RQ}$,   (c) $\overline{PQ}$,   (d) PQ,
(e) $\angle POR$,   (f) $\overset{\frown}{PR}$,   (g) $\overset{\frown}{QPR}$,   (h) $\overset{\frown}{PQ}$,   (i) region I.

Ans. (a) radius   (f) minor arc
     (b) chord   (g) major arc
     (c) diameter   (h) semicircle
     (d) length of diameter   (i) sector
     (e) central angle

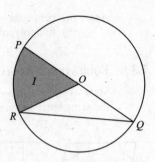

Fig. 10-50

9. TRIANGLES

A *triangle* is a polygon of three sides. The symbol for triangle is $\triangle$. Thus, "triangle ABC" is denoted by $\triangle ABC$.

Classifying Triangles by Number of Equal Sides

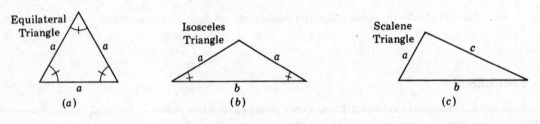

Fig. 10-51

1. An *equilateral triangle* is a triangle having three congruent sides, Fig. 10-51(a). An equilateral triangle is also equiangular. Each of its angles is an angle of 60°.

2. An *isosceles triangle* is a triangle having two congruent sides, Fig. 10-51(b). An isosceles triangle has two congruent angles. The two congruent angles are referred to as *base angles* and lie along the *base*, b. The two equal sides are referred to as *legs*.

3. A *scalene triangle* has no congruent sides, Fig. 10-51(c).

Classifying Triangles by Kinds of Angles

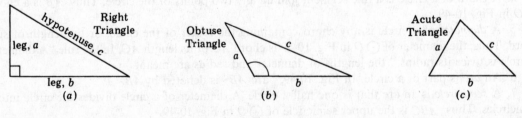

Fig. 10-52

4. A *right triangle* is a triangle having one right angle, Fig. 10-52(a). The *hypotenuse* is the side opposite the right angle and is the greatest side of the right triangle. The other two sides are referred to as the *legs* or *arms* of the right triangle.

5. An *obtuse triangle* is a triangle having one obtuse angle (more than 90° and less than 180°), Fig. 10.52(b).

6. An *acute triangle* is a triangle having all acute angles (less than 90°), Fig. 10-52(*c*).

9.1 RIGHT AND OBTUSE TRIANGLES

In Fig. 10-53, name   (*a*) each right triangle,   (*b*) each obtuse triangle.

Ans.   (*a*) △I or △ABC, △ACD, △ACE
(*b*) △II or △ABD, △III or △ADE, △ABE

9.2 ISOSCELES AND SCALENE TRIANGLES

In Fig. 10-54, *a*, *b*, and *s* are measures of lengths of segments. Name (*a*) each isosceles triangle,   (*b*) each scalene triangle.

Ans.   (*a*) △ABD, △BCD, △ABC, △ADC
(*b*) △I or △ABE, △II or △BCE, △III or △ECD, △IV or △AED

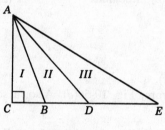

Fig. 10-53

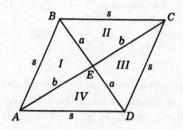

Fig. 10-54

10. QUADRILATERALS

A *quadrilateral* is a polygon of four sides. We shall consider two important types of quadrilaterals: trapezoids and parallelograms.

Trapezoids

1. A *trapezoid* is a quadrilateral having one and only one pair of parallel sides, Fig. 10-55(*a*). The parallel sides are the *bases* of the trapezoid. An *altitude* of the trapezoid is a segment perpendicular to the bases whose length, *h*, is the distance between the bases.

2. An *isoceles trapezoid* is a trapezoid having two congruent nonparallel sides, Fig. 10-55(*b*). The nonparallel sides of the trapezoid are its *legs*. Note that the isosceles trapezoid has two pairs of congruent angles. Either pair of congruent angles has a base of the trapezoid in common.

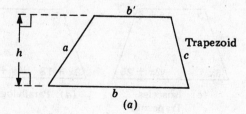

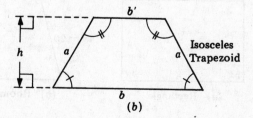

Fig. 10-55

Parallelograms

3. A *parallelogram* is a quadrilateral having two pairs of parallel sides, Fig. 10-56(*a*). Opposite angles of a parallelogram are congruent. Opposite sides of a parallelogram are congruent. The symbol for parallelogram is ⟏. Thus, "parallelogram *ABCD*" is denoted by ⟏*ABCD*.

4. A *rectangle* is a parallelogram all of whose angles are right angles, Fig. 10-56(*b*). A rectangle may also be defined as an equiangular parallelogram. Any side of a rectangle may be referred to as the base, in which case an adjacent side becomes the altitude.

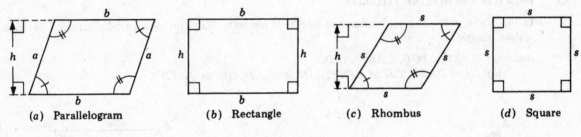

(*a*) **Parallelogram**   (*b*) **Rectangle**   (*c*) **Rhombus**   (*d*) **Square**

Fig. 10-56

5. A *rhombus* is an equilateral parallelogram, Fig. 10-56(*c*).

6. A *square* is an equilateral and equiangular parallelogram, Fig. 10-56(*d*). Also, a square is an equilateral rectangle, or an equiangular rhombus.

Important Note: Since rectangles, rhombuses, and squares are parallelograms, they have all the properties of parallelograms. Since the square is both equilateral and equiangular, it is a regular quadrilateral, the only quadrilateral to be regular.

10.1   NAMING QUADRILATERALS

In each, name the most general type of quadrilateral having the property: (*a*) regular polygon,   (*b*) equilateral,   (*c*) equiangular,   (*d*) two parallel sides and two congruent nonparallel sides,   (*e*) two and only two parallel sides,   (*f*) congruent opposite angles,   (*g*) congruent opposite sides.

Ans.   (*a*) square   (*c*) rectangle   (*e*) trapezoid   (*g*) parallelogram
             (*b*) rhombus   (*d*) isosceles trapezoid   (*f*) parallelogram

10.2   APPLYING ALGEBRA TO QUADRILATERALS

In each part of Fig. 10-57, find *x* and *y*.

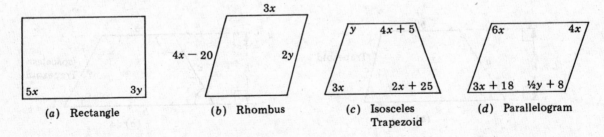

(*a*) **Rectangle**   (*b*) **Rhombus**   (*c*) **Isosceles Trapezoid**   (*d*) **Parallelogram**

Fig. 10-57

Solutions

(a) $5x = 90$, $x = 18$ (c) $3x = 2x + 25$, $x = 25$
     $3y = 90$, $y = 30$ $y = 4x + 5 = 105$

(b) $4x - 20 = 3x$, $x = 20$ (d) $3x + 18 = 4x$, $x = 18$
     $2y = 3x = 60$, $y = 30$ $\frac{1}{2}y + 8 = 6x = 108$
                                                $\frac{1}{2}y = 100$, $y = 200$

11. ANGLE-SUMS OF TRIANGLES AND QUADRILATERALS

Rule 1: The angle-sum of the measures of three angles of a triangle is 180.

Thus, if the angles of a triangle contain $a°$, $b°$, and $c°$, then $a + b + c = 180$.

See-for-Yourself Test

You can see that the measures of the three angles of a triangle total 180 by simply tearing off each of the three angles as shown in Fig. 10-58(a) and putting them together as shown in Fig. 10-58(b). The angle that results when the three angles are placed together is a straight angle of 180°.

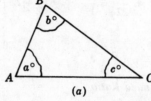
(a)

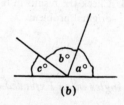
(b)

Fig. 10-58

Rule 2: The angle-sum of the measures of the acute angles of a right triangle is 90.

Thus, in Fig. 10-59, if the acute angles contain $a°$ and $b°$, then $a + b = 90$.

Rule 3: The angle-sum of the measures of the angles of an isosceles triangle equals twice the measure of a base angle plus the measure of the vertex angle.

Thus, in Fig. 10-60, if each base angle contains $b°$ and the vertex angle contains $a°$, then $a + 2b = 180$.

Rule 4: The angle-sum of the measures of the four angles of a quadrilateral is 360.

Thus, if the angles of a quadrilateral contain $a°$, $b°$, $c°$, and $d°$, then $a + b + c + d = 360$.

Rule 5: The angle-sum of the measures of any two consecutive angles of a parallelogram is 180.

Thus, if the angles of a parallelogram, Fig. 10-61, contain $a°$ or $b°$, then $a + b = 180$.

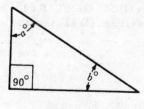

Fig. 10-59

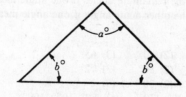

Fig. 10-60

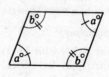

Fig. 10-61

Angles in a Given Ratio

For simplicity, we say "angles in a given ratio" instead of "angles whose measures in degrees are in a given ratio."

To Solve Problems Involving Angles in a Given Ratio

(a) Find the acute angles of a right triangle if the angles are in the ratio of $3:1$.

(b) Find the angles of a parallelogram if consecutive angles are in the ratio of $2::3$.

PROCEDURE	SOLUTIONS	
1. Represent the measures of the angles using x as a common factor:	1. Let $3x$ and x represent the measures of angles.	Let $2x$ and $3x$ represent the measures of the angles.
2. Translate the relationship between the angles into an equation:	2. The angle-sum is 90 (Rule 2) Hence, $3x + x = 90$.	The angle-sum is 180 (Rule 5). Hence, $2x + 3x = 180$.
3. Solve the equation:	3. $\begin{aligned} 4x &= 90 \\ x &= 22\tfrac{1}{2}, \\ 3x &= 67\tfrac{1}{2} \end{aligned}$ *Ans.* $22\tfrac{1}{2}°,\ 67\tfrac{1}{2}°$	$\begin{aligned} 5x &= 180 \\ x &= 36,\ 2x = 72,\ 3x = 108 \end{aligned}$ *Ans.* $72°,\ 108°$
4. Check the results in the original problem:	4. $\dfrac{67\tfrac{1}{2}}{22\tfrac{1}{2}} \overset{?}{=} \dfrac{3}{1}$ $67\tfrac{1}{2} + 22\tfrac{1}{2} \overset{?}{=} 90$ $\dfrac{3}{1} = \dfrac{3}{1}$ $90 = 90$	$\dfrac{72}{108} \overset{?}{=} \dfrac{2}{3}$ $72 + 108 \overset{?}{=} 180$ $\dfrac{2}{3} = \dfrac{2}{3}$ $180 = 180$

Angles in an Extended or Continued Ratio

The angles of a triangle may be in the ratio of $2:3:4$, in which case the angles are $40°$, $60°$, and $80°$. To find the measures of these angles, let $2x$, $3x$, and $4x$ represent the measures. By Rule 1,

$$2x + 3x + 4x = 180$$
$$9x = 180$$
$$x = 20,\ 2x = 40,\ 3x = 60,\ 4x = 80$$

A ratio such as $2:3:4$ is an *extended* or *continued ratio*. Actually, a continued ratio is a combination of ratios. Thus, the ratio $2:3:4$ is a combination of three ratios: $2:3$, $2:4$, and $3:4$. Keep in mind that a ratio of $2:4$ is equivalent to a ratio of $1:2$.

11.1 NUMERICAL ANGLE PROBLEMS

(a) Find the third angle of a triangle if the other angles (1) each measure $60°$, (2) measure $35°$ and $65°$, (3) have an angle-sum of 115.

(b) Find the fourth angle of a quadrilateral if the other angles (1) each measure $90°$; (2) measure $70°$, $100°$, and $110°$; (3) have an angle-sum of 288.

(c) In an isosceles triangle, find the vertex angle if each base angle measures (1) $40°$, (2) $75°$, (3) $89°$.

(d) In a parallelogram, find the remaining angles if one angle measures (1) $50°$, (2) $80°$, (3) $90°$.

(e) In a right triangle, find the other acute angle if one angle measures (1) $30°$, (2) $45°$, (3) $89°$.

Solutions

(a) Apply Rule 1:   (1) $60°$ (2) $80°$ (3) $65°$

(b) Apply Rule 4:   (1) $90°$ (2) $80°$ (3) $72°$

(c) Apply Rule 3:   (1) $100°$ (2) $30°$ (3) $2°$

(d) Apply Rule 5:   (1) $50°$, $130°$, $130°$ (2) $80°$, $100°$, $100°$ (3) $90°$, $90°$, $90°$

(e) Apply Rule 2:   (1) $60°$ (2) $45°$ (3) 1

11.2 SOLVING PROBLEMS INVOLVING ANGLES IN A GIVEN RATIO

Using x as a common factor in their representation, find the required angles if the angles are in the ratio of (a) $2:3$ and the angles are the acute angles of a right triangle, (b) $8:1$ and the angles are the consecutive angles of a parallelogram, (c) $1:2:3$ and the angles are the three angles of a triangle, (d) $1:2:3:4$ and the angles are the four angles of a quadrilateral,   (e) 1 : 2 and the greater is the vertex angle of an isosceles triangle of which the smaller is a base angle.

Solutions

(a) $2x + 3x = 90$ (Rule 2)
$5x = 90$
$x = 18, 2x = 36, 3x = 54$
Ans. 36°, 54°

(b) $8x + x = 180$ (Rule 5)
$9x = 180$
$x = 20, 8x = 160$
Ans. 160°, 20°

(c) $x + 2x + 3x = 180$ (Rule 1)
$6x = 180$
$x = 30, 2x = 60, 3x = 90$
Ans. 30°, 60°, 90°

(d) $x + 2x + 3x + 4x = 360$ (Rule 4)
$10x = 360$
$x = 36, 2x = 72, 3x = 108, 4x = 144$
Ans. 36°, 72°, 108°, 144°

(e) $x + x + 2x = 180$ (Rule 3)
$4x = 180$
$x = 45, 2x = 90$
Ans. 45°, 90°

11.3 SOLVING PROBLEMS INVOLVING THREE ANGLES IN A GIVEN RATIO

Using $2x$, $3x$ and $5x$ as their representation, find the three required angles if   (a) the angles are the angles of a triangle,   (b) the angles are the angles of a quadrilateral whose remaining angle measures 90°,   (c) the two greatest angles are the acute angles of a right triangle,   (d) the largest angle is the vertex angle of an isosceles triangle whose base angle is the smallest   (e) the smallest angle is one of two consecutive angles of a parallelogram and the other consecutive angle measures 47° more than the largest angle of the set.

Solutions

(a) $2x + 3x + 5x = 180$ (Rule 1)
$10x = 180$
$x = 18, 2x = 36, 3x = 54, 5x = 90$
Ans. 36°, 54°, 90°

(b) $10x + 90 = 360$ (Rule 4)
$10x = 270$
$x = 27, 2x = 54, 3x = 81, 5x = 135$
Ans. 54°, 81°, 135°

(c) $3x + 5x = 90$ (Rule 2)
$8x = 90$
$x = 11\frac{1}{4}, 2x = 22\frac{1}{2}, 3x = 33\frac{3}{4}, 5x = 56\frac{1}{4}$
Ans. $22\frac{1}{2}°, 33\frac{3}{4}°, 56\frac{1}{4}°$

(d) $2x + 2x + 5x = 180$ (Rule 3)
$9x = 180$
$x = 20, 2x = 40, 3x = 60, 5x = 100$
Ans. 40°, 60°, 100°

(e) $2x + (5x + 47) = 180$ (Rule 5)
$7x = 133$
$x = 19, 2x = 38, 3x = 57, 5x = 95$
Ans. 38°, 57°, 95°

11.4 SOLVING ANGLE PROBLEMS

The measure of the first of three angles exceeds the measure of the second by 12°. The third angle measures 24° less than the sum of the other two. Find the measure of the three required angles if   (a) the angles are the angles of a triangle,   (b) the first two are the acute angles of a right triangle,   (c) the first and third are the consecutive angles of a parallelogram,   (d) the second is one of the base angles of an isosceles triangle whose vertex angle is the third angle,   (e) the angles are three angles of a quadrilateral whose remaining angle is a right angle.

Solutions Let x = no. of degrees in the measure of the second angle
$x + 12$ = no. of degrees in the measure of the first angle
$2x - 12$ = no. of degrees in the measure of the third angle

(a) Since the angle-sum is 180, $4x = 180$. Thus $x = 45$, $x + 12 = 57$, $2x - 12 = 78$.
Ans. 57°, 45°, 78°

(b) Since the angle-sum is 90, $2x + 12 = 90$. Thus $x = 39$, $x + 12 = 51$, $2x - 12 = 66$.
Ans. 51°, 39°, 66°

(c) Since the angle-sum is 180, $3x = 180$. Thus $x = 60$, $x + 12 = 72$, $2x - 12 = 108$.
   Ans. 72°, 60°, 108°

(d) Since the angle-sum is 180, $x + x + 2x - 12 = 180$. Thus $x = 48$, $x + 12 = 60$, $2x - 12 = 84$.
   Ans. 60°, 48°, 84°

(e) Since the angle-sum is 360, $4x = 270$. Thus $x = 67\frac{1}{2}$, $x + 12 = 79\frac{1}{2}$, $2x - 12 = 123$.
   Ans. $79\frac{1}{2}$°, $67\frac{1}{2}$°, 123°

12. PERIMETERS AND CIRCUMFERENCES: LINEAR MEASURE FORMULAS

Perimeters of Polygons

The *perimeter of a polygon* is the sum of the lengths of its sides. In Fig. 10-62, the perimeter of equilateral triangle *ADC* is 3s, the perimeter of isosceles triangle *ABC* is $2a + s$, and the perimeter of quadrilateral *ABCD* is $2a + 2s$.

The *perimeter of an equilateral or regular polygon* is the product of the measure of a side, *s*, and the number of sides, *n*. Hence, if *p* represents the perimeter of a regular polygon, then

$$p = ns$$

Thus, the perimeter of an equiliateral or regular hexagon is 6s, and the perimeter of an equilateral or regular decagon is 10s.

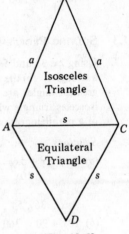

Important Notes:
1. In perimeter formulas, the unit of measurement for the perimeter must be the same as that used for the sides. Thus, if $p = 3s$ and *s* is in centimeters, then *p* is in centimeters.
2. For the sake of simplicity in future discussions, "a side of" or "a side is" means "a side whose measure is."

Fig. 10-62

Circumferences of Circles

The *circumference of a circle* is the distance around it. For any circle, the ratio of the circumference to the diameter equals π. To 9 decimal places, $\pi = 3.141592654$. To 4 decimal places, the approximation to use is 3.1416. To 2 decimal places, use 3.14. A fractional approximation is $3\frac{1}{7}$ or 22/7.

If *c* represents the circumference, *d* the diameter, and *r* the radius, then

$$c = \pi d \quad \text{or} = 2\pi r$$

Hence, if $r = 50$, then $d = 100$, and, using $\pi = 3.14$, $c = (3.14)(100) = 314$.

Length of an Arc of a Circle

Since an arc is part of a circle, the length of an arc is a fraction of the circumference of a circle. For example, the length of an arc whose central angle is 90° is 90/360 or 1/4 of the circumference of the circle, as in Fig. 10-63. The length of an arc whose central angle is 60° is 60/360 or 1/6 of the circumference of the circle.

Rule: The length of an arc of n° is n/360 of the circumference.

Important Note: For the sake of simplicity in future discussions, "an arc of n°" or "an n° arc" means "an arc whose central angle measures n°."

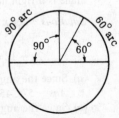

Fig. 10-63

12.1   PERIMETER OF A REGULAR POLYGON

(*a*) State the formula to use to find the perimeter of (1) a square, (2) a regular pentagon, (3) a regular dodecagon.

(*b*) In each, name the regular polygon to which the perimeter formula applies: (1) $p = 6s$, (2) $p = 8s$, (3) $p = 9s$.

(*c*) If its perimeter is 45 yd, what is the side of (1) an equilateral triangle, (2) a regular decagon, (3) a regular *n*-gon where $n = 15$?

Ans.   (*a*) (1) $p = 4s$ (2) $p = 5s$ (3) $p = 12s$
        (*b*) (1) regular hexagon   (2) regular octagon   (3) regular nonagon
        (*c*) (1) 15 yd   (2) 4.5 yd   (3) 3 yd

12.2   PERIMETER OF AN EQUILATERAL *n*-GON

(*a*) Find the perimeter of an equilateral *n*-gon, given *n* and given the side, *s*, in feet: (1) $n = 6$ and $s = 10$, (2) $n = 10$ and $s = 24.5$, (3) $n = 8$ and $s = 12\frac{1}{2}$.

(*b*) Find the side of an equilateral *n*-gon, given *n* and given the perimeter, *p*, in inches: (1) $n = 5$ and $p = 12$, (2) $n = 7$ and $p = 24.5$, (3) $n = 20$ and $p = 75$.

(*c*) Find the number of sides of an equilateral *n*-gon, given the perimeter, *p*, and the side, *s*, both in meters: (1) $p = 45$ and $s = 9$, (2) $p = 6.8$ and $s = 1.7$, (3) $p = 51$ and $s = 4.25$.

Solutions

(*a*) Apply $p = ns$:   (1) 60 ft   (2) 245 ft   (3) 100 ft   (*c*) Apply $n = \dfrac{p}{s}$: (1) 5   (2) 4   (3) 12

(*b*) Apply $s = \dfrac{p}{n}$:   (1) 2.4 in.   (2) 3.5 in.   (3) 3.75 in.

12.3   PERIMETER OF AN ISOSCELES TRIANGLE

In Fig. 10-64, find:

(*a*) perimeter, *p*, if   (1) $a = 7$, $b = 10$;   (2) $a = 7\frac{1}{2}$, $b = 10\frac{1}{2}$;   (3) $a = 7.2$, $b = 10.4$.

(*b*) base, *b*, if   (1) $p = 20$, $a = 5$;   (2) $p = 22\frac{1}{2}$; $a = 5\frac{1}{2}$;   (3) $p = 21.4$, $a = 5.7$.

(*c*) leg, *a*, if   (1) $p = 30$, $b = 20$;   (2) $p = 30\frac{1}{4}$, $b = 15\frac{1}{4}$;   (3) $p = 30.4$, $b = 15.6$.

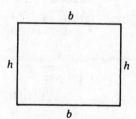

Fig. 10-64

Illustrative Solution   (*b*) (2) Since $p = 2a + b$,   then $b = p - 2a$. Since $p = 22\frac{1}{2}$ and $a = 5\frac{1}{2}$, then $b = 22\frac{1}{2} - 11 = 11\frac{1}{2}$ *Ans.*

Ans.   (*a*) (1) 24   (2) $25\frac{1}{2}$ (3) 24.8
        (*b*) (1) 10   · (3) 10
        (*c*) (1) 5   (2) $7\frac{1}{2}$ (3) 7.4

12.4   PERIMETER OF A RECTANGLE

In Fig. 10-65, find:

(*a*) perimeter, *p*, if (1) $b = 10$, $h = 8$;   (2) $b = 10\frac{1}{4}$, $a = 5\frac{1}{2}$;   (3) $b = 5.2$, $h = 4.7$.

(*b*) base, *b*, if   (1) $p = 30$, $h = 9$;   (2) $p = 20\frac{1}{2}$, $h = 5\frac{3}{4}$;   (3) $p = 21.8$, $h = 6.9$.

(*c*) altitude, *h*, if (1) $p = 25$, $b = 8$;   (2) $p = 30$, $b = 5\frac{1}{4}$;   (3) $p = 36$, $b = 14.3$.

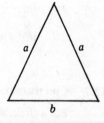

Fig. 10-65

Illustrative Solution   (*b*) (3) Since $p = 2b + 2h$, then $b = (p - 2h)/2$. Since $p = 21.8$ and $h = 6.9$, then

$$b = \frac{1}{2}(p - 2h) = \frac{1}{2}(21.8 - 13.8) = \frac{1}{2}(8) = 4 \quad \textit{Ans.}$$

Ans.   (*a*) (1) 36   (2) $31\frac{1}{2}$ (3) 19.8   (*c*) (1) $4\frac{1}{2}$ (2) $9\frac{3}{4}$ (3) 3.7
        (*b*) (1) 6   (2) $4\frac{1}{2}$

12.5   SOLVING PROBLEMS INVOLVING THE PERIMETER OF A RECTANGLE

The base of a rectangle exceeds three times the height by 8, Fig. 10-66. Find the dimensions of the rectangle if

(*a*) the semiperimeter is 32
(*b*) the perimeter is 100
(*c*) the semiperimeter is 10 less than five times the height
(*d*) the perimeter is 20 more than twice the base.

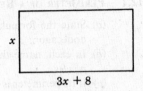

Fig. 10-66

Solutions

(*a*) $3x + 8 + x = 32$
$\qquad 4x = 24$
$\qquad x = 6,\ 3x + 8 = 26$
   Ans.   6, 26

(*b*) $2(4x + 8) = 100$
$\qquad 8x = 84$
$\qquad x = 10\frac{1}{2},\ 3x + 8 = 39\frac{1}{2}$
   Ans. $10\frac{1}{2}$, $39\frac{1}{2}$

(*c*) $4x + 8 = 5x - 10$
$\qquad x = 18,\ 3x + 8 = 62$
   Ans.   18, 62

(*d*) $8x + 16 = 2(3x + 8) + 20$
$\qquad 2x = 20$
$\qquad x = 10,\ 3x + 8 = 38$
   Ans.   10, 38

12.6   CIRCUMFERENCE AND ARC FORMULAS

For any circle, state a formula in which   (*a*) the circumference, *c*, is expressed in terms of the diameter, *d*;   (*b*) the diameter, *d*, is expressed in terms of the circumference, *c*;   (*c*) an arc of 90°, *a*, is expressed in terms of the circumference, *c*;   (*d*) an arc of 90°, *a*, is expressed in terms of the radius, *r*;   (*e*) an arc of 60°, *a'*, is expressed in terms of the circumference, *c*;   (*f*) an arc of 60°, *a'*, is expressed in terms of the diameter, *d*.

Ans.   (*a*) $c = \pi d$ (*c*) $a = \dfrac{c}{4}$ (*e*) $a' = \dfrac{c}{6}$

$\qquad\quad$ (*b*) $d = \dfrac{c}{\pi}$ (*d*) $a = \dfrac{\pi r}{2}$ (*f*) $a' = \dfrac{\pi d}{6}$

12.7   CIRCUMFERENCE OF A CIRCLE AND LENGTH OF AN ARC

For a circle, if *c*, *d*, and *r* are in inches, using $\pi = 3.14$, find   (*a*) the circumference if $d = 10$,   (*b*) the length of a 90° arc if $r = 4$,   (*c*) the length of a 60° arc if $d = 24$,   (*d*) the radius if $c = 157$,   (*e*) the diameter if $c = 7.85$,   (*f*) the circumference if an arc of 45° is $10\frac{7}{8}$ in. long.

Solutions

(*a*) $c = \pi d$
$\qquad = (3.14)(10) = 31.4$ in.

(*b*) $a = \dfrac{\pi r}{2}$
$\qquad = \dfrac{(3.14)(4)}{2} = 6.28$ in.

(*c*) $a' = \dfrac{\pi d}{6}$
$\qquad = \dfrac{3.14(24)}{6} = 12.56$ in.

(*d*) $\qquad c = 2\pi r$
$\qquad 157 = 2(3.14)r$
$\qquad 25$ in. $= r$

(*e*) $\qquad c = \pi d$
$\qquad 7.85 = (3.14)d$
$\qquad 2.5$ in. $= d$

(*f*) An arc of 45° is 1/8 of the circumference. Hence,
$\qquad c = 8a = 8(10\frac{7}{8}) = 87$ in.

13.   AREAS: SQUARE MEASURE FORMULAS

A *square unit* is a square having a side of 1 unit, as in Fig. 10-67. The *area* of the region enclosed by a polygon or a circle is the number of square units contained in the region.

Thus, in Fig. 10-68, if the base of a rectangle is 6 units and the altitude is 5 units, then the number of square units in the region enclosed by the rectangle is 6×5 or 30 square units.

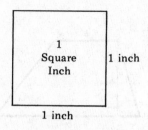

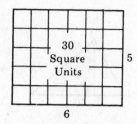

Fig. 10-67   Fig. 10-68

Important Notes:

1. In area formulas, the unit of measurement for the area must be the square of the unit for the dimensions. Thus, if $A = s^2$ and s is in meters (m), then A is in square meters (m^2).
2. For the sake of simplicity in future discussions, "area of a polygon (circle)" will be used to mean "area of the region enclosed by a polygon (circle)."

Areas of Polygons

1. The *area of a rectangle* equals the product of its base and altitude. If, as in Fig. 10-69, b is the measure of the base of the rectangle, h its altitude, and A its area, then $A = bh$.

Thus, if $b = 10$ in. and $h = 5$ in., then $A = 10 \cdot 5 = 50$ in^2.

2. The *area of a square* equals the square of a side. If, as in Fig. 10-70, s is the measure of the side of the square and A the area, then $A = s^2$.

Thus, if the side of a square is 10 ft, then the area of the square is 100 ft^2.

3. The *area of a parallelogram* equals the product of a side (base) and the altitude to that side. If, as in Fig. 10-71, b is the measure of the base of the parallelogram, h its altitude, and A the area, then $A = bh$.

Thus, if $b = 7$ m and $h = 5$ m, then $A = 7 \times 5 = 35$ m^2.

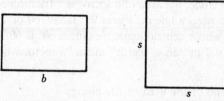

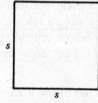

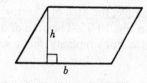

Fig. 10-69   Fig. 10-70   Fig. 10-71

4. The *area of a triangle* is equal to one-half the product of a side (base) and the altitude to that side. If, as in Fig. 10-72, b is the measure of the base of the triangle, h its altitude, and A the area, then $A = bh/2$.

Thus, if $b = 8$ cm and $h = 10$ cm, then $A = 8 \times 10/2 = 40$ cm^2.

5. The *area of a trapezoid* is equal to one-half the product of the altitude and the sum of the bases. If, as in Fig. 10-73, b and b' are the measures of the bases of the trapezoid, h its altitude, and A the area, then $A = h(b + b')/2$.

Thus, if $b = 20$ mi, $b' = 10$ mi, and $h = 6$ mi, then $A = 6(20 + 10)/2 = 90$ mi^2.

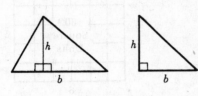

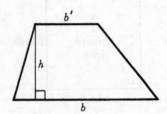

Fig. 10-72   Fig. 10-73

Area of a Circle

The *area of a circle* equals the product of π and the square of the radius, or one-quarter the product of π and the square of the diameter. If, as in Fig. 10-74, r is the measure of the radius of the circle, d its diameter, and A the area, then $A = \pi r^2 = \pi d^2/4$.

Thus, if $d = 10\,\text{mm}$, then $r = 5\,\text{mm}$ and $A = \pi(25)$ or $25\pi\,\text{mm}^2$.

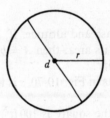

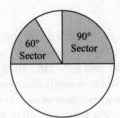

Fig. 10-74   Fig. 10-75

Area of a Sector

Since a sector is part of the inner region of a circle, the area of a sector is a fraction of the area of the circle. For example, the area of a sector whose central angle measures $90°$ is $90/360$ or $1/4$ of the area of the circle, as in Fig. 10-75. The area of a sector whose central angle measures $60°$ is $60/360$ or $1/6$ of the area of the circle. For simplicity, let "a sector of $n°$" or "an $n°$ sector" mean "a sector whose central angle measures $n°$."

Rule: The area of a sector of $n°$ equals $n/360$ of the area of the circle.

13.1 RELATING SQUARE UNITS

Complete each:

(a) $1\,\text{ft}^2 = ?\,\text{in}^2$ (c) $1\,\text{mi}^2 = ?\,\text{ft}^2$ (e) $1\,\text{m}^2 = ?\,\text{km}^2$

(b) $1\,\text{yd}^2 = ?\,\text{ft}^2$ (d) $1\,\text{m}^2 = ?\,\text{cm}^2$ (f) $1\,\text{mm}^2 = ?\,\text{cm}^2$

Solutions Refer to Table 10-2. Use the square of the linear conversion factor when converting from a greater unit of area measurement to a smaller one. Use the reciprocal of this square when converting from a smaller unit of area measurement to a greater one.

Table 10-2.   Linear Conversion Factors

Between English Units	Between Metric Units
1 ft = 12 in.	1 m = 10 dm = 100 cm = 1000 mm
1 yd = 3 ft (= 36 in.)	1 hm = 100 m
1 mi = 5280 ft (= 1760 yd)	1 km = 1000 m

(a) $12^2 = 144$ *Ans.*   (c) $5,280^2 = 27,878,400$ *Ans.*   (e) $\dfrac{1}{1,000^2} = \dfrac{1}{1,000,000}$ *Ans.*

(b) $3^2 = 9$ *Ans.*   (d) $100^2 = 10,000$ *Ans.*   (f) $\dfrac{1}{10^2} = \dfrac{1}{100}$ *Ans.*

13.2   AREAS OF POLYGONS

(a) In a rectangle, apply $A = bh$ to find (1) A if $b = 8$ and $h = 4.5$, (2) b if $A = 126$ and $h = 18$, (3) h if $A = 1,575$ and $b = 45$.

(b) In a triangle, apply $A = bh/2$ to find (1) A if $b = 22$ and $h = 15$, (2) h if $A = 75$ and $b = 7.5$, (3) b if $A = 44$ and $h = 10$.

(c) In a trapezoid, apply $A = h(b + b')/2$ to find (1) A if $b = 23$, $b' = 17$, and $h = 13$; (2) h if $b = 29$, $b' = 21$, and $A = 250$; (3) b if $b' = 20$, $h = 14$, and $A = 210$.

Solutions   When b, h, or b' is to be found, it is recommended that the method of first transforming the formula and then substituting given values be used.

(a) (1) $A = 8(4.5) = 36$ (2) $b = \dfrac{A}{h} = \dfrac{126}{18} = 7$ (3) $h = \dfrac{A}{b} = \dfrac{1,575}{45} = 35$

(b) (1) $A = \dfrac{(22)(15)}{2} = 165$ (2) $h = \dfrac{2A}{b} = \dfrac{2(75)}{7.5} = 20$ (3) $b = \dfrac{2A}{h} = \dfrac{2(44)}{10} = 8.8$

(c) (1) $A = \dfrac{(13)(23 + 17)}{2} = 260$ (2) $h = \dfrac{2A}{b + b'} = \dfrac{2(250)}{29 + 21} = 10$ (3) $b = \dfrac{2A}{h} - b' = \dfrac{2(210)}{14} - 20 = 10$

13.3   CIRCLE AND SECTOR FORMULAS

For a circle, state a formula in which   (a) the area, A, is expressed in terms of the radius, r;   (b) a sector of $90°$, S, is expressed in terms of the area, A;   (c) a sector of $90°$, S, is expressed in terms of the radius, r;   (d) a sector of $60°$, S', is expressed in terms of the area, A;   (e) a sector of $60°$, S', is expressed in terms of the radius, r.

Ans.   (a) $A = \pi r^2$ (b) $S = \dfrac{A}{4}$ (c) $S = \dfrac{\pi r^2}{4}$ (d) $S' = \dfrac{A}{6}$ (e) $S' = \dfrac{\pi r^2}{6}$

13.4   AREA OF A CIRCLE AND SECTOR

For a circle, if d and r are in feet, using $\pi = 3.14$, find, to the nearest whole number, the area of   (a) a circle if $r = 5$,   (b) a circle if $d = 20$,   (c) a $90°$ sector if $r = 12$,   (d) a $60°$ sector if $r = 6$,   (e) a circle if the area of a $30°$ sector is $35\,\text{ft}^2$,   (f) a $45°$ sector if the area of a circle is $120\,\text{ft}^2$.

Solutions

(a) $A = \pi r^2$
$= (3.14)(25) = 78.5$
Ans. $79\,\text{ft}^2$

(b) $A = \pi r^2$
$= (3.14)(100) = 314$
Ans. $314\,\text{ft}^2$

(c) $S = \dfrac{\pi r^2}{4}$

$\quad = \dfrac{(3.14)(144)}{4} = 113.04$

$\quad$ Ans. $113\,\text{ft}^2$

(d) $S = \dfrac{\pi r^2}{6}$

$\quad = \dfrac{(3.14)(36)}{6} = 18.84$

$\quad$ Ans. $19\,\text{ft}^2$

(e) $A = 12S$

$\quad = 12(35) = 420$

$\quad$ Ans. $420\,\text{ft}^2$

(f) $S' = \dfrac{A}{8}$

$\quad = \dfrac{120}{8} = 15$

$\quad$ Ans. $15\,\text{ft}^2$

14. UNDERSTANDING SQUARES OF NUMBERS

Square of a Whole Number

The *square of a whole number* is the product obtained by multiplying the number by itself. Thus, the square of 4 is 16, the product obtained by multiplying 4 by 4.

Area of a Square

The *area of a square* equals the square of a side. Thus, if a side is 4 units, Fig. 10-76, the area of the square is 16 square units.

Recall that an exponent may be used when a factor is repeated. Hence, 4×4 may be written as 4^2 using 2 as an exponent to show that 4 is a factor twice. Read "4^2" as "four squared."

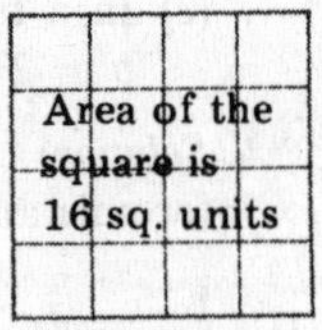

Area of the square is 16 sq. units

Side is 4 units

Fig. 10-76

Table 10-3. **Frequently Used Squares**

Number, N	Square of Number, N^2	Number, N	Square of Number, N^2	Number, N	Square of Number, N^2
0	$0^2 = 0$	7	$7^2 = 49$	20	$20^2 = 400$
1	$1^2 = 1$	8	$8^2 = 64$	25	$25^2 = 625$
2	$2^2 = 4$	9	$9^2 = 81$	30	$30^2 = 900$
3	$3^2 = 9$	10	$10^2 = 100$	35	$35^2 = 1225$
4	$4^2 = 16$	11	$11^2 = 121$	40	$40^2 = 1600$
5	$5^2 = 25$	12	$12^2 = 144$	45	$45^2 = 2025$
6	$6^2 = 36$	15	$15^2 = 225$	50	$50^2 = 2500$

Square of a Whole Number Terminating in Zero Digits

Rule 1: If a whole number terminates in one or more zero digits, the square of the number terminates in twice as many zero digits.

Thus, $50^2 = 2,500$, $500^2 = 250,000$, and $5,000^2 = 25,000,000$.

Square of a Fraction

The *square of a fraction* is the product obtained by multiplying the fraction by itself. Thus, the square of 3/4 is 9/16, the product obtained by multiiplying 3/4 by 3/4.

Area of a Square Whose Side Is a Fraction

Note in Fig. 10-77 that the area of a square is 9/16 if its side is 3/4. Observe that the side of the shaded square is 3/4 of the side of the entire square, while the area of the shaded square is 9/16 of the area of the entire square; that is, the shaded square contains 9 of the 16 small squares.

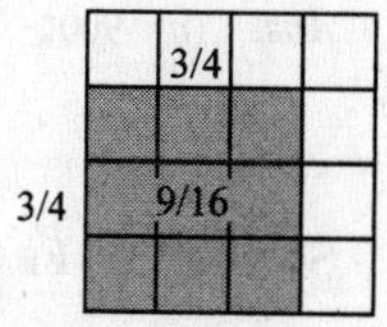

Rule 2: To square a fraction, square its numerator and its denominator.

Thus, the square of $\frac{5}{7}$ is $\frac{25}{49}$ or $\frac{5^2}{7^2}$. Also, $\left(\frac{1}{12}\right)^2 = \frac{1}{144}$ or $\frac{1^2}{12^2}$. In general,

$$\left(\frac{n}{d}\right)^2 = \frac{n^2}{d^2}$$

Fig. 10-77

Square of a Decimal

The *square of a decimal* is the product obtained by multiplying the decimal by itself. Thus, the square of .3 is .09, the product obtained by multiplying .3 by .3.

Area of a Square Whose Side Is a Decimal

Note in Fig. 10-78 that the area of a square is .09 if its side is .3. Observe that the side of the shaded square is 3/10 or .3 of the side of the large square, while the area of the shaded square is .09 of the area of the large square; that is, the shaded square contains 9 of the 100 small squares.

Rule 3: The square of a decimal is a decimal having twice as many decimal places as the decimal being squared.

Thus, the square of .003, which has 3 places, is .000009, which has 6 places.

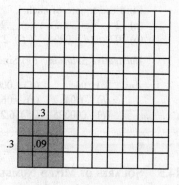

Fig. 10-78

Square of a Mixed Number

A mixed number may be squared by converting it to a fraction or a decimal and then squaring the result.

Thus, square $1\frac{1}{2}$ by changing it to 3/2 or to 1.5; then square the result to obtain 9/4 or to obtain $2.25 = 2\frac{1}{4}$.

14.1 SQUARES OF WHOLE NUMBERS

Find each square:

(a) 0^2	(c) 9^2	(e) 14^2	(g) 21^2	(i) 42^2	(k) 55^2	(m) 75^2	(o) 101^2
(b) 1^2	(d) 10^2	(f) 18^2	(h) 33^2	(j) 49^2	(l) 70^2	(n) 83^2	(p) $1,001^2$

Illustrative Solution (p) $1,001^2 = (1,001)(1,001) = 1,002,001.$

Ans. (a) 0	(c) 81	(e) 196	(g) 441	(i) 1,764	(k) 3,025	(m) 5,625	(o) 10,201
(b) 1	(d) 100	(f) 324	(h) 1,089	(j) 2,401	(l) 4,900	(n) 6,889

14.2 Squares of Whole Numbers Terminating in Zeros

Find each square:
(a) 30^2 (b) 300^2 (c) $3,000^2$ (d) 100^2 (e) $10,000^2$ (f) 120^2 (g) $12,000^2$

Illustrative Solution   (g) Since 12,000 terminates in 3 zeros, its square terminates in 6 zeros (Rule 1).

Ans.   144,000,000

Ans.   (a) 900   (b) 90,000   (c) 9,000,000   (d) 10,000   (e) 100,000,000   (f) 14,400

14.3 Squares of Fractions

Find each square:
(a) $\left(\dfrac{2}{5}\right)^2$ (b) $\left(\dfrac{1}{7}\right)^2$ (c) $\left(\dfrac{3}{11}\right)^2$ (d) $\left(\dfrac{11}{3}\right)^2$ (e) $\left(\dfrac{3}{80}\right)^2$ (f) $\left(\dfrac{17}{100}\right)^2$ (g) $\left(\dfrac{100}{17}\right)^2$
(h) $\left(\dfrac{1}{1,000}\right)^2$

Illustrative Solution   (g) Square numerator and denominator (Rule 2):

$$\frac{100^2}{17^2} = \frac{10,000}{289} \quad Ans.$$

Ans.   (a) $\dfrac{4}{25}$ (b) $\dfrac{1}{49}$ (c) $\dfrac{9}{121}$ (d) $\dfrac{121}{9}$ (e) $\dfrac{9}{6,400}$ (f) $\dfrac{289}{10,000}$ (h) $\dfrac{1}{1,000,000}$

14.4 Squares of Decimals

Find the square of each:
(a) .2   (c) .005   (e) .25   (g) .025   (i) .0001   (k) 1.001   (m) .011   (o) 1.01
(b) .02   (d) .0004   (f) 2.5   (h) .001   (j) 1.1   (l) 100.1   (n) 10.1   (p) 3.0303

Illustrative Solution   (g) Since .025 has 3 places, its square, $.025^2$, has 6 places (Rule 3).   *Ans.* .000625

Ans.   (a) .04   (d) .00000016   (h) .000001   (k) 1.002,001   (n) 102.01
         (b) .0004   (e) .0625   (i) .00000001   (l) 10,020.01   (o) 1.0201
         (c) .000025   (f) 6.25   (j) 1.21   (m) .000121   (p) 9.18271809

14.5 Squares of Mixed Numbers

Find the square of each:
(a) $1\frac{1}{2}$ (b) $1\frac{1}{4}$ (c) $2\frac{1}{2}$ (d) $3\frac{1}{3}$ (e) $4\frac{4}{9}$ (f) $1\frac{2}{3}$ (g) $2\frac{3}{5}$ (h) $5\frac{3}{4}$ (i) $3\frac{4}{5}$ (j) $12\frac{1}{5}$

Illustrative Solutions

(e) Square $4\frac{4}{9}$ by converting it to the fraction 40/9. Hence,

$$(4\tfrac{4}{9})^2 = \frac{40^2}{9^2} = \frac{1,600}{81} \quad \text{or} \quad 19\tfrac{61}{81} \quad Ans.$$

(g) Square $2\frac{3}{5}$ by converting it to the decimal 2.6. Hence,

$$(2\tfrac{3}{5})^2 = 2.6^2 = 6.76 \quad \text{or} \quad 6\tfrac{19}{25} \quad Ans.$$

Ans.   (a) 2.25 or $2\frac{1}{4}$ (c) 6.25 or $6\frac{1}{4}$ (f) $2\frac{7}{9}$ (i) 14.44 or $14\frac{11}{25}$

         (b) 1.5625 or $1\frac{9}{16}$ (d) $11\frac{1}{9}$ (h) 33.0625 or $33\frac{1}{16}$ (j) 148.84 or $148\frac{21}{25}$

14.6   AREAS OF SQUARES

Find the area of a square whose side is

(*a*) 6 in.   (*c*) 15 m   (*e*) 2.5 cm   (*g*) .03 mi   (*i*) 7/12 ft   (*k*) 15/8 dm   (*m*) 11/5 km

(*b*) 10 ft   (*d*) 500 mi   (*f*) 3.5 m   (*h*) 2/5 in.   (*j*) 11/90 mi   (*l*) $1\frac{1}{3}$ mi   (*n*) $3\frac{1}{4}$ yd

Illustrative Solution   (*g*) Square .03 and use square measure. Hence, $(.03\,\text{mi})^2 = .0009\,\text{mi}^2$ *Ans.*

Ans. (*a*) $36\,\text{in}^2$ (*d*) $250{,}000\,\text{mi}^2$ (*h*) $4/25\,\text{in}^2$ (*k*) $225/64\,\text{dm}^2$ (*n*) 10.5625 or $10\frac{9}{16}\,\text{yd}^2$

     (*b*) $100\,\text{m}^2$ (*e*) $6.25\,\text{cm}^2$ (*i*) $49/144\,\text{ft}^2$ (*l*) $1\frac{7}{9}\,\text{mi}^2$

     (*c*) $225\,\text{yd}^2$ (*f*) $12.25\,\text{m}^2$ (*j*) $121/8{,}100\,\text{mi}^2$ (*m*) $121/25\,\text{km}^2$

15.   UNDERSTANDING SQUARE ROOTS OF NUMBERS

Square Root of a Number

A *square root* of a number is one of its two equal factors. Thus, 7 is a square root of 49 because $7 \times 7 = 49$. Also, .7 is a square root of .49, and 1/7 is a square root of 1/49.

Note: In addition, -7 is a square root of 49 since $(-7) \times (-7) = 49$. Each whole number has two square roots, one positive and the other negative. Thus, 9 has two square roots, $+3$ and -3. It is the positive square root that we study in this chapter.

Radical and Radicand

The symbol $\sqrt{}$ denotes the square root of a number. Thus, $7 = \sqrt{49}$ denotes that 7 is the square root of 49. Also, $.7 = \sqrt{.49}$ and $1/7 = \sqrt{1/49}$.

In $\sqrt{N}$, N, the number within the square root symbol, is the *radicand* and $\sqrt{N}$ is the *radical*. Thus, in the radical $\sqrt{49}$, 49 is the radicand. Also, .49 is the radicand of $\sqrt{.49}$.

Exact and Approximate Square Roots of Whole Numbers

There are four exact square roots in Table 10-4, namely, $0 = \sqrt{0}$, $1 = \sqrt{1}$, $2 = \sqrt{4}$, and $3 = \sqrt{9}$. Hence, 3 is the *exact square root* of 9 and, in turn, 9 is the *perfect square* of 3.

However, the remaining square roots in the table are *approximate square roots*. The symbol $\approx$ means "is approximately equal to." The statement $\sqrt{2} \approx 1.414$ means that 1.414 is the best three-place approximation of $\sqrt{2}$. $\sqrt{2}$ cannot be an exact square root. For example, to 6 places, $\sqrt{2} \approx 1.414214$ and to 9 places, $\sqrt{2} \approx 1.414213562$.

Table 10-4.   Square Roots of Whole Numbers from 0 to 10

Number, N	Square Root of Number, $\sqrt{N}$	Number, N	Square Root of Number, $\sqrt{N}$
0	$\sqrt{0} = 0$	6	$\sqrt{6} \approx 2.449$
1	$\sqrt{1} = 1$	7	$\sqrt{7} \approx 2.646$
2	$\sqrt{2} \approx 1.414$	8	$\sqrt{8} \approx 2.828$
3	$\sqrt{3} \approx 1.732$	9	$\sqrt{9} = 3$
4	$\sqrt{4} = 2$	10	$\sqrt{10} \approx 3.162$
5	$\sqrt{5} \approx 2.236$		

15.1 **RELATING SQUARES AND SQUARE ROOTS OF NUMBERS**

Relate each pair of numbers, using both the square and square root symbols: (*a*) 5 and 25,   (*b*) 400 and 20,   (*c*) .2 and .04,   (*d*) 1/100 and 1/10.

Solutions

(*a*) Since $25 = 5^2$, then $5 = \sqrt{25}$.

(*c*) Since $.04 = .2^2$, then $.2 = \sqrt{.04}$.

(*b*) Since $400 = 20^2$, then $20 = \sqrt{400}$.

(*d*) Since $\dfrac{1}{100} = \left(\dfrac{1}{10}\right)^2$, then $\dfrac{1}{10} = \sqrt{\dfrac{1}{100}}$.

15.2 **EVALUATING EXPRESSIONS CONTAINING SQUARE ROOTS**

Evaluate, expressing each decimal answer to the nearest hundredth:

(*a*) $\sqrt{1} + \sqrt{4} + \sqrt{9}$ (*c*) $\sqrt{0} + 5\sqrt{1} - \sqrt{4}$ (*e*) $3\sqrt{2}$

(*b*) $2\sqrt{9} + 3\sqrt{4} - 10\sqrt{1}$ (*d*) $\sqrt{7} + \sqrt{9}$ (*f*) $\sqrt{5}/2$

Solutions

(*a*) $1 + 2 + 3 = 6$

(*b*) $2 \cdot 3 + 3 \cdot 2 - 10 \cdot 1 = 6 + 6 - 10 = 2$

(*c*) $0 + 5 \cdot 1 - 2 = 5 - 2 = 3$

(*d*) $2.646 + 3 = 5.646 \approx 5.65$

(*e*) $3(1.414) = 4.242 \approx 4.24$

(*f*) $2.236/2 = 1.118 \approx 1.12$

16. FINDING THE SQUARE ROOT OF A NUMBER BY USING A CALCULATOR

More comprehensive calculations involving square roots should be accomplished using a calculator. The calculator is used to find square roots by scientists, machinists, engineers, and mathematicians.

To find the square root of a number using a calculator, simply enter the number and press the $(\sqrt{\ })$ button.

Henceforth, for the sake of uniformity, we shall use the equal sign, =, for approximate as well as exact square roots. Thus, $(\sqrt{5} = 2.236)$ will be used instead of $(\sqrt{5} \approx 2.236)$.

16.1 **FINDING SQUARE ROOTS USING A CALCULATOR**

Using a calculator, find each square root to the nearest thousandth if the result is not exact.

(*a*) $\sqrt{8}$ *Ans.* 2.828   (*g*) $\sqrt{109}$ *Ans.* 10.440   (*m*) $\sqrt{8,281}$ *Ans.* 91

(*b*) $\sqrt{18}$ *Ans.* 4.243   (*h*) $\sqrt{118}$ *Ans.* 10.863   (*n*) $\sqrt{9,409}$ *Ans.* 97

(*c*) $\sqrt{35}$ *Ans.* 5.916   (*i*) $\sqrt{144}$ *Ans.* 12   (*o*) $\sqrt{10,609}$ *Ans.* 103

(*d*) $\sqrt{75}$ *Ans.* 8.660   (*j*) $\sqrt{484}$ *Ans.* 22   (*p*) $\sqrt{12,769}$ *Ans.* 113

(*e*) $\sqrt{93}$ *Ans.* 9.644   (*k*) $\sqrt{1,849}$ *Ans.* 43   (*q*) $\sqrt{17,161}$ *Ans.* 131

(*f*) $\sqrt{104}$ *Ans.* 10.198   (*l*) $\sqrt{2,916}$ *Ans.* 54   (*r*) $\sqrt{20,736}$ *Ans.* 144

16.2 **EVALUATING EXPRESSIONS INVOLVING SQUARE ROOTS USING A CALCULATOR**

Using a calculator, find each to the nearest hundredth:

(*a*) $3\sqrt{2}$ (*b*) $2\sqrt{30} - 10$ (*c*) $\dfrac{\sqrt{45}}{3} + \dfrac{1}{\sqrt{2}}$

Solutions

(a) Since $\sqrt{2} = 1.414$,

$$3\sqrt{2} = 3(1.414)$$

$$= 4.242$$

Ans. 4.24

(b) Since $\sqrt{30} = 5.477$,

$$2\sqrt{30} - 10 = 2(5.477) - 10$$

$$= .954$$

Ans. .95

(c) Since $\sqrt{45} = 6.708$ and $\sqrt{2} = 1.414$,

$$\frac{\sqrt{45}}{3} + \frac{1}{\sqrt{2}} = \frac{6.708}{3} + \frac{1}{1.414}$$

$$= 2.236 + .707$$

$$= 2.943$$

Ans. 2.94

17. SIMPLIFYING THE SQUARE ROOT OF A PRODUCT

Rule: The square root of a product of two or more numbers equals the product of the separate square roots of these numbers.

Thus, $\sqrt{3600} = \sqrt{(36)(100)} = \sqrt{36}\sqrt{100} = (6)(10) = 60$, and in general, $\sqrt{ab} = \sqrt{a}\,\sqrt{b}$.

To Simplify the Square Root of a Product

Simplify $\sqrt{72}$.

PROCEDURE	SOLUTION
1. Factor the radicand, choosing perfect square factors:	**1.** $\sqrt{(4)(9)(2)}$
2. Form a separate square root for each factor:	**2.** $\sqrt{4}\,\sqrt{9}\,\sqrt{2}$
3. Extract the square roots of each perfect square:	**3.** $(2)(3)\sqrt{2}$
4. Multiply the factors outside the radical:	**4.** $6\sqrt{2}$ *Ans.*

17.1 FINDING EXACT SQUARE ROOTS THROUGH SIMPLIFICATION

Find each square root through simplification: (a) $\sqrt{1600}$,   (b) $\sqrt{256}$,   (c) $\sqrt{225}$,   (d) $\sqrt{324}$,   (e) $\frac{1}{2}\sqrt{576}$.

Solutions

(a) $\sqrt{(16)(100)}$

$\sqrt{16}\,\sqrt{100}$

$(4)(10)$

Ans. 40

(b) $\sqrt{(4)(64)}$

$\sqrt{4}\,\sqrt{64}$

$(2)(8)$

Ans. 16

(c) $\sqrt{(25)(9)}$

$\sqrt{25}\,\sqrt{9}$

$(5)(3)$

Ans. 15

(d) $\sqrt{(9)(36)}$

$\sqrt{9}\,\sqrt{36}$

$(3)(6)$

Ans. 18

(e) $\frac{1}{2}\sqrt{(4)(144)}$

$\frac{1}{2}\sqrt{4}\,\sqrt{144}$

$\frac{1}{2}(2)(12)$

Ans. 12

17.2 SIMPLIFYING SQUARE ROOTS OF NUMBERS

Simplify: (a) $\sqrt{75}$,   (b) $\sqrt{90}$,   (c) $\sqrt{112}$,   (d) $5\sqrt{288}$,   (e) $\frac{1}{3}\sqrt{486}$,   (f) $\frac{1}{2}\sqrt{1200}$.

Solutions

(a) $\sqrt{(25)(3)}$

$\sqrt{25}\,\sqrt{3}$

Ans. $5\sqrt{3}$

(b) $\sqrt{(9)(10)}$

$\sqrt{9}\,\sqrt{10}$

Ans. $3\sqrt{10}$

(c) $\sqrt{(16)(7)}$

$\sqrt{16}\,\sqrt{7}$

Ans. $4\sqrt{7}$

(d) $5\sqrt{(144)(2)}$

$5\sqrt{144}\,\sqrt{2}$

Ans. $60\sqrt{2}$

(e) $\frac{1}{3}\sqrt{(81)(6)}$

$\frac{1}{3}\sqrt{81}\,\sqrt{6}$

Ans. $3\sqrt{6}$

(f) $\frac{1}{2}\sqrt{(400)(3)}$

$\frac{1}{2}\sqrt{400}\,\sqrt{3}$

Ans. $10\sqrt{3}$

18. LAW OF PYTHAGORAS

In the rest of this chapter, when measures are involved, the terms *hypotenuse*, *leg*, *side*, and *angle* should be taken to mean their respective measures. We shall also understand that, in a triangle, the small letter for a side should agree with the capital letter for the vertex of the angle opposite that side. A right triangle labeled in this way is shown in Fig. 10-79.

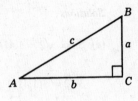

Fig. 10-79

Law of Pythagoras

In a right triangle, the square of the hypotenuse equals the sum of the squares of the two legs.

In terms of Fig. 10-79, the Law of Pythagoras reads: $c^2 = a^2 + b^2$.

Test Rule for a Right Triangle: If $c^2 = a^2 + b^2$ applies to the three sides of a triangle, then the triangle is a right triangle; but if $c^2 \neq a^2 + b^2$, where c is the longest side, then the triangle is not a right triangle.

18.1 FINDING THE HYPOTENUSE OF A RIGHT TRIANGLE

Find hypotenuse c in Fig. 10-79 when   (a) $a = 12$, $b = 9$;   (b) $a = 3$, $b = 7$,   (c) $a = 3$, $b = 6$; (d) $a = 3$, $b = 3\sqrt{3}$. Express your answers in simplified form.

Solutions

(a) $c^2 = a^2 + b^2$
$c^2 = 12^2 + 9^2$
$c^2 = 144 + 81$
$c^2 = 225$
$c = \sqrt{225}$
$c = 15$ *Ans.*

(b) $c^2 = a^2 + b^2$
$c^2 = 3^2 + 7^2$
$c^2 = 9 + 49$
$c^2 = 58$
$c = \sqrt{58}$ *Ans.*

(c) $c^2 = a^2 + b^2$
$c^2 = 3^2 + 6^2$
$c^2 = 9 + 36$
$c^2 = 45$
$c = \sqrt{45}$
$c = 3\sqrt{5}$ *Ans.*

(d) $c^2 = a^2 + b^2$
$c^2 = 3^2 + (3\sqrt{3})^2$
$c^2 = 9 + 27$
$c^2 = 36$
$c = \sqrt{36}$
$c = 6$ *Ans.*

18.2 FINDING A LEG OF A RIGHT TRIANGLE

In Fig. 10-79, find the missing leg when   (a) $b = 5$, $c = 13$;   (b) $a = 24$, $c = 25$;   (c) $b = 6$, $c = 8$; (d) $a = 4\sqrt{3}$, $c = 8$. Express your answers in simplified form.

Solutions

(a) $a^2 = c^2 - b^2$
$a^2 = 13^2 - 5^2$
$a^2 = 169 - 25$
$a^2 = 144$
$a = \sqrt{144}$
$a = 12$ *Ans.*

(b) $b^2 = c^2 - a^2$
$b^2 = 25^2 - 24^2$
$b^2 = 625 - 576$
$b^2 = 49$
$b = \sqrt{49}$
$b = 7$ *Ans.*

(c) $a^2 = c^2 - b^2$
$a^2 = 8^2 - 6^2$
$a^2 = 64 - 36$
$a^2 = 28$
$a = \sqrt{28}$
$a = 2\sqrt{7}$ *Ans.*

(d) $b^2 = c^2 - a^2$
$b^2 = 8^2 - (4\sqrt{3})^2$
$b^2 = 64 - 48$
$b^2 = 16$
$b = \sqrt{16}$
$b = 4$ *Ans.*

18.3 RATIOS IN A RIGHT TRIANGLE

In a right triangle whose hypotenuse is 20, the ratio of the two legs is $3:4$. Find each leg.

Solution   Let $3x$ and $4x$ represent the two legs of the right triangle.

$$(3x)^2 + (4x)^2 = 20^2 \qquad \text{If} \quad x = 4,$$
$$9x^2 + 16x^2 = 400 \qquad\qquad 3x = 12 \quad \textit{Ans.}$$
$$25x^2 = 400 \qquad\qquad 4x = 16 \quad \textit{Ans.}$$
$$x^2 = 16$$
$$x = 4$$

18.4   APPLYING LAW OF PYTHAGORAS TO A RECTANGLE

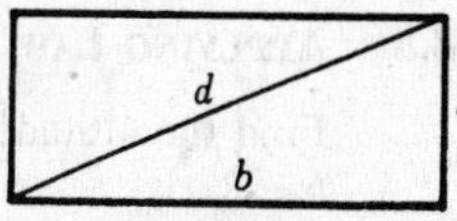

In a rectangle, find   (a) the diagonal if its sides are 9 and 40,   (b) one side if the diagonal is 30 and the other side is 24.

Fig. 10-80

Solution   The diagonal of the rectangle is the hypotenuse of a right triangle (Fig. 10-80).

(a) $d^2 = 9^2 + 40^2$, $d^2 = 1681$, $d = 41$ *Ans.*
(b) $h^2 = 30^2 - 24^2$, $h^2 = 324$, $h = 18$ *Ans.*

18.5   TESTING FOR RIGHT TRIANGLES

Using the three sides given, which triangles are right triangles?

$\triangle I$: 8, 15, 17 $\triangle II$: 6, 9, 11 $\triangle III$: $1\frac{1}{2}$, 2, $2\frac{1}{2}$

Solutions   Test whether or not $c^2 = a^2 + b^2$, where c is the longest side in each case.

$\triangle I$: $8^2 + 15^2 \overset{?}{=} 17^2$ $\triangle II$: $6^2 + 9^2 \overset{?}{=} 11^2$ $\triangle III$: $(1\frac{1}{2})^2 + 2^2 \overset{?}{=} (2\frac{1}{2})^2$

$64 + 225 \overset{?}{=} 289$ $36 + 81 \overset{?}{=} 121$ $2\frac{1}{4} + 4 \overset{?}{=} 6\frac{1}{4}$

$289 = 289$ $117 \neq 121$ $6\frac{1}{4} = 6\frac{1}{4}$

$\triangle I$ is a right $\triangle$ *Ans.* $\triangle II$ is not a right $\triangle$ *Ans.* $\triangle III$ is a right $\triangle$ Ans.

18.6   USING LAW OF PYTHAGORAS TO DERIVE A FORMULA

Derive a formula for the altitude h of any equilateral triangle in terms of any side s (Fig. 10-81).

Solution   The altitude h of the equilateral triangle bisects the base s.

$$h^2 = s^2 - \left(\frac{s}{2}\right)^2$$

$$h^2 = s^2 - \frac{s^2}{4} = \frac{3s^2}{4}$$

$$h = \frac{s}{2}\sqrt{3} \ \ Ans.$$

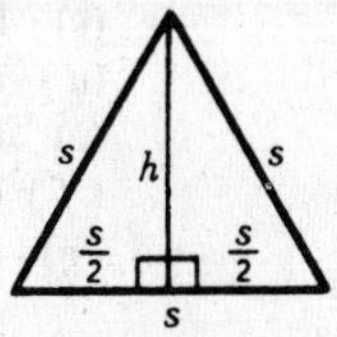

Fig. 10-81

18.7   APPLYING LAW OF PYTHAGORAS TO AN INSCRIBED SQUARE

The largest possible square is to be cut from a circular piece of cardboard having a diameter of 10 in. Find the side of the square to the nearest inch.

Solution   The diameter of the circle will be the diagonal of the square (Fig. 10-82). Hence,

$$s^2 + s^2 = 100$$

$$2s^2 = 100$$

$$s^2 = 50$$

$$s = 5\sqrt{2} = 7.07 \approx 7 \text{ in.} \ \ Ans.$$

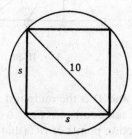

Fig. 10-82

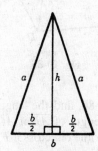

Fig. 10-83

18.8 **APPLYING LAW OF PYTHAGORAS TO AN ISOSCELES TRIANGLE**

Find the altitude of the isosceles triangle shown in Fig. 10-83 if (*a*) $a = 25$ and $b = 30$, (*b*) $a = 12$ and $b = 8$.

Solution The altitude h of the isosceles triangle bisects the base b. Hence,

$$h^2 = a^2 - \left(\frac{b}{2}\right)^2$$

Supplementary Problems

10.1. In Fig. 10-84, state the meaning of each: (1.2)

(*a*) $\overline{AB}$ (*b*) AB (*c*) $\overline{ABC}$ (*d*) $\overrightarrow{AC}$ (*e*) $\overrightarrow{BA}$ (*f*) $\overleftrightarrow{AC}$ (*g*) $AB + BC$

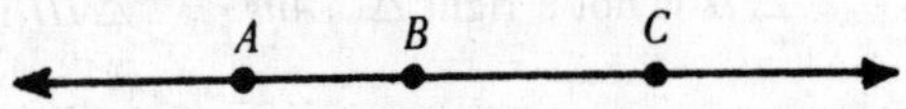

A B   C

Fig. 10-84

Ans. (*a*) line segment $\overline{AB}$
  (*b*) measure or length of line segment $\overline{AB}$
  (*c*) line segment $\overline{AC}$ with point B lying between A and C
  (*d*) ray with endpoint A and passing through C
  (*e*) ray with endpoint B and passing through A
  (*f*) line passing through or determined by points A and C
  (*g*) the sum of the lengths of $\overline{AB}$ and $\overline{BC}$

10.2. (*a*) Name the line segments in Fig. 10-85 that intersect at E. (1.3)
  (*b*) Name the line segments that intersect at D.
  (*c*) What other line segments can be drawn?
  (*d*) Name the point of intersection of $\overline{AC}$ and $\overline{BD}$.

Ans. (*a*) $\overline{AE}$ $\overline{DE}$ (*c*) $\overline{AD}$ $\overline{BE}$ $\overline{CE}$ $\overline{EF}$
   (*b*) $\overline{ED}$ $\overline{CD}$ $\overline{BD}$ $\overline{FD}$ (*d*) F

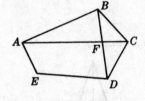

Fig. 10-85

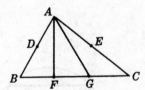

Fig. 10-86

10.3. (*a*) In Fig. 10-86, find the length of $\overline{AB}$ if AD is 8 and D is the midpoint of $\overline{AB}$. (1.4)
  (*b*) Find the length of $\overline{AE}$ if AC is 21 and E is the midpoint of $\overline{AC}$.
  (*c*) Name two line bisectors if F and G are the trisection points of $\overline{BC}$ (that is, $BF = FG = GC$).

Ans. (*a*) $AB = 16$ (*b*) $AE = 10\frac{1}{2}$ (*c*) $\overline{AF}$ bisects $\overline{BG}$, $\overline{AG}$ bisects $\overline{FC}$

10.4. (*a*) In Fig. 10-87(*a*), using three capital letters, name (1) ∠*a* (2) ∠*b* (3) ∠*c*, (4) a **(2.1)**
straight angle of which ∠*a* and ∠*b* are part.
(*b*) In Fig. 10-87(*b*), name (1) ∠*d*, (2) ∠*e*, (3) ∠*f*, (4) an angle of which ∠*d* and ∠*e* are parts.

Ans. (In each of the answers, the letters may be reversed.)
    (*a*) (1) ∠*ABC*   (2) ∠*CBE*   (3) ∠*BCD*   (4) ∠*ABE*
    (*b*) (1) ∠*RTS*   (2) ∠*QTR*   (3) ∠*PTQ*   (4) ∠*QTS* or ∠*PTS*

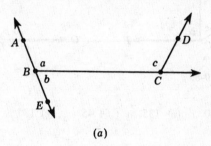

(*a*)   (*b*)

Fig. 10-87

10.5. What turn or rotation is made (*a*) by an hour hand in 3 hours, (*b*) by the minute **(2.2)**
hand in 1/3 hour? What rotation is needed to turn from (*c*) west to northeast in a clockwise
direction, (*d*) east to south in a counterclockwise direction, (*e*) southwest to northeast in
either direction?

Ans. (*a*) 90°   (*b*) 120°   (*c*) 135°   (*d*) 270°   (*e*) 180°

10.6. Find the angle formed by the hands of a clock at (*a*) 3 o'clock, (*b*) 10 o'clock, **(2.3)**
(*c*) 5:30 o'clock, (*d*) 11:30 o'clock.

Ans. (*a*) 90°   (*b*) 60°   (*c*) 15°   (*d*) 165°

10.7. **(2.4)**

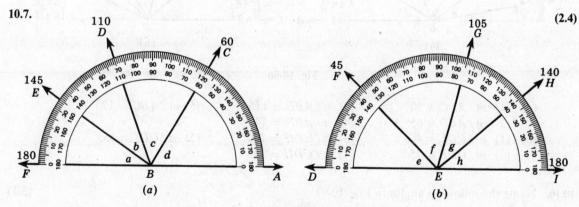

(*a*)   (*b*)

Fig. 10-88

(*a*) In Fig. 10-88, using the inner scale of the protractor, find the measures of (1) ∠*a*,
(2) ∠*b*, (3) ∠*c*, (4) ∠*d*, (5) ∠*ABD*, (6) ∠*CBE*, (7) ∠*ABE*, (8) ∠*CBF*.
(*b*) In Fig. 10-88, using the outer scale of the protractor, find the measures of (1) ∠*e*,
(2) ∠*f*, (3) ∠*g*, (4) ∠*h*, (5) ∠*DEH*, (6) ∠*FEH*, (7) ∠*GEI*, (8) ∠*FEI*.

Ans. (*a*) (1) 35° (2) 35° (3) 50° (4) 60° (5) 110° (6) 85° (7) 145° (8) 120°
     (*b*) (1) 45° (2) 60° (3) 35° (4) 40° (5) 140° (6) 95° (7) 75° (8) 135°

10.8. Use the protractor to find the measure of each angle indicated in Fig. 10-89:   **(2.5)**

  (*a*) ∠*a*   (*b*) ∠*b*   (*c*) ∠*c*   (*d*) ∠*ACB*   (*e*) ∠*p*   (*f*) ∠*q*   (*g*) ∠*QPS*

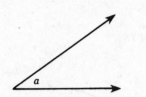
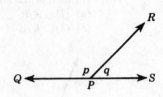

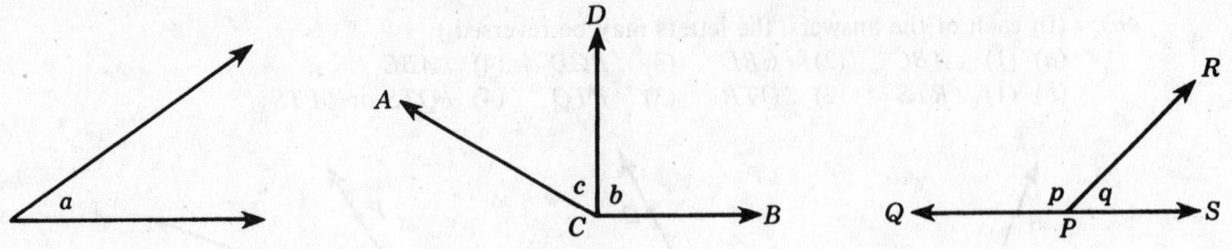

Fig. 10-89

Ans.   (*a*) 35   (*b*) 90   (*c*) 60   (*d*) 150   (*e*) 135   (*f*) 45   (*g*) 180

10.9. (*a*) Draw the following angles using the inner (counterclockwise) scale of the   **(2.6)**
           protractor:
           (1) 30°   (2) 65°   (3) 115°   (4) 143°   (5) 170°
       (*b*) Draw the following angles using the outer (clockwise) scale of the protractor:
           (1) 40°   (2) 75°   (3) 120°   (4) 138°   (5) 155°

Ans.   Refer to Fig. 10-90.

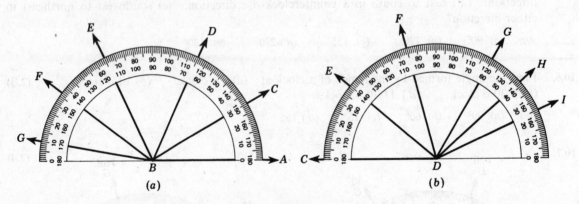

Fig. 10-90

  (*a*) (1) $m\angle ABC = 30°$ (3) $m\angle ABE = 115°$ (5) $m\angle ABG = 170°$
         (2) $m\angle ABD = 65°$ (4) $m\angle ABF = 143°$
  (*b*) (1) $m\angle CDE = 40°$ (3) $m\angle CDG = 120°$ (5) $m\angle CDI = 155°$
         (2) $m\angle CDF = 75°$ (4) $m\angle CDH = 138°$

10.10. Name the following angles in Fig. 10-91:   **(3.1)**

  (*a*) an acute angle at *B*.
  (*b*) an acute angle at *E*
  (*c*) a right angle
  (*d*) three obtuse angles
  (*e*) a straight angle

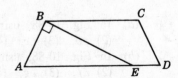

Ans.   (*a*) ∠*CBE*   (*d*) ∠*ABC*, ∠*BCD*, ∠*BED*
         (*b*) ∠*AEB*   (*e*) ∠*AED*
         (*c*) ∠*ABE*

Fig. 10-91

10.11. Find the number of degrees in   (*a*) 1/6 of a right angle,   (*b*) 30% of a right   **(3.2)**
angle,   (*c*) 150% of a right angle,   (*d*) 4/9 of a straight angle,   (*e*) 40% of a straight
angle,   (*f*) 150% of a straight angle.

Ans. (*a*) 15°   (*b*) 27°   (*c*) 135°   (*d*) 80°   (*e*) 72°   (*f*) 270°

10.12.   **(3.3)**

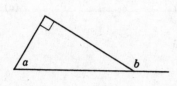   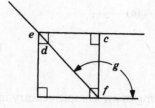

Fig. 10-92

In Fig. 10-92, what kind of angle is   (*a*) ∠*a*,   (*b*) ∠*b*,   (*c*) ∠*c*,   (*d*) ∠*d*,   (*e*) ∠*e*,   (*f*) ∠*f*,
(*g*) ∠*g*?

Ans. (*a*) acute   (*b*) obtuse   (*c*) right   (*d*) acute   (*e*) obtuse   (*f*) right   (*g*) obtuse

10.13. Complete Tables 10-5 and 10-6.   **(4.1)**

Table 10-5

Angle	Complement	Supplement
5°	?	?
25°	?	?
32°	?	?
51°	?	?
88°	?	?

Table 10-6

Angle	Supplement
95°	?
104°	?
125°	?
152°	?
175°	?

Ans. See Tables 10-7 and 10-8

Table 10-7

Angle	Complement	Supplement
5°	85°	175°
25°	65°	155°
32°	58°	148°
51°	39°	129°
88°	2°	92°

Table 10-8

Angle	Supplement
95°	85°
104°	76°
125°	55°
152°	28°
175°	5°

10.14.   **(4.2)**

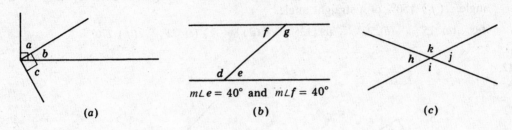

Fig. 10-93

(*a*)   Which angles are complementary in Fig. 10-93?   (*b*) Which angles are supplementary in Fig. 10-93?   (*c*) Which angles are vertical in Fig. 10-93?

Ans.   (*a*) ∠*a*, and ∠*b*, ∠*b* and ∠*c*
         (*b*) ∠*d* and ∠*e*, ∠*f* and ∠*g*, ∠*f* and ∠*d*, ∠*e* and ∠*g*
         (*c*) ∠*h* and ∠*j*, ∠*k* and ∠*i*

10.15. Find the angles in each: **(4.3)**
   (*a*)   The angles are complementary and the smaller is 40° less than the larger.
   (*b*)   The angles are complementary and the larger is four times the smaller.
   (*c*)   The angles are supplementary and the smaller is one-half of the larger.
   (*d*)   The angles are supplementary and the larger is 58° more than the smaller.
   (*f*)   The angles are adjacent and form an angle of 140°. The smaller is 28° less than the larger.
   (*g*)   The angles are vertical and supplementary.

Ans.   (*a*) 25°, 65°   (*d*) 61°, 119°   (*f*) 56°, 84°
         (*b*) 18°, 72°   (*e*) 50°, 130°   (*g*) 90°, 90°
         (*c*) 60°, 120°

10.16. In Fig. 10-94, find the number of degrees in   **(4.4)**
   (*a*) ∠*a*   (*b*) ∠*b*   (*c*) ∠*c*   (*d*) ∠*d*   (*e*) ∠*e*   (*f*) ∠*f*   (*g*) ∠*g*

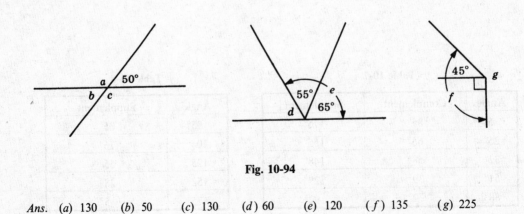

Fig. 10-94

Ans. (*a*) 130   (*b*) 50   (*c*) 130   (*d*) 60   (*e*) 120   (*f*) 135   (*g*) 225

10.17.   **(5.1)**

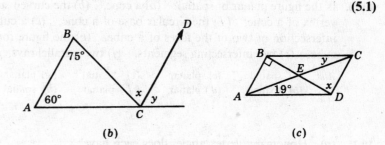

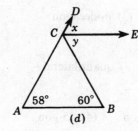   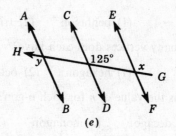

Fig. 10-95

Find $m\angle x$ and $m\angle y$ if   (*a*) in Fig. 10-95(*a*), $\overleftrightarrow{AB} \parallel \overleftrightarrow{CD}$;   (*b*) in Fig. 10-95(*b*), $\overleftrightarrow{AB} \parallel \overleftrightarrow{CD}$;   (*c*) in Fig. 10-95(*c*), $\overleftrightarrow{AB} \parallel \overleftrightarrow{CD}$, $\overleftrightarrow{BC} \parallel \overleftrightarrow{AD}$;   (*d*) in Fig. 10-95(*d*), $\overleftrightarrow{AB} \parallel \overrightarrow{CE}$;   (*e*) in Fig. 10-95(*e*), $\overleftrightarrow{AB} \parallel \overleftrightarrow{CD} \parallel \overleftrightarrow{EF}$.

Ans. (*a*) $m\angle x = 112$ by Rule 1; $m\angle y = 112$ by Rule 3   (*d*) $m\angle x = 58$ by Rule 1; $m\angle y = 60$ by Rule 3
      (*b*) $m\angle x = 75$ by Rule 3; $m\angle y = 60$ by Rule 1   (*e*) $m\angle x = 125$ by Rule 1; $m\angle y = 125$ by
      (*c*) $m\angle x = 90$ by Rule 3; $m\angle y = 19$ by Rule 3   congruent vertical angles and Rule 1

10.18. In Fig. 10-96, $\overleftrightarrow{AB} \parallel \overleftrightarrow{CD}$, $\overleftrightarrow{EF} \parallel \overleftrightarrow{GH}$.   **(5.2)**

(*a*) If $d = 110$,
      find (1) f, (2) h, (3) g.

(*b*) If $f = 105$,
      find (1) j, (2) l, (3) k.

(*c*) If $i = 76$,
      find (1) g, (2) c, (3) b.

Ans. (*a*) (1) 110   (2) 110   (3) 70
      (*b*) (1) 105   (2) 105   (3) 75
      (*c*) (1)   76   (2)   76   (3) 104

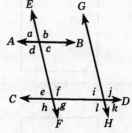

Fig. 10-96

10.19. Referring to Fig. 10-97, state the rule or rules that justify
each statement: **(5.3)**

(*a*) If $m\angle p = 88$ and $m\angle q = 88$, then $\overleftrightarrow{AB} \parallel \overleftrightarrow{CD}$.
(*b*) If $m\angle p = 87$ and $m\angle r = 87$, then $\overleftrightarrow{EF} \parallel \overleftrightarrow{GH}$.
(*c*) If $m\angle p = 90$ and $m\angle q = 90$, then $\overleftrightarrow{AB} \parallel \overleftrightarrow{CD}$.
(*d*) If $m\angle p = 90$ and $m\angle r = 90$, then $\overleftrightarrow{EF} \parallel \overleftrightarrow{GH}$.

Ans. (*a*) Rule 4   (*c*) Rule 4 or 5
      (*b*) Rule 2   (*d*) Rule 2 or 5

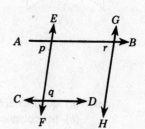

Fig. 10-97

10.20. Is the figure planar or spatial? (*a*) a cone, (*b*) the curved surface of a cone, (*c*) the **(6.1)** vertex of a cone, (*d*) the circular base of a cone, (*e*) a cube, (*f*) a face of a cube, (*g*) the intersection of two of the faces of a cube, (*h*) the figure formed by two intersecting faces of a cube, (*i*) two intersecting segments, (*j*) two parallel rays, (*k*) a circular region.

Ans. (*a*) spatial   (*c*) planar   (*e*) spatial   (*g*) planar   (*i*) planar   (*k*) planar
      (*b*) spatial   (*d*) planar   (*f*) planar   (*h*) spatial   (*j*) planar

10.21. (*a*) How many vertex angles does each have? **(7.1)**

          (1) pentagon   (2) triangle   (3) dodecagon

(*b*) How many vertices does each have?

          (1) hexagon   (2) octagon   (3) quadrilateral

(*c*) What is the value of *n* for each *n*-gon?

          (1) decagon   (2) nonagon   (3) septagon   (4) 25-gon

Ans. (*a*) (1) 5   (2) 3   (3) 12
      (*b*) (1) 6   (2) 8   (3) 4
      (*c*) (1) 10   (2) 9   (3) 7   (4) 25

10.22. In Fig. 10-98, name each figure that is a polygon and, for each figure that is not a **(7.2)** polygon, state the reason why it is not a polygon.

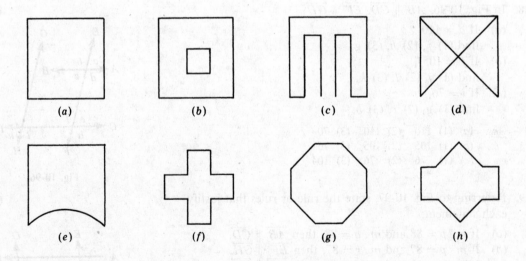

Fig. 10-98

Ans. (*a*) quadrilateral   (*e*) nonpolygon (curved side)
      (*b*) nonpolygon (encloses more than 1 region)   (*f*) dodecagon
      (*c*) nonpolygon (open figure)   (*g*) octagon
      (*d*) nonpolygon (crossing segments)   (*h*) decagon

10.23. State whether each polygon in Fig. 10-99 is equilateral, equiangular, regular, or none   **(7.3)**
of these.

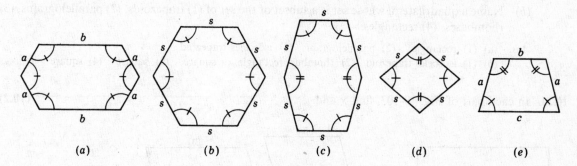

Fig. 10-99

Ans.   (a) equiangular   (b) regular   (c) equilateral   (d) equilateral   (e) none of these

10.24. In Fig. 10-100, find an example of   (a) diameter,   (b) radius,   **(8.1)**
(c) sector,   (d) minor arc,   (e) semicircle,   (f) chord,
(g) central angle,   (h) major arc.

Ans.   (a) $\overline{AB}$
      (b) $\overline{OA}, \overline{OB}, \overline{OC}$
      (c) regions I, II, III and the sum of any two of these
      (d) $\overset{\frown}{AC}, \overset{\frown}{CB}$
      (e) $\overset{\frown}{ACB}, \overset{\frown}{ADB}$
      (f) $\overline{AC}, \overline{AB}$
      (g) $\angle AOC, \angle BOC, \angle AOB$
      (h) $\overset{\frown}{ABC}, \overset{\frown}{BAC}$

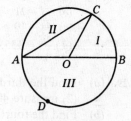

Fig. 10-100

10.25. In Fig. 10-101, name   (a) each right triangle,   (b) each obtuse triangle.   **(9.1)**

Ans.   (a) △I or △AEF,   △II or △BCF,   △ABE
      (b) △ABC,   △AFB

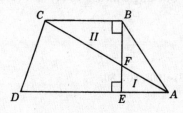

Fig. 10-101

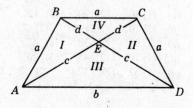

Fig. 10-102

10.26. In Fig. 10-102, a, b, c, and d are measures of lengths of segments. Name   (a) each   **(9.2)**
isosceles triangle,   (b) each scalene triangle.

Ans.   (a) △III or △AED,   △IV or △BEC,   △ABC,   △BCD
      (b) △,I or △ABE,   △II or △ECD,   △ABD,   △ACD

10.27. (*a*)　Name the most general type of quadrilateral having the property: (1) all angles　　**(10.1)**
are right angles, (2) there are two pairs of parallel sides, (3) each of its two pairs of equal
angles has a side in common.

(*b*)　Name a quadrilateral whose set is a subset of the set of (1) trapezoids, (2) parallelograms, (3)
rhombuses, (4) rectangles.

Ans.　(*a*) (1) rectangle　(2) parallelogram　(3) isosceles trapezoid

(*b*) (1) isosceles trapezoid　(2) rhombus, rectangle, or square　(3) square　(4) square

10.28. In each part of Fig. 10-103, find *x* and *y*.　　　　　　　　　　　　　　　　**(10.2)**

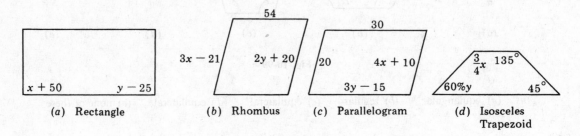

(*a*)　Rectangle　　　(*b*)　Rhombus　　　(*c*)　Parallelogram　　(*d*)　Isosceles
　　　　　　　　　　　　　　　　　　　　　　　　　　　　　　　　　　　　　　Trapezoid

Fig. 10-103

Ans.　(*a*)　$x + 50 = 90$, $x = 40$; $y - 25 = 90$, $y = 115$

(*b*)　$3x - 21 = 54$, $x = 25$; $2y + 20 = 54$, $y = 17$

(*c*)　$4x + 10 = 20$, $x = 2\frac{1}{2}$; $3y - 15 = 30$, $y = 15$

(*d*)　$\frac{3}{4}x = 135$, $x = 180$; $60\% y = \frac{3}{5}y = 45$, $y = 75$

10.29. (*a*)　Find the third angle of a triangle if the other angles (1) each measure 70°,　　**(11.1)**
(2) measure 40° and 50°, (3) have an angle-sum of 168.

(*b*)　Find the fourth angle of a quadrilateral if the other angles (1) each measure 85°; (2) measure
85°, 90° and 100°; (3) have an angle-sum of 287.

(*c*)　In an isosceles triangle, find the vertex angle if each base angle measures (1) 42°, (2) 71°, (3)
88°.

(*d*)　In an isosceles triangle, find each base angle if the vertex angle measures (1) 40°, (2) 90°, (3)
140°.

(*e*)　In a parallelogram, find the remaining angles if one angle measures (1) 52°, (2) 78°, (3) 100°.

(*f*)　In a right triangle, find the other acute angle if one angle measures (1) 43°, (2) 67°, (3) 86°.

Ans.　(*a*)　(1) 40°　(2) 90°　(3) 12°

(*b*)　(1) 105°　(2) 85°　(3) 73°

(*c*)　(1) 96°　(2) 38°　(3) 4°

(*d*)　(1) 70°　(2) 45°　(3) 20°

(*e*)　(1) 52°, 128°, 128°　(2) 78°, 102°, 102°　(3) 100°, 80°, 80°

(*f*)　(1) 47°　(2) 23°　(3) 4°

10.30. Using *x* as a common factor in their representation, find the required angles if the　　**(11.2)**
angles are in the ratio of　(*a*) 4 : 5 and the angles are the acute angles of a right triangle,　(*b*) 7 : 3
and the angles are consecutive angles of a parallelogram,　(*c*) 2 : 3 : 4 and the angles are the three
angles of a triangle,　(*d*) 3 : 4 : 5 : 6 and the angles are the four angles of a quadrilateral,　(*e*) 1 : 2
and the greater is a base angle of an isosceles triangle of which the smaller is the vertex angle.

Ans.　(*a*)　40°, 50°　　　(*c*)　40°, 60°, 80°　　　(*e*)　36°, 72°

(*b*)　126°, 54°　　　(*d*)　60°, 80°, 100°, 120°

10.31. Using $3x$, $4x$, and $5x$ as their representation, find the required three angles if   **(11.3)**
(a) the angles are the angles of a triangle,   (b) the angles are angles of a quadrilateral whose remaining angle is 120°,   (c) the two greatest are the acute angles of a right triangle,   (d) the smallest angle is a base angle of an isosceles triangle of which the middle-sized angle is the vertex angle,   (e) the smallest and greatest are the consecutive angles of a parallelogram.

Ans.   (a) 45°, 60°, 75°   (c) 30°, 40°, 50°   (e) $67\frac{1}{2}$°, 90°, $112\frac{1}{2}$°
        (b) 60°, 80°, 100°   (d) 54°, 72°, 90°

10.32. The first of three angles measures 10° less than the second. The third angle   **(11.4)**
measures 20° more than the sum of the other two. Find the three angles if   (a) the angles are the angles of a triangle,   (b) the two smallest angles are the acute angles of a right triangle,   (c) the smallest and the greatest are consecutive angles of a parallelogram,   (d) the greatest is a base angle of an isosceles triangle of which the smallest is the vertex angle,   (e) the angles are angles of a quadrilateral whose remaining angle measures 20° more than the greatest of the three angles.

Ans.   (a) 35°, 45°, 100°   (c) 50°, 60°, 130°   (e) 45°, 55°, 120
        (b) 40°, 50°, 110°   (d) 24°, 34°, 78°

10.33. (a) State the formula to use to find the perimeter of (1) an equilateral triangle,   **(12.1)**
(2) an equilateral hexagon, (3) a regular dodecagon, (4) a regular centagon, (5) an equilateral 40-gon.

(b) In each, name the regular polygon to which the formula applies: (1) $p = 4s$, (2) $p = 5s$, (3) $p = 9s$, (4) $p = 10s$, (5) $p = 15s$.

(c) If the perimeter is 60 m, find the side of a (1) square, (2) regular octagon, (3) regular decagon, (4) regular n-gon if $n = 50$.

Ans.   (a) (1) $p = 3s$ (2) $p = 6s$ (3) $p = 12s$ (4) $p = 100s$ (5) $p = 40s$
        (b) (1) square   (2) regular pentagon   (3) regular nonagon   (4) regular decagon
            (5) regular 15-gon
        (c) (1) 15 m   (2) 7.5 m   (3) 6 m   (4) 1.2 m

10.34. (a) Find the perimeter of an equilateral n-gon, given n and given the side, s, in yards: **(12.2)**
            (1) $n = 5$ and $s = 8\frac{1}{5}$ (2) $n = 20$ and $s = 13.7$ (3) $n = 12$ and $s = 7\frac{3}{4}$

(b) Find the side of an equilateral n-gon, given n and given the perimeter, p, in feet:
            (1) $n = 8$ and $p = 92$ (2) $n = 6$ and $p = 44$ (3) $n = 25$ and $p = 110$

(c) Find the number of sides of an equilateral n-gon, given the perimeter, p, and the side, s, both in centimeters:
            (1) $p = 117$ and $s = 9$ (2) $p = 36$ and $s = 1.2$ (3) $p = 51$ and $s = 17$

Ans.   (a) (1) 41 yd   (2) 274 yd   (3) 93 yd
        (b) (1) $11\frac{1}{2}$ ft   (2) $7\frac{1}{3}$ ft   (3) 4.4 ft
        (c) (1) 13   (2) 30   (3) 3

10.35. In Fig. 10-104, find: **(12.3)**
(a) perimeter, p, if (1) $a = 7\frac{1}{2}$, $b = 11$; (2) $a = 7.3$, $b = 12.4$; (3) $a = 3\frac{1}{4}$, $b = 6.2$.
(b) base, b, if (1) $p = 28$, $a = 10$; (2) $p = 30$, $a = 9\frac{1}{4}$; (3) $p = 46.2$, $a = 13.7$.
(c) leg, a, if (1) $p = 24$, $b = 7$; (2) $p = 36\frac{1}{2}$, $b = 16$; (3) $p = 40.1$, $b = 13.1$.

Ans.   (a) (1) 26   (2) 27   (3) 12.7
        (b) (1) 8   (2) $11\frac{1}{2}$ (3) 18.8
        (c) (1) $8\frac{1}{2}$ (2) $10\frac{1}{4}$ (3) 13.5

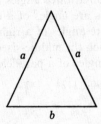

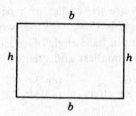

Fig. 10-104　　　　　　　　　　　　　　　　　**Fig. 10-105**

10.36. In Fig. 10-105, find: **(12.4)**

    (a) perimeter, p, if (1) $b = 7$, $h = 13$; (2) $b = 10\frac{1}{4}$, $h = 5\frac{1}{2}$; (3) $b = 6.8$, $h = 3.1$.

    (b) base, b, if (1) $p = 20$, $h = 7$; (2) $p = 25$, $h = 5\frac{1}{4}$; (3) $p = 35.6$, $h = 10.8$.

    (c) altitude, h, if (1) $p = 30$, $b = 9$; (2) $p = 31\frac{1}{2}$, $b = 11$; (3) $p = 31.4$, $b = 6.9$.

    Ans. (a) (1) 40   (2) $31\frac{1}{2}$ (3) 19.8

           (b) (1) 3   (2) $7\frac{1}{4}$ (3) 7

           (c) (1) 6   (2) $4\frac{3}{4}$ (3) 8.8

10.37. The base of a rectangle is 10 more than twice the height. Find the dimensions of the **(12.5)** rectangle if (a) the semiperimeter is 34, (b) the perimeter is 74, (c) the semiperimeter is 5 less than six times the height, (d) the perimeter is 20 more than twice the base.

    Ans. (a) 8, 26   (b) 9, 28   (c) 5, 20   (d) 10, 30

10.38. For any circle, state a formula in which (a) the circumference, c, is expressed in **(12.6)** terms of the radius, r, (b) the radius, r, is expressed in terms of the circumference, c, (c) an arc of 120° a, is expressed in terms of the circumference, c, (d) an arc of 120°, a, is expressed in terms of the diameter, d, (e) an arc of 45°, a', is expressed in terms of the circumference, c, (f) an arc of 45°, a', is expressed in terms of the radius, r.

    Ans. (a) $c = 2\pi r$ (b) $r = \dfrac{c}{2\pi}$ (c) $a = \dfrac{c}{3}$ (d) $a = \dfrac{\pi d}{3}$ (e) $a' = \dfrac{c}{8}$ (f) $a' = \dfrac{\pi r}{4}$

10.39. For a circle, if c, d, and r are in feet, using $\pi = 3.14$, find (a) the circumference if **(12.7)** $d = 20$, (b) a 90° arc if $r = 20$, (c) a 45° arc if $d = 12$, (d) the circumference if an arc of 120° has length 21.5 ft, (e) the diameter if $c = 785$, (f) the radius if $c = 188.4$.

    Ans. (a) 62.8 ft   (b) 31.4 ft   (c) 4.71 ft   (d) 64.5 ft   (e) 250 ft   (f) 30 ft

10.40. Complete each: **(13.1)**

    (a) $1\,\text{in}^2 = ?\,\text{ft}^2$ (c) $1\,\text{mi}^2 = ?\,\text{yd}^2$ (e) $1\,\text{km}^2 = ?\,\text{dm}^2$

    (b) $1\,\text{yd}^2 = ?\,\text{in}^2$ (d) $1\,\text{cm}^2 = ?\,\text{m}^2$ (f) $1\,\text{dm}^2 = ?\,\text{cm}^2$

    Ans. (a) $\dfrac{1}{144}$ (b) 1296   (c) 3,097,600   (d) $\dfrac{1}{10,000}$ (e) 100,000,000   (f) 100

10.41. (a) In a rectangle, apply $A = bh$ to find (1) A if $b = 6$ and $h = 7.5$, (2) b if $A = 11.48$ **(13.2)**
and $h = 1.4$, (3) h if $A = 104$ and $b = 12$.

(b) In a triangle, apply $A = bh/2$ to find (1) A if $b = 10$ and $h = 8.3$, (2) b if $A = 11.2$ and
$h = 3.5$, (3) h if $b = 16$ and $A = 26$.

(c) In a trapezoid, apply $A = h(b + b')/2$ to find (1) A if $b = 9$, $b' = 6$ and $h = 4$; (2) h if
$A = 47.5$, $b = 12$ and $b' = 7$; (3) b if $A = 70$, $h = 7$ and $b' = 14$.

Ans. (a) (1) 45   (2) 8.2   (3) $8\frac{2}{3}$

(b) (1) 41.5   (2) 6.4   (3) $3\frac{1}{4}$

(c) (1) 30   (2) 5   (3) 6

10.42. For any circle, state a formula in which (a) the area, A, is expressed in terms of **(13.3)**
the diameter, d;   (b) a sector of $36°$, S, is expressed in terms of the area, A;   (c) a sector of $36°$, S,
is expressed in terms of the radius, r;   (d) a sector of $72°$, S', is expressed in terms of the area,
A;   (e) a sector of $72°$, S', is expressed in terms of the diameter, d.

Ans. (a) $A = \dfrac{\pi d^2}{4}$ (b) $S = \dfrac{A}{10}$ (c) $S = \dfrac{\pi r^2}{10}$ (d) $S' = \dfrac{A}{5}$ (e) $S' = \dfrac{\pi d^2}{20}$

10.43. For a circle, if d and r are in inches, using $\pi = 3.14$, find, to the nearest whole **(13.4)**
number, the area of (a) a circle if $d = 8$, (b) a $90°$ sector if $d = 12$, (c) a $36°$ sector if
$d = 20$, (d) a circle if the area of a $36°$ sector is $37.28\,\text{in}^2$, (e) a circle if the area of a $72°$
sector is $17.3\,\text{in}^2$, (f) a $40°$ sector if the area of a semicircle is $120\,\text{in}^2$.

Ans. (a) $50\,\text{in}^2$ (b) $28\,\text{in}^2$ (c) $31\,\text{in}^2$ (d) $373\,\text{in}^2$ (e) $87\,\text{in}^2$ (f) $27\,\text{in}^2$

10.44. Find each square: **(14.1)**

(a) 7^2 (e) 32^2 (i) 85^2 (m) 130^2 (q) 450^2 (u) 515^2
(b) 11^2 (f) 45^2 (j) 91^2 (n) 203^2 (r) 455^2 (v) $1,111^2$
(c) 16^2 (g) 60^2 (k) 99^2 (o) 230^2 (s) 501^2 (w) $11,011^2$
(d) 19^2 (h) 65^2 (l) 105^2 (p) 405^2 (t) 510^2 (x) $23,456^2$

Ans. (a) 49   (e) 1,024   (i) 7,225   (m) 16,900   (q) 202,500   (u) 265,225
(b) 121   (f) 2,025   (j) 8,281   (n) 41,209   (r) 207,025   (v) 1,234,321
(c) 256   (g) 3,600   (k) 9,801   (o) 52,900   (s) 251,001   (w) 121,242,121
(d) 361   (h) 4,225   (l) 11,025   (p) 164,025   (t) 260,100   (x) 550,183,936

10.45. Find each square: **(14.2)**

(a) 40^2 (c) $4,000^2$ (e) $1,000^2$ (g) 150^2 (i) $15,000^2$ (k) $13,000^2$ (m) $500,000^2$
(b) 400^2 (d) 10^2 (f) $10,000^2$ (h) $1,500^2$ (j) 130^2 (l) 500^2

Ans. (a) 1,600   (e) 1,000,000   (i) 225,000,000   (m) 250,000,000,000
(b) 160,000   (f) 100,000,000   (j) 16,900
(c) 16,000,000   (g) 22,500   (k) 169,000,000
(d) 100   (h) 2,250,000   (l) 250,000

10.46. Find each square: **(14.3)**

(a) $\left(\dfrac{3}{5}\right)^2$ (c) $\left(\dfrac{6}{11}\right)^2$ (e) $\left(\dfrac{7}{30}\right)^2$ (g) $\left(\dfrac{21}{19}\right)^2$ (i) $\left(\dfrac{23}{250}\right)^2$ (k) $\left(\dfrac{70}{17}\right)^2$ (m) $\left(\dfrac{1,001}{1,000}\right)^2$

(b) $\left(\dfrac{1}{9}\right)^2$ (d) $\left(\dfrac{11}{6}\right)^2$ (f) $\left(\dfrac{19}{21}\right)^2$ (h) $\left(\dfrac{23}{25}\right)^2$ (j) $\left(\dfrac{230}{25}\right)^2$ (l) $\left(\dfrac{7}{170}\right)^2$

Ans. (a) $\dfrac{9}{25}$ (c) $\dfrac{36}{121}$ (e) $\dfrac{49}{900}$ (g) $\dfrac{441}{361}$ (i) $\dfrac{529}{62,500}$ (k) $\dfrac{4,900}{289}$ (m) $\dfrac{1,002,001}{1,000,000}$

(b) $\dfrac{1}{81}$ (d) $\dfrac{121}{36}$ (f) $\dfrac{361}{441}$ (h) $\dfrac{529}{625}$ (j) $\dfrac{52,900}{625}$ (l) $\dfrac{49}{28,900}$

10.47. Find the square of each: (a) .3, (b) .03, (c) .001, (d) .0001, (e) .15,   **(14.4)**
(f) 1.5, (g) .015, (h) 10.5, (i) 1.2, (j) .12, (k) .0012, (l) 10.2, (m) 1.02, (n) 8.5,
(o) .85, (p) 80.5, (q) 80.05.

Ans. (a) .09   (d) .00000001   (g) .000225   (j) .0144   (m) 1.0404   (p) 6480.25
(b) .0009   (e) .0225   (h) 110.25   (k) .00000144 (n) 72.25   (q) 6408.0025
(c) .000001   (f) 2.25   (i) 1.44   (l) 104.04   (o) .7225

10.48. Find the square of each: (a) $1\frac{1}{3}$, (b) $1\frac{2}{3}$, (c) $2\frac{1}{4}$, (d) $3\frac{1}{5}$, (e) $4\frac{3}{10}$, (f) $7\frac{1}{2}$,   **(14.5)**
(g) $8\frac{3}{4}$,   (h) $10\frac{1}{4}$,   (i) $12\frac{3}{8}$.

Ans. (a) $1\frac{7}{9}$ (c) $5\frac{1}{6}$ (e) $18\frac{49}{100}$ (g) $76\frac{9}{16}$ (i) $153\frac{9}{64}$

(b) $2\frac{7}{9}$ (d) $10\frac{6}{25}$ (f) $56\frac{1}{4}$ (h) $105\frac{1}{16}$

10.49. Find the area of a square whose side is (a) 8 in., (b) 12 ft, (c) 25 yd, (d) 60 mi,   **(14.6)**
(e) 600 yd, (f) 1.5 dm, (g) .25 km, (h) 3.5 hm, (i) .07 ft, (j) .85 in., (k) $\frac{3}{5}$ cm,
(l) $1\frac{3}{4}$ yd, (m) $3\frac{1}{2}$ mi, (n) $5\frac{1}{4}$ m.

Ans. (a) 64 in^2 (e) 360,000 yd^2 (i) .0049 ft^2 (m) 12.25 or $12\frac{1}{4}$ mi^2

(b) 144 ft^2 (f) 2.25 dm^2 (j) .7225 in^2 (n) 27.5625 or $27\frac{9}{16}$ m^2

(c) 625 yd^2 (g) .0625 km^2 (k) .36 or $\frac{9}{25}$ cm^2

(d) 3,600 mi^2 (h) 12.25 hm^2 (l) $3\frac{1}{16}$ yd^2

10.50. Relate each pair of numbers using both the square and square root symbols: (a) 7   **(15.1)**
and 49, (b) 900 and 30, (c) .25 and .5, (d) .1 and .01, (e) 3/5 and 9/25, (f) $1\frac{1}{2}$ and $2\frac{1}{4}$,
(g) .0225 and .15, (h) 1.1 and 1.21.

Ans. (a) $49 = 7^2$ and $7 = \sqrt{49}$ (e) $\dfrac{9}{25} = \left(\dfrac{3}{5}\right)^2$ and $\dfrac{3}{5} = \sqrt{\dfrac{9}{25}}$

(b) $900 = 30^2$ and $30 = \sqrt{900}$ (f) $2\frac{1}{4} = (1\frac{1}{2})^2$ and $1\frac{1}{2} = \sqrt{2\frac{1}{4}}$

(c) $.25 = .5^2$ and $.5 = \sqrt{.25}$ (g) $.0225 = .15^2$ and $.15 = \sqrt{.0225}$

(d) $.01 = .1^2$ and $.1 = \sqrt{.01}$ (h) $1.21 = 1.1^2$ and $1.1 = \sqrt{1.21}$

10.51. Evaluate: **(15.2)**

(a) $\sqrt{9} + \sqrt{4} - \sqrt{1}$ (d) $\dfrac{1}{3}\sqrt{9} + \dfrac{3}{2}\sqrt{4}$ (g) $\sqrt{8} - \sqrt{7}$

(b) $2\sqrt{9} + 10\sqrt{1}$ (e) $.2\sqrt{4} - .3\sqrt{1}$ (h) $\sqrt{4} + \sqrt{5} - \sqrt{10}$

(c) $8\sqrt{0} + \dfrac{1}{2}\sqrt{4}$ (f) $\sqrt{2} + \sqrt{3}$ (i) $10\sqrt{10} - 5\sqrt{8}$

Ans. (a) 4   (b) 16   (c) 1   (d) 4   (e) .1   (f) 3.146   (g) .182   (h) 1.074   (i) 17.48

10.52. Using a calculator, find each square root:　　　　　　　　　　　　**(16.1)**

(a) $\sqrt{7}$　　Ans. 2.646　　　　(g) $\sqrt{124}$　　Ans. 11.136　　　(m) $\sqrt{8,464}$　　Ans. 92

(b) $\sqrt{37}$　　Ans. 6.083　　　(h) $\sqrt{142}$　　Ans. 11.916　　(n) $\sqrt{11,449}$　　Ans. 107

(c) $\sqrt{46}$　　Ans. 6.782　　　(i) $\sqrt{196}$　　Ans. 14　　　　(o) $\sqrt{15,876}$　　Ans. 126

(d) $\sqrt{61}$　　Ans. 7.810　　　(j) $\sqrt{841}$　　Ans. 29　　　　(p) $\sqrt{18,225}$　　Ans. 135

(e) $\sqrt{89}$　　Ans. 9.434　　　(k) $\sqrt{1,936}$　　Ans. 44　　　(q) $\sqrt{19,881}$　　Ans. 141

(f) $\sqrt{105}$　　Ans. 10.247　　(l) $\sqrt{5,776}$　　Ans. 76　　　(r) $\sqrt{22,500}$　　Ans. 150

10.53. Using a calculator, find each to the nearest tenth:　　　　　　　**(16.2)**

(a) $2\sqrt{3}$　　　(c) $\frac{1}{2}\sqrt{140}$　　　(e) $100 - 10\sqrt{83}$　　　(g) $\frac{\sqrt{3}+\sqrt{2}}{5}$

(b) $10\sqrt{50}$　　(d) $\frac{2}{5}\sqrt{150}$　　　(f) $\frac{3+\sqrt{2}}{5}$　　　　(h) $\frac{10-\sqrt{2}}{3}$

Ans.　(a) 3.5　(b) 70.7　(c) 5.9　(d) 4.9　(e) 8.9　(f) .9　(g) .6　(h) 2.9

10.54. Find each square root through simplification:　　　　　　　　　　**(17.1)**

(a) $\sqrt{2500}$　　(c) $\sqrt{2025}$　　(e) $\sqrt{729}$　　(g) $\sqrt{48400}$

(b) $\sqrt{19600}$　(d) $\sqrt{441}$　　(f) $\sqrt{784}$　　(h) $\sqrt{562500}$

Ans.　(a) 50　(b) 140　(c) 45　(d) 21　(e) 27　(f) 28　(g) 220　(h) 750

10.55. Simplify:　　　　　　　　　　　　　　　　　　　　　　　　　　**(17.2)**

(a) $\sqrt{63}$　　(d) $\frac{1}{2}\sqrt{32}$　　(g) $\sqrt{17500}$　　(j) $\frac{3}{7}\sqrt{392}$

(b) $\sqrt{96}$　　(e) $\frac{1}{5}\sqrt{300}$　　(h) $\sqrt{4500}$　　(k) $\frac{2}{5}\sqrt{250}$

(c) $\sqrt{448}$　　(f) $\frac{7}{8}\sqrt{320}$　　(i) $\sqrt{21600}$　　(l) $\frac{5}{3}\sqrt{999}$

Ans.　(a) $3\sqrt{7}$　　(c) $8\sqrt{7}$　　(e) $2\sqrt{3}$　　(g) $50\sqrt{7}$　　(i) $60\sqrt{6}$　　(k) $2\sqrt{10}$

　　　(b) $4\sqrt{6}$　　(d) $2\sqrt{2}$　　(f) $7\sqrt{5}$　　(h) $30\sqrt{5}$　　(j) $6\sqrt{2}$　　(l) $5\sqrt{111}$

10.56. In a right triangle whose legs are a and b, find the hypotenuse c when:　　**(18.1)**

(a) $a = 15$, $b = 20$　　(c) $a = 5$, $b = 4$　　(e) $a = 7$, $b = 7$

(b) $a = 15$, $b = 36$　　(d) $a = 5$, $b = 5\sqrt{3}$　　(f) $a = 8$, $b = 4$

Ans.　(a) 25　　(b) 39　　(c) $\sqrt{41}$　　(d) 10　　(e) $7\sqrt{2}$　　(f) $4\sqrt{5}$

10.57. Find the missing leg of the right triangle, if one leg and the hypotenuse are　　**(18.2)**

(a) $a = 12$, $c = 20$　　(c) $b = 15$, $c = 17$　　(e) $a = 5\sqrt{2}$, $c = 10$

(b) $b = 6$, $c = 8$　　(d) $a = 2$, $c = 4$　　(f) $a = \sqrt{5}$, $c = 2\sqrt{2}$

Ans.　(a) $b = 16$　　(b) $a = 2\sqrt{7}$　　(c) $a = 8$　　(d) $b = 2\sqrt{3}$　　(e) $b = 5\sqrt{2}$　　(f) $b = \sqrt{3}$

10.58. Find the legs of a right triangle whose hypotenuse is c if these legs have a ratio of　　**(18.3)**

(a) 3:4 and $c = 15$,　(b) 5:12 and $c = 26$,　(c) 8:15 and $c = 170$,　(d) 1:2 and $c = 10$.

Ans.　(a) 9, 12　(b) 10, 24　(c) 80, 150　(d) $2\sqrt{5}$, $4\sqrt{5}$

10.59. In a rectangle, find the diagonal if its sides are   (*a*) 30 and 40,   **(18.4)**
(*b*) 9 and 40,   (*c*) 5 and 10,   (*d*) 2 and 6.

Ans.   (*a*) 50   (*b*) 41   (*c*) $5\sqrt{5}$ (*d*) $2\sqrt{10}$

10.60. In a rectangle, find one side if the diagonal is 15 and the other side is   (*a*) 9,   (*b*) 5,   **(18.4)**
(*c*) 10,   (*d*) 12.

Ans.   (*a*) 12   (*b*) $10\sqrt{2}$ (*c*) $5\sqrt{5}$ (*d*) 9

10.61. Using the three sides given, which triangles are right triangles?   **(18.5)**

(*a*)   33, 55, 44   (*c*) 4, $7\frac{1}{2}$, $8\frac{1}{2}$ (*e*) 5 in., 1 ft, 1 ft 1 in.   (*g*) 11 mi, 60 mi, 61 mi
(*b*)   120, 130, 50   (*d*) 25, 7, 24   (*f*) 1 yd, 1 yd 1 ft, 1 yd 2 ft   (*h*) 5 m, 5 m, 7 m

Ans.   Only (*h*) is not a right triangle, since $5^2 + 5^2 \neq 7^2$. In all the other cases, the square of the largest side equals the sum of the squares of the other two.

10.62. Using the formula $h = s\sqrt{3}/2$, express in radical form the altitude of an equilateral   **(18.6)**
triangle whose side is   (*a*) 6,   (*b*) 20,   (*c*) 11,   (*d*) 90 in.,   (*e*) 4.6 yd.

Ans.   (*a*) $3\sqrt{3}$ (*b*) $10\sqrt{3}$ (*c*) $11\sqrt{3}/2$ (*d*) $45\sqrt{3}$ in.   (*e*) $2.3\sqrt{3}$ yd

10.63. The largest possible square is to be cut from a circular piece of wood. Find the side of   **(18.7)**
the square, to the nearest inch, if the diameter of the circle is   (*a*) 30 in.,   (*b*) 14 in.,   (*c*) 17 in.

Ans.   (*a*) 21 in.   (*b*) 10 in.   (*c*) 12 in.

10.64. Find the altitude of an isosceles triangle if one of its two equal sides is 10 and its base   **(18.8)**
is   (*a*) 12,   (*b*) 16,   (*c*) 18,   (*d*) 10.

Ans.   (*a*) 8   (*b*) 6   (*c*) $\sqrt{19}$ (*d*) $5\sqrt{3}$

Index